T0185842

ANALYTICAL PYROLYSIS HANDBOOK

ANALYTICAL PYROLYSIS
HANDBOOK
Third Edition

Edited by
Karen D. Sam and Thomas P. Wampler

CRC Press
Taylor & Francis Group
Boca Raton London New York

CRC Press is an imprint of the
Taylor & Francis Group, an **informa** business

Third edition published 2021
by CRC Press
6000 Broken Sound Parkway NW, Suite 300, Boca Raton, FL 33487-2742

and by CRC Press
2 Park Square, Milton Park, Abingdon, Oxon, OX14 4RN

© 2021 Taylor & Francis Group, LLC

[First edition published by CRC Press 1995]
[Second edition published by CRC Press 2006]

CRC Press is an imprint of Taylor & Francis Group, LLC

The right of Karen D. Sam to be identified as the author of the editorial material, and of the authors for their individual chapters, has been asserted in accordance with sections 77 and 78 of the Copyright, Designs and Patents Act 1988.

Reasonable efforts have been made to publish reliable data and information, but the author and publisher cannot assume responsibility for the validity of all materials or the consequences of their use. The authors and publishers have attempted to trace the copyright holders of all material reproduced in this publication and apologize to copyright holders if permission to publish in this form has not been obtained. If any copyright material has not been acknowledged please write and let us know so we may rectify in any future reprint.

Except as permitted under the US Copyright Law, no part of this book may be reprinted, reproduced, transmitted, or utilized in any form by any electronic, mechanical, or other means, now known or hereafter invented, including photocopying, microfilming, and recording, or in any information storage or retrieval system, without written permission from the publishers.

For permission to photocopy or use material electronically from this work, access www.copyright.com or contact the Copyright Clearance Center, Inc. (CCC), 222 Rosewood Drive, Danvers, MA 01923, 978-750-8400. For works that are not available on CCC please contact mpkbookspermissions@tandf.co.uk

Trademark notice: Product or corporate names may be trademarks or registered trademarks and are used only for identification and explanation without intent to infringe.

Library of Congress Cataloging in Publication Data
Names: Sam, Karen D., editor. | Wampler, Thomas P., editor.
Title: Analytical pyrolysis handbook / edited by Karen D. Sam and Thomas P. Wampler.
Other titles: Applied pyrolysis handbook.
Description: Third edition. | Boca Raton : CRC Press, 2021. | Revised edition of: Applied pyrolysis handbook / edited by Thomas P. Wampler. 2nd ed. c2007. | Includes bibliographical references and index.
Identifiers: LCCN 2020055154 (print) | LCCN 2020055155 (ebook) | ISBN 9780367192327 (hardback) | ISBN 9780429201202 (ebook)
Subjects: LCSH: Pyrolysis--Handbooks, manuals, etc.
Classification: LCC TP156.P9 A67 2021 (print) | LCC TP156.P9 (ebook) | DDC 660/.296--dc23
LC record available at https://lccn.loc.gov/2020055154
LC ebook record available at https://lccn.loc.gov/2020055155

ISBN: 978-0-367-19232-7 (hbk)
ISBN: 978-0-367-74054-2 (pbk)
ISBN: 978-0-429-20120-2 (ebk)

Typeset in Times New Roman
by MPS Limited, Dehradun

Contents

Preface

Analytical pyrolysis is the study of materials by adding adequate thermal energy to cause bond cleavage and then analyzing the resulting fragments by gas chromatography, mass spectrometry, or infrared spectroscopy. Pyrolysis has been used for the analysis of organic molecules. Initially connected with investigations of vapor-phase hydrocarbons, it later became a routine technique for analyzing fuel sources and polymers, both natural and synthetic. Current applications include analysis of trace evidence in forensic laboratories, evaluation of new composite formulations, authentication and conservation of artworks, identification of microorganisms, and the study of complex biological and ecological systems.

Pyrolysis instrumentation has evolved into versatile front-end instruments, capable of not just pyrolysis, but also thermal desorption, pyrolysis in reactive atmospheres, and even elevated pressures. The advancement of more sensitive detectors has also allowed analysis at trace levels, to investigate additives or contamination. Since the second edition of this book, there has been the development of new applications, like analysis of personal lubricants in forensic applications, the study of polyurethanes in the conservation field, determination of sludge pollution in ocean sediments, and the study of plastic pollution in the ecosystem.

This book serves as a starting point for analysts who are adding pyrolysis to their collection of analytical techniques, by providing concrete examples and suggesting additional reading. General and theoretical considerations written by the original book editor—including instrumentation and degradation mechanisms—are contained in the first few chapters. The remainder of the book describes the use of pyrolysis as a tool in specific areas of study. These areas include the conservation of cultural materials, forensic analysis, and environmental studies. The chapters examine the scope of work based on pyrolysis in these specific fields of analysis, giving specific examples of methods currently used for the examination of representative samples.

I am thankful to all the authors for their contributions. They each updated their respective chapters, adding recent examples and references. In addition, new authors provided valuable information in their respective fields. The time each one has spent to contribute to this book is tremendously appreciated. I am extremely grateful to the founding editor, Thomas P. Wampler, who has updated many of the chapters contained within and provided treasured guidance for the development and continuation of this book.

Editors

Karen D. Sam is an Application Scientist and Lab Manager at CDS Analytical. She has over 20 years of experience in sample preparation and analysis, including 15 years in application development experience, supporting the design and testing of analytical pyrolysis instruments. She is the author or co-author of several professional papers on the use of analytical pyrolysis. She holds an A.A. degree from Cottey College in Nevada, MO, and a B.S. degree in Chemistry from Bloomsburg University, in Bloomsburg, PA.

Thomas P. Wampler earned his B.S. degree (1970) in chemistry and an M.Ed. degree (1973) in natural science from the University of Delaware in Newark, Delaware. Currently retired, he was actively engaged in the field of analytical pyrolysis for over 25 years. He was the director of science and technology at CDS Analytical, LLC, in Oxford, Pennsylvania. He is the author and coauthor of numerous professional papers on the use of analytical pyrolysis and other thermal sampling techniques.

List of Contributors

Nathalie Balcar
Center for Research and Restoration of the
Museums of France – C2RMF
Paris, France

Céline Burnier
Ecole des Sciences Criminelles,
Université de Lausanne
Switzerland

John M. Challinor
ChemCentre
Bentley, Western Australia

David A. DeTata
ChemCentre, Bentley
Western Australia

Randolph C. Galipo
The University of South Carolina
Columbia, SC 29208

C. J. Maddock
Horizon Instruments, Ltd.,
Heathfield, East Sussex, England

Stephen L. Morgan
The University of South Carolina
Columbia, SC 29208

T. O. Munson
Department of Math/Science,
Concordia University
Portland, Oregon

Hajime Ohtani
Nagoya Institute of Technology
Nagoya, Japan

T. W. Ottley
Horizon Instruments, Ltd.,
Heathfield
East Sussex, England

Kari M. Pitts
ChemCentre, Bentley
Western Australia

Alexandra ter Halle
Paul Sabatier University
Toulouse, France

Shin Tsuge
Nagoya University
Nagoya, Japan

Bruce E. Watt
The University of South Carolina
Columbia, SC 29208

Charles Zawodny
CDS Analytical
Oxford, Pennsylvania

1 Analytical Pyrolysis: An Overview

Thomas P. Wampler
CDS Analytical, Inc., Oxford, Pennsylvania

I INTRODUCTION

Pyrolysis, simply put, is the breaking apart of chemical bonds only by thermal energy. Analytical pyrolysis is the technique of studying molecules either by observing their behavior during pyrolysis or by studying the resulting molecular fragments. The analysis of these processes and fragments tells us much about the nature and identity of the original larger molecule. The production of a variety of smaller molecules from some larger original molecule has fostered the use of pyrolysis as a sample preparation technique, extending the applicability of instrumentation designed for the analysis of gaseous species to solids, especially polymeric materials. As a result, gas chromatography, mass spectrometry, and Fourier-transform infrared (FTIR) spectrometry may be used routinely for the analysis of samples such as synthetic polymers, biopolymers, composites, and complex industrial materials.

The fragmentation that occurs during pyrolysis is analogous to the processes that occur during the production of a mass spectrum. Energy is put into the system, and as a result, the molecule breaks apart into stable fragments. If the energy parameters (temperature, heating rate, and time) are controlled in a reproducible way, the fragmentation is characteristic of the original molecule, based on the relative strengths of the bonds between its atoms. The same distribution of smaller molecules will be produced each time an identical sample is heated in the same manner, and the resulting fragments carry with them much information concerning the arrangement of the original macromolecule.

The application of pyrolysis techniques to the study of complex molecular systems covers a wide and diversified field. Several books have been published that present theoretical as well as practical aspects of the field, including a good introductory text by Irwin [1] and a compilation of gas chromatographic applications by Liebman and Levy [2]. Books dealing with the analysis of synthetic polymers have been published by Moldoveanu [3] and Watanabe, and one on biomass by Basu [4]. A 1989 bibliography [5] lists approximately 500 papers in areas as diverse as food and environmental and geochemical analysis; an excellent review by Blazsó [6] lists over 150 papers just for the analysis of polymers, while the application to microorganisms has been examined by Morgan et al. [7]; and a 1998 bibliography by Haken [8] concentrates on pyrolysis-GC of synthetic polymers. This chapter will include only a few representational examples of the kinds of applications being pursued, with references for further reading. Specific areas of analysis are detailed in subsequent chapters.

II DEGRADATION MECHANISMS

The degradation of a molecule that occurs during pyrolysis is caused by the dissociation of a chemical bond and the production of free radicals. The general processes employed to explain the behavior of these molecules are based on free radical degradation mechanisms. The way a molecule fragments during pyrolysis, and the identity of the fragments produced, depend on the types of chemical bonds involved and on the stability of the resulting smaller molecules. If the

1

subject molecule is based on a carbon chain backbone, such as that found in many synthetic polymers, it may be expected that the chain will break apart in a fairly random fashion to produce smaller molecules chemically similar to the parent molecule. Some of the larger fragments produced will preserve intact structural information snipped out of the polymer chain, and the kinds and relative abundances of these specific smaller molecules give direct evidence of macro-molecular structure. The traditional degradation mechanisms generally applied to explain the pyrolytic behavior of macromolecules will now be reviewed, followed by some general comments on degradation via free radicals.

A   RANDOM SCISSION

Breaking apart a long-chain molecule—such as the carbon backbone of a synthetic polymer—into a distribution of smaller molecules is referred to as random scission. If all of the C–C bonds are of about the same strength, there is no reason for one to break more than another, and consequently, the polymer fragments to produce a wide array of smaller molecules. The poly-olefins are good examples of materials that behave in this manner. When poly(ethylene) (shown as structure I with hydrogen atoms left off for simplicity) is heated sufficiently to cause pyrolysis, it breaks apart into hydrocarbons, which may contain any number of carbons, including methane, ethane, propane, etc.

I   —C—C—C—C—C—C—

II   —C—C— C •   • C—C— C—

Chain scission produces hydrocarbons with terminal free radicals (structure II) that may be stabilized in several ways. If the free radical abstracts a hydrogen atom from a neighboring molecule, it becomes a saturated end, and creates another free radical in the neighboring molecule (structure III), which may stabilize in a number of ways. The most likely of these is beta scission, which accounts for most of the polymer backbone degradation by producing an unsaturated end and a new terminal free radical.

III   —C—C—C—C—C—C—

Beta scission   ↓

IV   —C—C—C = CH2   +   •C—C—

This process continues, producing hydrocarbon molecules that are saturated and have one terminal double bond or a double bond at each end. When analyzed by gas chromatography, the resulting pyrolysate looks like the bottom chromatogram in Figure 1.1. Each triplet of peaks represents the diene, alkene, and alkane containing a specific number of carbons and eluting in that order. The next set of three peaks contain one more carbon. etc. It is typical to see all chain lengths—from methane to compounds—containing 35–40 carbons, limited only by the upper temperature of the gas chromatography (GC) column.

When poly(propylene) is pyrolyzed, it behaves in much the same manner, producing a series of hydrocarbons that have methyl branches indicative of the structure of the original polymer. The center pyrogram in Figure 1.1 shows poly(propylene) revealing again a recurring pattern of peaks, with each group now containing three more carbons than the preceding group. Likewise, when a

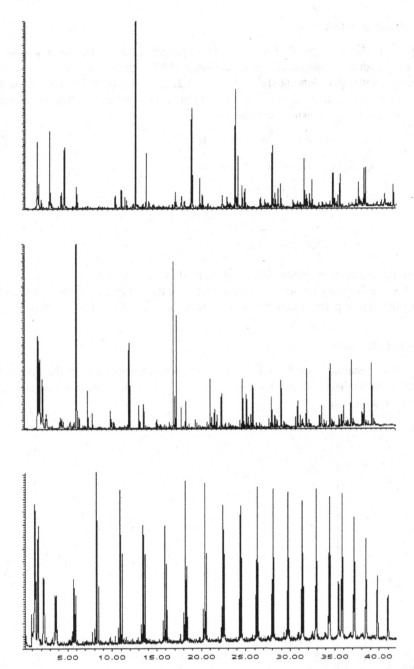

FIGURE 1.1   Pyrograms of poly(1-butene) (top), poly(propylene) (center), and poly(ethylene) (bottom).

polymer made from a four-carbon monomer such as 1-butene is pyrolyzed, it produces yet another pattern of peaks, with oligomers differing by four carbons, as seen in the top pyrogram in Figure 1.1. The relationships of specific compounds produced in the pyrolysate to the original polymer structure have been extensively studied by Tsuge et al. [9], for example, in the case of poly(propylenes). The effects of temperature and heating rate have also been studied [10].

B SIDE GROUP SCISSION

When poly(vinyl chloride) is pyrolyzed, no oligomeric pattern occurs. Instead of undergoing random scission to produce chlorinated hydrocarbons, PVC produces aromatics, especially benzene, toluene and naphthalene, as shown in Figure 1.2. This is the result of a two-step degradation mechanism that begins with the elimination of HCl from the polymer chain (structure V), leaving the polyunsaturated backbone shown in structure VI.

V

Cl   H   Cl   H   Cl   H
| | | | | |
—C — C— C — C — C — C—

↓ — HCl

VI   —C = C — C = C — C = C—

Upon further heating, this unsaturated backbone produces the characteristic aromatics, as seen in Figure 1.2 This mechanism has been well characterized and the occurrence of chlorinated aromatics is used as an indication of polymer defect structures, as in the work of Lattimer and Kroenke [11].

C MONOMER REVERSION

A third pyrolysis behavior is evidenced by polymers such as poly(methyl methacrylate). Because of the structure of methacrylate polymers (structure VII), the favored degradation is essentially a reversion to the monomer.

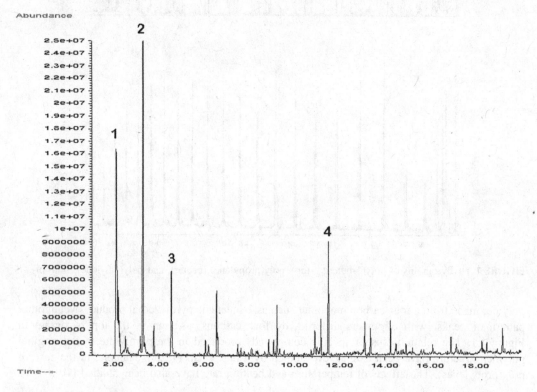

FIGURE 1.2 Pyrogram of poly(vinyl chloride) at 750 °C for 15 seconds. Peak 1 = HCl, 2 = Benzene, 3 = Toluene, 4 = Naphthalene.

VII
$$
\begin{array}{ccccc}
CH_3 & & CH_3 & & CH_3 \\
| & H & | & H & | \\
-C & -C & -C & -C & -C\, \bullet \\
| & H & | & H & | \\
CO_2R & & CO_2R & & CO_2R
\end{array}
$$

↓ Beta Scission

$$
\begin{array}{ccccc}
CH_3 & & CH_3 & & CH_3 \\
| & H & | & & | \\
-C & -C & -C\, \bullet & + \quad CH_2 & =C \\
| & H & | & & | \\
CO_2R & & CO_2R & & CO_2R \\
& & & & \text{Monomer}
\end{array}
$$

Monomer production is, for the most part, unaffected by the R group, so poly(methyl methacrylate) will revert to methyl methacrylate, and poly(ethyl methacrylate) will produce ethyl methacrylate, etc. This proceeds in copolymers as well, with the production of both monomers in roughly the original polymerization ratio. Figure 1.3 shows a pyrogram of poly(butyl methacrylate), with the butyl methacrylate monomer peak by far the predominant product. A pyrogram of a copolymer of two or more methacrylate monomers would contain a peak for each of the monomers in the polymer.

D RELATIVE BOND STRENGTHS

The question of which degradation mechanism a particular polymer will be subjected to—random scission, side group scission, or monomer reversion, or a combination of these—is simplified by

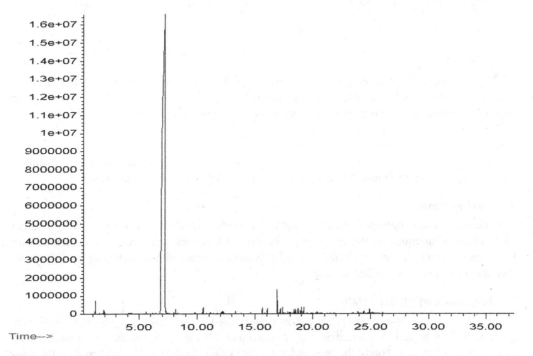

FIGURE 1.3 Pyrogram of poly(butyl methacrylate), showing large monomer peak (750 °C for 15 seconds).

considering the nature of thermal degradation as a free radical process. All of the degradation products—as well as minor constituents, and deviations to the simplified rules—are consistent with the following general statements:

> Pyrolysis degradation mechanisms are free radical processes and are initiated by breaking the weakest bonds first.

> The composition of the pyrolysate will be based on the stability of the free radicals involved and on the stabilities of the product molecules.

> Free radical stability follows the usual order of $3° > 2° > 1° > CH_3$, and intramolecular rearrangements, which produce more stable free radicals, play an important role—particularly the shift of a hydrogen atom.

A quick review of the previous degradation examples will help show how each of the above categories is, in reality, just one aspect of the general rule of free radical processes.

1 Polyolefins

Poly(ethylene) and the other polyolefins contain only C–C bonds and C–H bonds. Since an average C–C bond is about 83 kcal/mol and a C–H bond 94 kcal/mol, the initiation step involves breaking the backbone of the molecule, with subsequent stabilization of the free radical. In the case of poly (ethylene), the original free radicals formed are terminal or primary. Hydrogen abstraction from a neighboring molecule creates a C–H bond (stable product) and a new, secondary free radical, which may then undergo beta scission to form an unsaturated end. In addition, transfer of a hydrogen atom from the carbon five removed from the free radical (via a six-membered ring) transforms a primary free radical to a secondary, increasing the free radical stability.

```
     H                              H
     /                              \
----C5      1C•        →     ----C•       C
     |       |                    |       |
    C4      2C                    C       C
     \       /                    \       /
      C3                              C
```

This new secondary free radical will either undergo beta scission or another 1–5 H shift. Beta scission produces a molecule of hexene, the trimer of ethylene, while a second 1–5 shift moves the unpaired electron to carbon number 9. Beta scission of this new free radical would generate a molecule of decene, the pentamer. This stabilization by 1–5 hydrogen shifting explains the increased abundance of products in the pyrolysate of poly(ethylene) containing 6, 10, and 14 carbons. These products are the result of performing a 1–5 hydrogen shift one, two, and three times, respectively (see again Figure 1.1 in which 10 and 14 carbon species are marked).

2 Vinyl Polymers

Poly(vinyl chloride) contains C–C, C–H, and C–Cl bonds, with the C–Cl bonds weakest at about 73 kcal/mol. Consequently, the first step is the loss of Cl•, which subsequently combines with hydrogen to form HCl, leaving the unsaturated polymer backbone with formation of the very stable aromatic products upon further heating.

3 Acrylates and Methacrylates

Carbons in the chain of a methacrylate polymer are bonded to the CO_2R side group, the CH_3 side group to hydrogens and to other chain carbons (structure VII), with C–C bonds being weaker than C–H bonds. Of the C–C bonds, the ones making up the chain are the weakest and produce the most stable free radicals, since breaking C–CH$_3$ produces CH$_3$•, the least stable free radical. In addition, the free radicals produced are already tertiary—the most stable—and there is no hydrogen atom on carbon number 5 to shift, so no additional pathway. Consequently, beta scission with an unzipping back to the monomer represents the most stable product formed by the most stable free radical.

With poly(acrylates) on the other hand, the methyl groups are absent, (structure VIII) so there are hydrogens available to shift. Bond dissociation produces a secondary free radical, which can be stabilized by the 1–5 H shift to a tertiary free radical.

$$VIII \quad \begin{array}{ccccccccccc} & & H & & & & H & & & & H \\ & & | & & H & & | & & H & & | \\ -C & - & C & - & C & - & C & - & C & - & C\; • \\ & & | & & H & & | & & H & & | \\ & & CO_2R & & & & CO_2R & & & & CO_2R \end{array}$$

⬇ 1-5 H transfer

$$\begin{array}{ccccccccccc} & & & & & & H & & & & H \\ • & & H & & | & & H & & | \\ -C & - & C & - & C & - & C & - & C & - & C\,H \\ & & | & & H & & | & & H & & | \\ & & CO_2R & & & & CO_2R & & & & CO_2R \end{array}$$

When this new free radical undergoes beta scission, the acrylate trimer is formed, with the three monomeric units connected in the same way they were in the polymer. Consequently, the poly(acrylates) pyrolyze to produce monomer, dimer, and most especially trimer, while the poly(methacrylates) produce mostly monomer.

III   EXAMPLES AND APPLICATIONS

Analytical pyrolysis is frequently considered to be a technique mainly applied to the analysis of polymers, which may at first seem fairly limited. However, when one considers that proteins, polysaccharides, plastics, adhesives, paints, etc. are included in this general category of "polymers,"

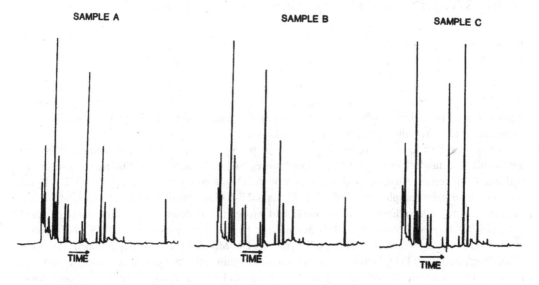

FIGURE 1.4   Comparison of paint pyrolyses, showing paints A and B matching, C being a different formulation.

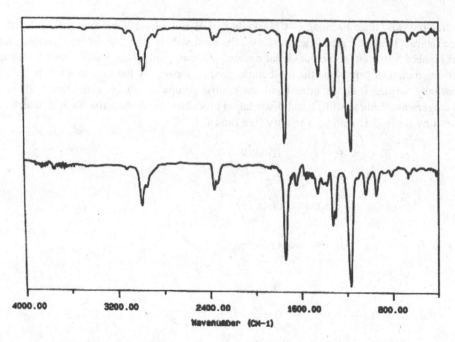

FIGURE 1.5 Comparison of poly(methyl methacrylate) (top) and poly(ethyl methacrylate) (bottom) by pyrolysis-FT-IR.

the list of applications becomes much longer. Natural and synthetic polymers—in the forms of textile fibers, wood products, foods, leather, paints, varnishes, plastic bottles and bags, and paper and cardboard—make up the bulk of what we come into contact with every day. In fact, it is difficult to sit in a room and touch something—paint, paneling, carpet, clothing, countertop, telephone, upholstery, books—that is not made of some sort of polymer. Consequently, the study of materials using pyrolysis has become a very broad field, including such diverse topics as soil nutrients, plastic recycling, criminal evidence, bacteria and fungi, fuel sources, oil paintings, and computer circuit boards. The examples in this chapter will review in only a very general way some of the applications of analytical pyrolysis. Subsequent chapters treat some of these areas in greater depth.

A FORENSIC MATERIALS

The application of pyrolysis techniques to the study of forensic samples has a long and well-documented history, including a review of pyrolysis-mass spectrometry as a forensic tool in 1977 by Saferstein and Manura [12] and a general review by Wheals [13]. A wide variety of sample types has been investigated—including chewing gum, rubber and plastic parts from automobiles, drugs, tapes [14,15], adhesives [16] and bloodstains.

Perhaps the best known forensic application of pyrolysis is the analysis of paint flakes from automobiles, a standard practice in many laboratories backed by substantial libraries of pyrograms and sample materials. Munson et al. [17] describe their work using pyrolysis-capillary GC-MS for the analysis of paint samples, and Fukuda [18] has published results on nearly 80 paints used in the Japanese automobile industry. Kochanowski and Morgan [19] describe a multivariate statistical approach to the discrimination of 100 automotive paints using pyrolysis-GC/MS, and a good evaluation of library searching using both pyrolysis-GC/MS and FT-IR was published by Chang et al. [20], and Burns [21]. Ways in which automotive paint formulations have changed, partly in response to environmental concerns, have also been studied via pyrolysis [22]. The same techniques may be applied to paint samples recovered from nonautomotive sources, including house paints and tool and machine coatings as well as varnishes from furniture and musical instruments. Armitage et al. [23] used a laser micropyrolysis system to characterize paint, photocopy toner, and fibers.

Most automotive finishes are applied in layers—which may be removed selectively and analyzed individually—or pyrolyzed intact. An advantage provided by pyrolysis is that the inorganic pigment material is left behind and only the organic pyrolysate is transferred to the analytical instrument. Because of the great variety of polymeric materials used as paints and coatings—acrylics, urethanes, styrenes, epoxies, etc.—the resulting pyrograms may be quite complex. It is not always necessary to identify all of the constituents involved, however, to compare related paints. Figure 1.4 shows a comparison of three paint samples of similar monomer composition. Although many of the peaks are very similar for all three formulations, the inversion on the relative peak height in the second and third largest peaks makes it relatively straightforward to see that paints A and B are of same formulation and paint C is not a match.

Frequently, samples like paint flakes present a problem to the analytical lab because they are small, nonvolatile, and opaque with inorganic pigments. Since pyrolysis prepares a volatile organic sample from a polymer or composite, it offers the ability to introduce these organics to an analytical instrument separate from the inorganics, using only a few micrograms of sample. This extends the use of analytical techniques such as mass spectrometry and FT-IR spectroscopy to the investigation of small complex samples. When an opaque paint is pyrolyzed, the organic constituents are volatilized and available for analysis apart from the pigment material. A paint formulation based on methacrylate monomers, for example, will pyrolyze to reveal the methacrylates despite the presence of the pigment, and techniques such as FT-IR—that were previously unable to provide good spectral information—may be applied to the pyrolysate only. Figure 1.5 shows the pyrolysis-FT-IR comparison of poly(methyl methacrylate) and poly(ethyl methacrylate). In each case, a 200 µg sample of the solid polymer was pyrolyzed for 5 s in a cell fitted directly into the sample compartment of the FT-IR. The cell was positioned so that the FT-IR beam passed directly over the platinum filament of the pyrolyzer. The samples were pyrolyzed, and the pyrolysate scanned for 10 s, producing the spectra shown. This system, details of which are published [24], permits the rapid scanning of polymer-based materials, requiring approximately one minute per sample.

B FIBERS AND TEXTILES

Almost all clothing is made from fibers of natural polymers such as the proteins in silk and wool, cellulose in cotton, synthetic polymers including the various nylons and polyesters, or blends of both natural and synthetic polymers. Since these polymers are all chemically different, the pyrolysates they generate are all distinctive and provide a ready means of fiber analysis. Significant work has been done in the analysis and comparison of the various nylons by Tsuge et al. [25] among others, as well as acrylate and methacrylate/acrylonitrile copolymer fibers by Saglam [26], Almer [27], and Causin et al. [28]. A good overview of fiber analysis by pyrolysis-MS was published by Hughes et al. [29].

Figure 1.6 shows a pyrogram of silk fibers that are made of the protein fibroin, which is nearly 50% glycene. Figure 1.7 is a pyrogram of the polyamide nylon 6/12, which is formulated using a diamine containing six carbons and a dicarboxylic acid containing 12 carbons. Although both silk and nylon are polyamides, the chemical differences between them make distinctions using pyrolysis gas chromatography relatively simple. The same techniques may be used to differentiate among the various nylon formulations, to distinguish silk from wool, etc.

Polymer blends used in clothing may be analyzed in the same manner. Because the degradation of a specific polymer is largely an intramolecular event, the presence of two different fibers being pyrolyzed simultaneously generally produces a pyrogram resembling the superimposition of the pyrograms of the two pure materials. A good example of this is shown in Figure 1.8, which

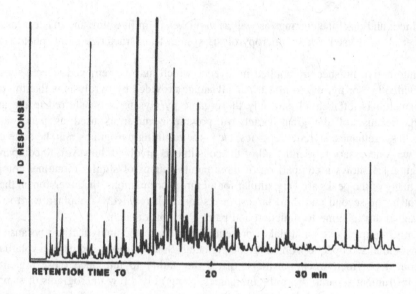

FIGURE 1.6   Pyrolysis of silk thread (675 °C for 10 s in a glass-lined system).

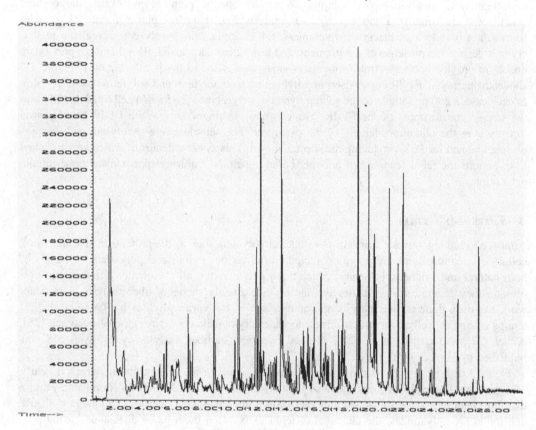

FIGURE 1.7   Pyrogram of nylon 6/12 (800 °C for 10 s).

compares pyrograms of cotton, polyester, and cotton/polyester blend threads. Each sample was a piece of thread 1 cm in length, pyrolyzed at 750 °C. The top pyrogram shows just cotton, which produces large amounts of CO_2 and H_2O, as well as some larger molecules, which appear in the

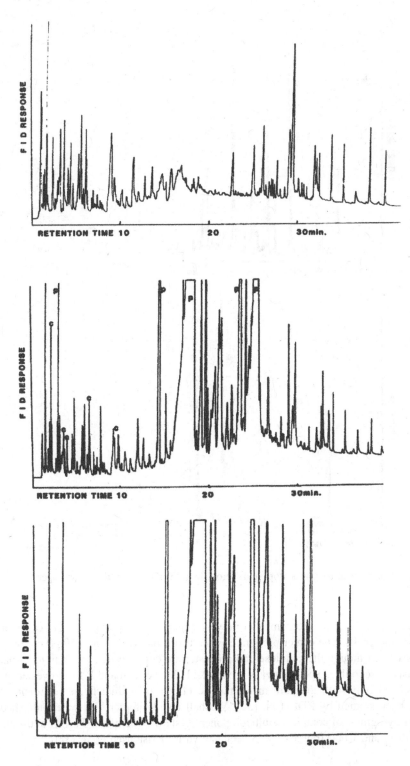

FIGURE 1.8 Comparison of pyrograms obtained from cotton (top), polyester (bottom), and cotton/polyester blend (center).

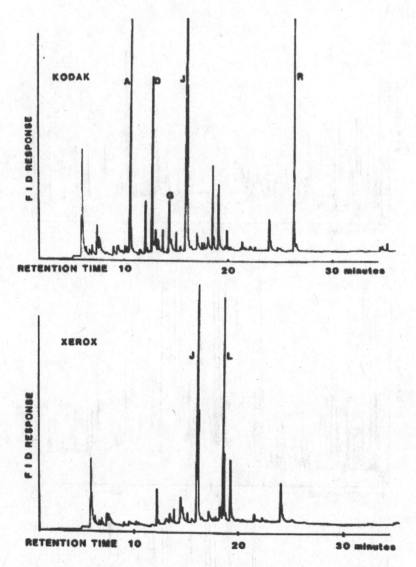

FIGURE 1.9   Comparison of Kodak (top) and Xerox (bottom) photocopy toners on paper.

chromatogram. The bottom pyrogram is of polyester only, which produces a larger abundance of chromatographic peaks. The center pyrogram is the result of pyrolyzing a 50/50 cotton-polyester blend thread. Even though the cellulose and polyester are copyrolyzed, the individual patterns of each are easily discernible. Specific peaks are marked C for cotton and P for polyester, indicating peaks associated only with the pure material. The effects of adding flame retardants to cotton fabrics has been studied by Zhu et al. [30]; Washall and Wampler [31] have published further examples of pyrograms of complex, multicomponent systems. The use of TMAH in the analysis of vegetable fibers and wood fragments has been shown by Kristensen et al. [32].

C   Paper, Ink, and Photocopies

Paper—primarily cellulose, printing ink, writing ink, and photocopy toners—have all been analyzed by pyrolysis, sometimes independently as pure materials, and sometimes intact as a fragment

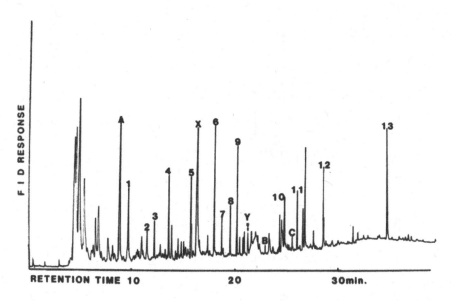

FIGURE 1.10   Pyrolysis of ballpoint pen black ink (NPC) on paper (650 °C for 10 s). Numbered peaks result from paper pyrolysis. Peaks A, B, and C are from pyrolysis of ink non-volatiles.

of a document. Zimmerman et al. [33] used pyrolysis-GC to extend the specificity of toner identification in photocopies. Ballpoint pen ink, various papers, and photocopy toners have been analyzed using a combination of headspace sampling and pyrolysis by Wampler and Levy [34], while Ohtani et al. [35] evaluated paper deterioration in library materials.

Photocopy toner materials are generally complex formulations of pigments and polymers unique to a specific manufacturer. The formulations of these toners frequently include poly (styrene), polyesters, acrylic polymers, methacrylic polymers of various side-chain lengths, and various other additives. The formulations vary from manufacturer to manufacturer and year to year, as improvements are made and new processes are introduced. Figure 1.9 shows a comparison of a Kodak and a Xerox photocopy made in 1985. The pyrograms were prepared by pyrolyzing a single letter punched from each photocopy. The punch, both paper and toner letter, was pyrolyzed at 650 °C in a small quartz tube and analyzed by capillary gas chromatography. The paper, which is essentially cellulose and produces a pyrogram like the one shown for cotton thread in Figure 1.8, contributes many of the smaller peaks, including the one marked G, which is furfural. The synthetic polymers in the toner materials pyrolyze to generate the larger peaks, including A, methyl methacrylate, and J, which is styrene. A multivariate statistical approach to the analysis of photocopy and printer toners is presented by Egan et al. [36].

When the same analysis is carried out using a sample of paper that was written upon with ballpoint pen, three sets of peaks are produced. Because the ink was applied as a fluid, there may be traces of the liquid or semiliquid vehicle remaining in the sample. In addition, there are peaks resulting from actual pyrolysis of nonvolatile ink constituents, and from the paper itself. Since less ink is applied than in the case for photocopy toner, the peaks from the paper generally make up a larger part of the chromatogram for ink analysis than for toner analysis.

Figure 1.10 shows the pyrogram resulting from pyrolyzing a 2 mm^2 piece of paper with ballpoint ink on it. All of the peaks labeled with numbers are cellulose pyrolysates from the paper. Peaks X and Y are from the ink vehicle and will show up in an independent headspace sampling at 200 °C and consequently, they leave the sample before pyrolysis takes place. The peaks labeled A, B, and C are pyrolysis products from the remaining nonvolatile residue of

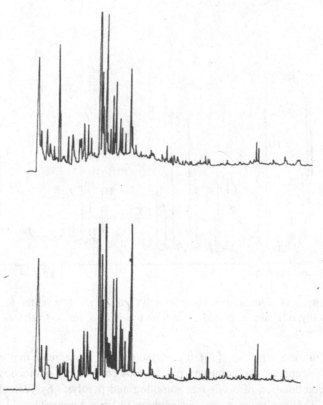

FIGURE 1.11 Pyrolysis (500 °C for 20 s) of Egyption sarcophagus ground material for binder content (bottom), compared to sample of ancient animal glue (top).

the ink. Just as with the cotton/polyester blend thread and the photocopy toner on paper, the presence of several components in the sample causes little effect in the pyrolytic behavior of each individual material, and peaks A, B, and C are seen if the ink is pyrolyzed alone or while it is on the paper.

D ART MATERIALS AND MUSEUM PIECES

Valuable art objects like paintings [37], furniture, and archeological artifacts [38] are frequently investigated via pyrolysis of the nonvolatile materials used in their construction. Paints, varnishes, glues, pigments, waxes, and organic binder formulations have been studied from the aspects of both conservation and authentication. The materials used for varnishes and other protective coatings, and glues used in assembling pieces and other polymeric species, change from region to region, time to time, even with specific artists and craftsmen, so that identification of these materials goes far to indicate the authenticity—and even age—of a particular object. Natural resins used to formulate varnishes for centuries have been analyzed by Shedrinsky et al. [39], showing good distinctions among dammar, mastic, sandarac, and copals. More recent media have been studied as well, including artists' acrylic paints [40] and twentieth century binding media [41]. Wright and Wheals [42] have published results on natural gums, waxes, and resins as they apply to Egyptian artifacts. Residues from mummies [43] and even tissue samples from preserved bodies [44] have been investigated using pyrolysis techniques.

Frequently, art and archeological samples must be investigated in layers, since protective coatings are generally applied over the original artwork, which itself may have been applied onto a

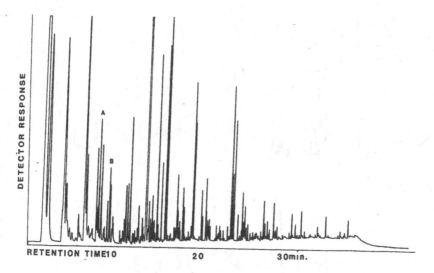

FIGURE 1.12 Pyrogram of a copolymer of poly(propylene) and 1-butene, with peaks marked A and B associated with the propylene monomer and 1-butene, respectively.

prepared surface. The identification may involve analysis of the binder used in a subsurface, the oil or resin in the artwork, and the natural or synthetic polymer present in the protective coating. An example of this is shown in Figure 1.11. The sample came from an Egyptian sarcophagus that was believed to originate from about the fourth century AD. The object was made of wood that was covered with a white layer (the "ground") of essentially inorganic material, used as a base for decorative paintings. It was decided to investigate the organic binder used in preparing the ground as a measure of the authenticity of the sarcophagus. If an ancient material had been used, the authenticity would be supported; if a modern adhesive was detected, the sarcophagus would be fraudulent. Various natural binder materials were proposed and investigated, including egg, wax, animal glue, and copal resin. When a sample of the ground layer was pyrolyzed at 500 °C, the pyrogram matched that produced from ancient animal glue which had been independently authenticated.

An extensive review of pyrolysis applications in the analysis of artwork and antiquities has been published by Shedrinsky et al. [45], with examples including glues and other adhesives, oil paints, varnishes, natural resins, and reference to the famous Van Meegeren case.

E SYNTHETIC POLYMERS

Perhaps the widest application of analytical pyrolysis is in the analysis of synthetic polymers [46,47], both from the standpoints of product analysis and quality control as well as polymer longevity, degradation dynamics, and thermal stability [48]. Several specific polymers have been discussed in Section II, and subsequent chapters treat specific families of polymers in detail.

Individual polymers may be distinguished from one another fairly readily. For example, telling poly(ethylene) from poly(propylene)—as in Figure 1.1—or poly(methyl methacrylate) from poly (butyl methacrylate). Pyrolysis generates fragments that retain molecular structures intact; however, much finer distinctions may be made. Defect structures, branching, head-to-head or tail-to-tail linking, extraneous substitutions, and efficiency of curing have all been investigated. Copolymers may be studied, revealing the monomers involved and even the relative amounts of each monomer present in the final polymer. Wang [49], for example, has explored the microstructure of thermoplastic copolymers formed using a variety of monomers. Branching, cross-linking [50], and even the stereochemistry of polymers may be investigated, since the positioning

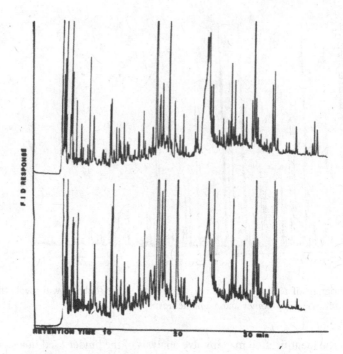

FIGURE 1.13 Comparison of two polymers of glucose pyrolyzed at 750 °C for 10 s, cellulose (top) and starch (bottom).

of side groups in different stereochemical orientations produces pyrolysates of different composition.

Copolymers of methyl methacrylate and ethyl acrylate (EA), varying from 2% to 32% EA, have been studied by Shen and Woo [51], showing a calibration curve for the determination of monomer ratios in unknowns. Similar work for copolymers of butadiene and acrylonitrile is reported by Weber [52]. When a copolymer is pyrolyzed, the resulting pyrogram may be quite complex—as in the case of polyolefins—or rather simple, if the polymer chain unzips to a monomer. If the monomers are all methacrylates, for example, the pyrogram will show major peaks for each of the corresponding monomers involved. Even for more complex materials, however, the pyrograms are characteristic of the original macromolecular system. Information is present indicating whether the sample is a physical blend of homopolymers, or resulting from the copolymerization of monomers. In the latter case, fragments that incorporate molecules of both monomers will be present. These fragments could not result from the pyrolysis of homopolymers, but rather indicate the position of monomer units relative to each other in the macromolecule. It is possible to identify small fragments resulting from only one or the other of the monomers and, thus, study the effect of relative monomer concentration on abundance of specific peaks in the pyrogram. Figure 1.12 shows the pyrogram of a copolymer of polypropylene and 1-butene, with peaks marked A and B associated with the propylene monomer and 1-butene, respectively. Similar work, comparing copolymers and polymer blends of ethylene and propylene, has been published by Tsuge et al. [53].

In addition to the polymer matrix itself, information may be obtained about additives such as tackifier resins in rubber [40,54], flame retardants in polystyrenes [55], plasticizers [56], anti-oxidants, etc. The study of polymers using pyrolysis in general has been broadened by techniques permitting hydrolysis and methylation, typically by the addition of tetramethylammonium hydroxide to the sample, as described in a review by Challinor [57].

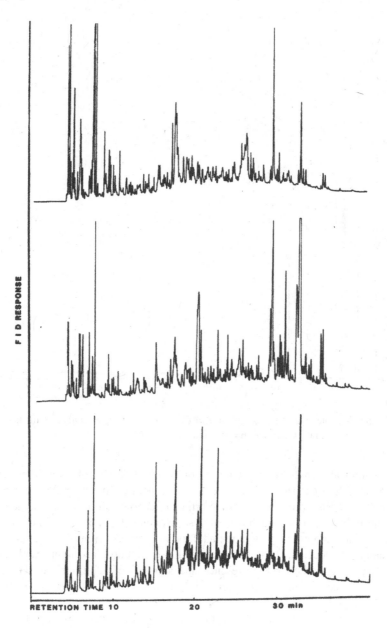

FIGURE 1.14   Comparison of three protein materials pyrolyzed at 750 °C for 10 s, gelatin (top), human hair (center), and fingernail (bottom).

F   NATURAL MATERIALS AND BIOLOGICALS

A rapidly growing area of pyrolytic investigation involves the study of natural materials [58] and biopolymers. Although fuel sources [59,60], kerogens [61,62], and coals [63,64] have been analyzed for decades, more recently, analysts are employing pyrolysis techniques in analyzing soil materials [65], microorganisms [66,67], and biomass [68–70]. Smith et al. [71] described a chemical marker for the differentiation of group A and group B streptococci, and *Salmonella* strain characterization has been demonstrated by Tas et al. [72]. Environmental samples are being analyzed as well, as in the case of the spruce needle analysis performed by Schulten [73] in conjunction with a study of forest death in Germany.

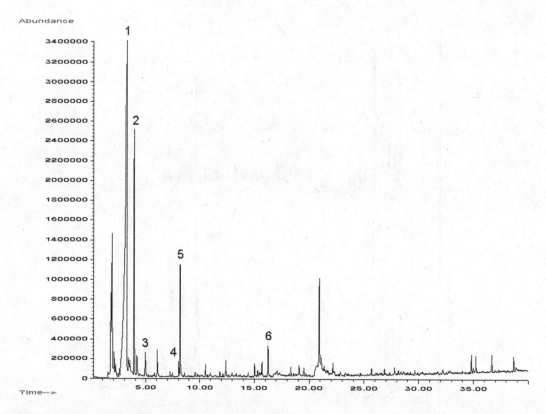

FIGURE 1.15 Interior latex paint pyrolyzed at 750°C. Peaks: 1 = acetic acid, 2 = methyl methacrylate, 3= toluene, 4 = styrene, 5= butyl acrylate, 6= naphthalene.

Although biological samples may be more complex than pure polymers, the degradation of biopolymers must follow the same kinds of chemical processes, producing a volatile distribution which, if complex, is also representative of the original material. Biopolymers include polyamides, as proteins, and polysaccharides, such as cellulose. Biological samples are likely to be complex systems based on such biopolymers with the addition of other, sometimes characteristic materials. Wood, for example, includes the basic macromolecules of cellulose and lignin, and different wood species differ from each other in the presence and amounts of additional substances, including terpenes. Microorganisms, including bacteria and fungi, have been studied in the intact state as well as in isolated parts, such as cell walls.

Figure 1.13 shows a comparison of two polymers of glucose—cellulose and starch. Since these materials are both comprised of the same monomer, it is understandable that the pyrograms are similar. Cellulose and starch differ, however, in the orientation of the linkage between the glucose units, and this difference affects the kinds and relative abundances of the pyrolysate products formed. Gelatin, hair, and nail are also similar in that they are all protein-based materials, but easily distinguishable in the pyrograms shown in Figure 1.14.

Besides plant materials, environmental applications include assays of soil and sediment contaminants, including Biogenic [74] as well as man-made [75]. Agricultural applications treat biomass as well as forages and crops [76].

G PAINTS AND COATINGS

In addition to automotive finishes and artistic media discussed above, there are many other polymeric products applied to materials as protective coatings. These include varnishes, using both

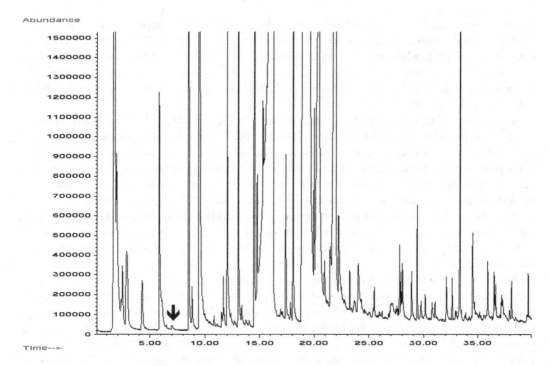

FIGURE 1.16 Pyrolysis of pharmaceutical showing trace amounts of styrene from polystyrene.

natural and synthetic materials, shellac [77], alkyd and latex architectural paints [78], decorative paints, spray paint, plastic coatings, laminates, and sealants. Many of these incorporate drying oils that contain long-chain fatty acids, which appear in the pyrogram along with pyrolysis fragments of the polymers used. Polyurethanes are frequently added to the formulations, and upon pyrolysis, regenerate the diisocyanate. Alkyd housepaints are glyceryl phthalate polyesters and produce a major peak for phthalic anhydride. Paints using acrylic polymers and copolymers with styrene, vinyl toluene, and methyl styrene produce at least the monomers and frequently dimers and trimers.

Figure 1.15 shows the pyrogram of a typical latex emulsion interior wall paint. These paints frequently contain substantial amounts of vinyl acetate. Like polyvinyl chloride, polyvinyl acetate loses the acetate side group to generate acetic acid, and then forms aromatics—, benzene, toluene, and naphthalene. In Figure 1.15, peak #1 is acetic acid, peak #3 is toluene and peak #6 is naphthalene. Benzene co-elutes with the large acetic acid peak, but may be determined by looking at ion 78. In addition to the polyvinyl acetate, the paint formulation includes methyl methacrylate (peak #2), butyl acrylate (#5), and even a small amount of styrene (peak #4). The inorganic pigments used in the paint (frequently a large amount of titanium dioxide) stay behind in the sample tube and do not interfere with the analysis.

H TRACE LEVEL ANALYSES

Pyrolysis techniques may be used to analyze minor constituents in a system in addition to looking at the polymeric matrix itself. Since the macromolecules involved generally degrade without interacting with each other, even a trace level of one polymer contained within another may be expected to produce characteristic pyrolysates and, thus, be identified and even quantified. The matrix need not be polymeric, or even organic, and analysts are using pyrolysis to look at coatings on minerals such as clays, pigments, and glass surfaces. Ezrin and Lavigne [79] published a technique for the analysis of silicone polymers in the parts per billion range in recycled papers. A method for quantitating various paper additives using internal standards has been shown by Odermatt et al. [80], while a variety of contamination products have been studied by del Rio et al. [81]. The contamination of soil with polymeric, non-volatile, or semi-volatile compounds—including trinitrotoluene [82]—has been addressed, as has the presence of trace levels of polymers in pharmaceuticals [83] been.

An example of this last application may be seen in Figure 1.16. When the pharmaceutical material was purified by elution from a bed of polymer beads, some of the polystyrene dissolved in the solvent used and became part of the dried finished product. To determine the amount of polystyrene present in the pharmaceutical powder, the product was pyrolyzed. Almost all of the peaks seen in the pyrogram (many of which are off-scale to show the styrene) came from pyrolyzing the pharmaceutical and, in this case, were of no interest. The polystyrene, even at such trace levels, provided a monomer peak in the pyrogram, marked with an arrow in Figure 1.16. The pure pharmaceutical was demonstrated not to produce styrene as a pyrolysate peak by itself, so this styrene monomer became a direct indication of the level of polystyrene that ended up in the finished product.

Polymeric materials, then—whether natural such as cellulose, resins, and proteins; or synthetic such as polyolefins, nylons, and acrylics—behave in reproducible ways when exposed to pyrolysis temperatures. This permits the use of pyrolysis as a sample preparation technique to allow the analysis of complex materials using routine laboratory instruments. Pyrolytic devices may now be interfaced easily to gas chromatographs, mass spectrometers, and FT-IR spectrometers, extending their use to solid, opaque, and multi-component materials. Laboratories have long made use of pyrolysis for the analysis of paint flakes, textile fibers, and natural and synthetic rubber and adhesives. The list of applications has been expanded to include documents, artwork, biological materials, antiquities, and other complex systems that may be analyzed with or without the separation of various layers and components involved.

REFERENCES

1. W. I. Irwin, *Analytical Pyrolysis: A Comprehensive Guide*. Marcel Dekker, New York (1982).
2. S. A. Liebman and E. J. Levy, *Pyrolysis and GC in Polymer Analysis*, Marcel Dekker, New York (1985).
3. S. C. Moldoveanu, *Analytical Pyrolysis of Synthetic Organic Polymers*, Elsevier, New York (2005).
4. Prabir Basu, *Biomass Gasification and Pyrolysis 1st Edition*, Practical Design and Theory Academic Press, New York (2010).
5. T. P. Wampler, *J. Anal. Appl. Pyrol.*, *16*: 291–322 (1989).
6. M. Blazsó, *J. Anal. Appl. Pyrol.*, *39*: 1–25 (1997).
7. S. L. Morgan, A. Fox, L. Larsson, and G. Odham, *Analytical Microbiology Methods, Chromatography and Mass Spectrometry*, Plenum Press, New York (1990).
8. J. K. Haken, *J. Chrom. A*, *825*, 2: 171–187 (1998).
9. S. Tsuge, Y. Sugimura, T. Nagaya, T. Murata, and T. Takeda, *Macromolecules*, *13*: 928–932 (1980).
10. T. P. Wampler, *J. Anal. Appl. Pyrol.*, *15*: 187–195 (1989).
11. R. P. Lattimer and W. J. Kroenke. *J. Appl. Polym. Sci.*, *25*: 101–110 (1980).
12. R. Saferstein and J. J. Manura, *J. Forens. Sci.*, *22*: 749–756 (1977).
13. B. B. Wheals, *J. Anal. Appl. Pyrolysis*, *2*: 277–292 (1981).
14. Y. Kumooka, *Forens. Sci. Inter.*, *206*: 1–3, 136–142 (2011).
15. M. J. Bradley, D. M. Wright, and A. H. Mehltretter, *J. Forens. Sci. 56*, 1: 82–94, (2011).
16. Y. Kumooka, *Forens. Sci. Inter.*, *163*: 1–2, 132–137 (2006).
17. T. T. Munson, D. G. McMinn, and T. L. Carlson, *J. Forens. Sci.*, *30*: 1064–1073 (1985).
18. K. Fukuda, *Forens. Sci. Inter.*, *29*: 227–236 (1985).
19. B. K. Kochanowski and S. L. Morgan, *J. Chrom. Sci.*, *38*, 3: 100–108 (2000).
20. W. Chang, C. Yu, C. Wang, and Y. Tsai, *Forens. Sci. J.*, *2*: 47–58 (2003).
21. D. T. Burns and K. P. Doolan, *Anal. Chim. Acta*, *539*: 1–2, 145–155 (2005).
22. T. P. Wampler, G. A. Bishea, and W. J. Simonsick, *J. Anal. Appl. Pyrol.*, *40–41*: 79–89 (1997).

23. S. Armitage, S. Saywell, C. Roux, C. Lennard, and P. Greenwood, *J. Forens. Sci. 46*, 5: 1043–1052 (2001).
24. J. W. Washall and T. P. Wampler, *Spectroscopy, 6,4*: 38–42 (1991).
25. S. Tsuge, H. Ohtani, T. Nagaya, and Y. Sugimura, *J. Anal. Appl. Pyrol., 4*: 117—131 (1982).
26. M. Saglam. *J. Appl. Polym. Sci.. 32*: 5717–5726 (1986).
27. J. Almer, *Can. Soc. Forens. Sci. J., 24, 1*: 51–64 (1991).
28. V. Causin, C. Marega, S. Schiavone, V. DellaGuarddia, and A. Marigo, *J. Anal. Appl. Pyrol., 75*: 43–48 (2006).
29. C. Hughes, B. B. Wheals, and M. J. Whitehouse. *J. Anal. Appl. Pyrol., 103*: 482—491 (1978).
30. P. Zhu, S. Sui, B. Wang and G. Sun, *J. Anal. Appl. Pyrol., 71*: 645–655 (2004).
31. J. W. Washall and T. P. Wampler. *J. Chrom. Sci., 27*: 144—148 (1989).
32. R. Kristensen, S. Coulson, and A. Gordon, *J. Anal. Appl. Pyrol., 86, 1*: 90–98 (2009).
33. J. Zimmerman, D. Mooney, and M. J. Kimmet, *JFSCA, 31*, 2: 489–493 (1986).
34. T. P. Wampler and E. J. Levy, *LC-GC, 4, 11*: 1112—1116 (1987).
35. H. Ohtani, T. Komura, N. Sonoda, and Y. Taguchi, *J. Anal. Appl. Pyrol., 85, 1–2*: 460–464 (2009).
36. W. J. Egan, R. C. Galipo, B. K. Kochanowski, S. L. Morgan, E. G. Bartick, M. L. Miller, D. C. Ward, and R. F. Mothershead, *J. Analyt. Bioanalyt. Chem., 376*: 1286–1297 (2003).
37. L. R. Ember, *C&E News, 79*, 31: 51–59 (2001).
38. J. J. Lucejko, F. Modugno, E. Ribechini, and J. del Rio, *Anal. Chim. Acta, 654, 1*: 26–34 (2009).
39. A. M. Shedrinsky, T. P. Wampler, and N. S. Baer, The identification of dammar, mastic, sandarac and copals by pyrolysis gas chromatography, *Wiener Berichie uber Naturwissenschaft in der Kunsi*, VWGO, Wien (1988).
40. D. Scalarone and O. Chiantore, *J. Sep. Sci., 27, 4*: 263–274 (2004).
41. F. Cappitelli, *J. Anal. Appl. Pyrol., 71*: 405–415 (2004).
42. M. M. Wright and B. B. Wheals, *J. Anal. Appl. Pyrol., 11*: 195–212(1987).
43. S. A. Buckley, A. W. Scott, and R. P. Evershed, *Analyst, 124, 4*: 443–452 (1999).
44. B. A. Stankiewicz, J. C. Hutchins, R. Thomson, D. E. Briggs, and R. P. Evershed, *Rapid Comm. Mass Spec., 11, 17*: 1884–1890 (1998).
45. A. M. Shedrinsky, T. P. Wampler, N. Indictor, and N. S. Baer, *J. Anal. Appl. Pyrol., 15*: 393–412 (1989).
46. F. Vilaplana, A. Ribes-Greus, and S. Karlsson, *Polym. Degrad. Stab., 95, 2*: 172–186 (2010).
47. J. A. Gonzalez-Perez, N. T. Jimenez-Morillo, J. M. de la Rosa, G. Almendros, and F. J. Gonzalez-Vila, *J. Chorm. A, 1388*: 236–243 (2015).
48. X. H. Li, Y. Z. Meng, Q. Zhu, and S. C. Tjong, *Polym. Degrad. Stab., 81, 1*: 157–165 (2003).
49. F. C. Wang, *J. Anal. Appl. Pyrol., 71*: 83–106 (2004).
50. K. Oba, Y. Ishida, H. Ohtani, and S. Tsuge, *Macromolecules, 33*, 22: 8173–8183 (2000).
51. J. J. Shen and E. Woo, *LC-GC. 6, 11*: 1020–1022 (1988).
52. D. Weber, *Int. Lab., 9*: 51–54 (1991).
53. S. Tsuge, Y. Sugimura, and T. Nagaya, *J. Anal. Appl. Pyrol., 1*: 222–229 (1980).
54. S. W. Kim, *Rapid Comm. Mass Spec., 13, 24*: 2518–2526 (1999).
55. E. Jakab, M. D. Uddin, T. Bhaskar, and Y. Sakata, *J. Anal. Appl. Pyrol., 68-69*: 83–99 (2003).
56. L. Bernard, D. Bourdeaux, B. Pereira, N. Azaroual, and C. Barthelemy, *Talanta, 16*: 604–611 (2017).
57. J. M. Challinor, *J. Anal. Appl. Pyrol., 61*: 3–34 (2001).
58. Z. Parsi, N. Hartog, T. Gorecki, and J. Poerschmann, *J. Anal. Appl. Pyrol., 79, 1-2*: 9–15 (2007).
59. R. P. Philp, *Org. Geochem., 6*: 489–501 (1984).
60. H. Bar, R. Ikan, and Z. Aizenshtat, *J. Anal. Appl. Pyrol., 10*: 153–162 (1986).
61. A. J. Barwise, A. L. Mann, G. Eglinton, A. P. Gowear, A. M. Wardroper, and C. S. Gutteridge, *Org. Geochem., 6*: 343—349 (1984).
62. A. K. Burnham, R. L. Braun, H. R. Gregg, A. M. Samoun, *Energy Fuels, 1*: 452–458 (1987).
63. J. J. Delpeuch, D. Nicole, D. Cagniat, P. Cleon, M. C. Foucheres, D. Dumay, J. P. Aune, and A. Genard, *Fuel Proc. Technol., 12*: 205–241 (1986).
64. A. M. Harper, H. L. C. Meuzelaar, and P. H. Given, *Fuel, 63*: 793–802 (1984).
65. J. M. Bracewell and G. W. Robertson, *Geoderma, 40*: 333–344 (1987).
66. H. Engman, H. T. Mayfield, T. Mar, and W. Bertsch, *J. Anal. Appl. Pyrol., 6*: 137–156 (1984).
67. C. Gutteridge, *Meth. Microbiol., 19*: 227–272 (1987).
68. T. A. Mime and M. N. Soltys, *J. Anal. Appl. Pyrol., 5*: 111–131 (1983).
69. A. Meng, H. Zhou, L. Qin, Y. Zhang, and Q. Li, *J. Anal. Appl. Pyrol., 104*: 28–37 (2013).
70. J. Cai, W. Wu, and R. Liu, *Renewable Sustainable Energy Rev., 36*: 236–246 (2014).

71. C. A. Smith, S. L. Morgan, C. D. Parks, A. Fox, and D. G. Pritchard, *Anal. Chem., 59*: 1410–1413 (1987).

72. A. C. Tas, J. DeWaart, and J. Van der Greef, *J. Anal. Appl. Pyrol., 11*: 329–340 (1987).

73. H.-R. Schulten, N. Simmleit, and H. H. Rump, *Int. J. Enriron. Anal. Chem., 27*: 241–260 (1986).

74. D. S. Garland, D. M. White, and C. R. Woolard, *J. Cold Regions Eng., 14, 1*: 1–12, (2000).

75. D. Fabbri, C. Trombini, and I. Vassura, *J. Chrom. Sci., 36, 12*: 600–604 (1998).

76. A. S. Fontaine, S. Bout, Y. Barriere and W. Vermerris, *J. Agric. Food Chem., 51, 27*: 8080–8087 (2003).

77. L. Wang, Y. Ishida, H. Ohtani, and S. Tsuge, *Anal. Chem., 71*: 1316–1322 (1999).

78. D. M. Wright, M. J. Bradley, and A. H. Mehltretter, *J. Forens, Sci., 58, 2*: 358–364 (2013).

79. M. Ezrin and G. Lavigne, *ANTEC 2002 Plastics: Ann. Tech Conf., 2*: 2046–2050 (2002).

80. J. Odermatt, D. Meier, K. Leicht, R. Meyer, and T. Runge, *J. Anal. Appl. Pyrol., 68–69*: 269–285 (2003).

81. J. del Rio, M. Hernando, P. Landin, A. Gutierrez, and J. Romero, *J. Anal. Appl. Pyrol., 68–69*: 251–268 (2003).

82. J. M. Weiss, A. J. McKay, C. DeRito, C. Watanabe, K. A. Thorn, and E. L. Madsen, *Environ. Sci. Technol., 38, 7*: 2167–2174 (2004).

83. S. Muguruma, S. Uchino, N. Oguri, and J. Kiji, *LC-GC Int., 12, 7*: 432–436 (1999).

2 Instrumentation and Analysis

Thomas P. Wampler
CDS Analytical, Inc., Oxford, Pennsylvania

I INTRODUCTION

To perform an analysis by pyrolytic techniques in the laboratory, it is necessary to assemble a system that is capable of heating small samples to pyrolysis temperatures in a reproducible way, interfaced with an instrument capable of analyzing the pyrolysis fragments produced. This chapter will discuss the various methods available for the convenient pyrolysis of laboratory samples and some general considerations regarding the interfacing of such units to analytical instruments. In addition, concerns about sample preparation, experimental reproducibility, and sources of error will be discussed.

A typical pyrolytic analysis involves sample preparation, pyrolysis, transfer of the pyrolysate to the analytical instrument, and then analysis. Pyrolysis-gas chromatography (GC) is still the most common technique, but pyrolysis-mass spectrometry (MS) and pyrolysis-Fourier-transform infrared (FT-IR) spectrometry are also common. In any system, the quality of the results will be no better than that permitted by any of its parts, and therefore, it is important not only to use a reliable pyrolysis technique, but also be aware of the effects of sample variations, mechanical and pneumatic connections, and instrument optimization. Although many analysts still design and use their own pyrolysis devices, variations in conditions and design frequently make it difficult to achieve reproducibility. The availability of a variety of pyrolysis instruments commercially has done much to improve the quality of pyrolysis experiments and reduce the frustrations common to early analysts.

Although there is a multitude of ways in which a sample could be heated sufficiently to break bonds, this chapter will treat only those ways that are readily available in the form of laboratory equipment. These instruments may be categorized as furnaces, both isothermal and programmable, inductively heated filaments, and resistively heated filaments. Interesting work has been done using lasers [1], solar radiation [2], electric arc, and other nonconventional heating units, but these experiments are frequently one-of-a-kind systems and are beyond the scope of this book.

II PYROLYSIS INSTRUMENTS

A GENERAL CONSIDERATIONS

In the typical analysis by pyrolysis, it is essential to heat a small sample to its final temperature as quickly as possible. Samples are generally small because of the sampling capacity of the analytical instrument. For example, most gas chromatographic columns and detectors cannot handle more than a few micrograms of sample. This works to the advantage of an analyst who is using pyrolysis as a sample introduction technique, since a small sample will heat to its endpoint temperature quicker—with less thermal gradient—than a large sample. It is important to heat the entire sample to the endpoint quickly, especially if one is studying the effects of pyrolysis temperature on the composition of the pyrolysate. Samples slowly heated will undergo considerable degradation while the instrument is heating to the final set-point temperature, and a large sample will degrade according to the temperature distribution across the sample as it is being heated. Consequently, reproducibility may depend heavily on the ability of the pyrolysis instrument to heat the sample uniformly, and to achieve the final temperature before the sample has already begun degradation. Pyrolysis instruments commercially available are capable of heating filaments to temperatures in excess of 1200 °C in milliseconds, producing rapid degradation of small, thin samples.

There are experiments and conditions, however, under which it is impossible or undesirable to pyrolyze the sample at such a fast rate. Large samples, used because they are non-homogeneous or low in organic content, may be analyzed by pyrolysis, but the effects of the slower heating rate and thermal gradient through a large sample must be considered. In addition, there are experiments and techniques in which slow heating is specifically the point (such as thermogravimetric analysis, or time-resolved spectroscopic analyses) and reproducible, slow-heating profiles or larger sample capacity (or both) are advantages rather than obstacles.

III ANALYTICAL PYROLYSIS AT RAPID RATES

The three most widespread techniques used to pyrolyze samples rapidly for analysis by GC, MS, or FT-IR are: isothermal furnaces, Curie-point (inductively heated) filaments, and resistively heated filaments. There are specific design advantages to each technique, depending on the sample to be analyzed and the physical requirements of the experiment conducted. Each way is capable of providing reproducible pyrolysis for small samples, and many laboratories use more than one type of instrument. The selection of one technique over another depends frequently on personal preference, experimental requirements, budget, or availability. Since each technique heats in its own unique fashion, it is important to keep the physical differences in mind when comparing pyrolysis results. Having chosen (or inherited) any specific instrument, it is important to understand its heating characteristics to capitalize upon its advantages and minimize the effects of its drawbacks.

A FURNACE PYROLYZERS

To pyrolyze samples rapidly for introduction to a gas chromatograph, furnace pyrolyzers are generally held isothermal at the desired pyrolysis temperature, and the samples introduced into the hot volume. Carrier gas is generally routed through the furnace to remove the pyrolysate quickly from the pyrolysis zone to minimize secondary pyrolysis.

1 Design

Isothermal furnace pyrolyzers are generally designed to be small enough to mount directly onto the inlet of a gas chromatograph. They are normally made of metal or quartz tube that is wrapped with a heating wire, then insulated (Figure 2.1). Carrier flow enters the top or front of the furnace, sweeping past a sample inlet or delivery system, then exits directly into the injection port of the chromatograph. They must contain enough mass to stabilize the temperature, which is generally held to within ±1 °C. A temperature sensor (thermocouple or resistance thermometer device) is usually installed between the heater and the furnace tube to indicate the wall temperature.

2 Sample Introduction

Care must be taken in introducing the sample for pyrolysis into the furnace without admitting air, since the pyrolysis zone is already hot and degradation begins immediately. The heating rate of the sample is dependent on the sample material itself, and on the composition of the sample introduction device. In the simplest configuration, a liquid sample is injected into the furnace using a standard syringe, the sample then vaporizes, and is followed by pyrolysis.

Solid samples present more of a problem, since they cannot be injected using a standard syringe. Some analysts dissolve soluble materials and inject the solution, pyrolyzing both the sample material and the solvent. Solid-injecting syringes have been designed that work on the principle of

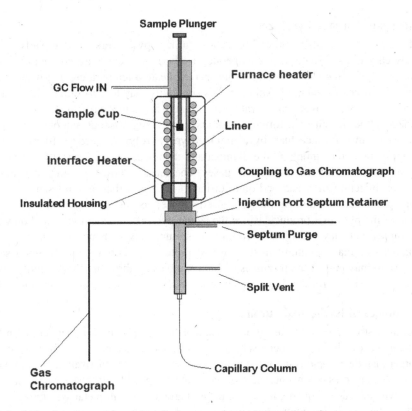

FIGURE 2.1   Microfurnace pyrolyzer installed on gas chromatograph injection port.

a needle inside a needle. The inside needle has a groove or slot into which a solid material may be placed. This needle then slides up into the outer needle for injection through the sample port of the furnace. Once inside the furnace, the inner needle is extended beyond the outer sheath, delivering the sample into the pyrolysis area.

Another approach has been developed by Shin Tsuge at the University of Nagoya in Japan [3]. His furnace pyrolyzer includes a cool chamber where samples are loaded into a small crucible above the hot zone. Once the sample is in place, the cup is rapidly lowered into the furnace for pyrolysis.

3   Temperature Control

The heating of a small, isothermal furnace is almost always achieved using a resistive electrical element wound around the central tube of the furnace. The temperature is monitored by a sensor that feeds back temperature data to the controller, where adjustments are made for deviations from set-point. It must be stressed that the temperature measured (and displayed) using this sensor is the temperature of the system at that specific location. Depending on the diameter, thickness, and mass of the furnace tube, the temperature experienced by the sample inside the tube may be quite different. It is possible to position a thermocouple inside the furnace to monitor temperature closer to the location of sample introduction. The temperature and rate of heating of the sample will also depend on the sample size and mass, and on the residence time inside the furnace. If the carrier gas is traveling too slowly through the furnace, the sample may degrade and the pyrolysate material may interact with the furnace wall, producing secondary pyrolysis products. In general, reasonable results may be duplicated on the same instrument using the same displayed wall temperature, assuming that the internal temperature caused by this wall temperature is also reproduced.

4   Advantages of Furnace Pyrolyzers

Because of their simple construction and operation, furnace pyrolyzers are frequently inexpensive and relatively easy to use. Since they are operated isothermally, there are no controls for heating ramp rate or pyrolysis time. The analyst simply sets the desired temperature and when the furnace is at equilibrium, inserts the sample. Although this simplicity may lose its attractiveness as soon as the analyst requires control over heating rate or time, there are some experiments and sample types that capitalize on the design of a furnace. Liquids—especially gaseous samples—are pyrolyzed much more easily in a furnace than by a filament-type pyrolyzer. Because filament pyrolyzers depend on applying a cold sample to the filament and then heating, it is very difficult to use them for any liquid that is readily volatile, or for gases. The furnace, however, since it is already hot, may receive an injection of a gas or liquid easier (or as easy) than a solid sample. Moreover, filament-type pyrolyzers are almost always housed or inserted into a heated zone, which prevents condensation of the pyrolysis products, since the actual pyrolysis filament is heated only for a few seconds. Samples that may be solids at room temperature may still melt or vaporize inside the heated chamber—or may denature, in the case of proteins. A furnace with a good sample introduction system may permit the pyrolysis of such samples by thrusting them rapidly into the hot zone of the pyrolyzer before they have a chance to undergo adverse lower-temperature changes.

5   Disadvantages of Isothermal Furnaces

Perhaps the greatest concern when using an isothermal furnace pyrolysis instrument is the size and construction of the pyrolysis chamber itself. To ensure thermal stability, the furnace tube is considerably larger than the sample being inserted into it. This produces a relatively large volume through which the sample must pass before entering the analytical device, with a large hot surface area. In some designs, this pyrolysis tube is quartz, in others, metal. Particularly with metal systems, the possibility exists that the initial pyrolysis will produce small organic fragments, which may then encounter the hot surface of the furnace tube and undergo secondary reactions. To counter this, furnaces are almost always operated with a high flow rate through the tube (e.g., 100 mL/min), generally necessitating split capillary analysis. This high flow rate reduces the residence time for the sample inside the hot zone, and the required high split ratio is generally not a problem unless one is sample limited, or unless the sample analyzed is low in organic content.

The nature of heating and sample introduction in a furnace precludes one from knowing the temperature rise time of the sample, which will depend on the nature of the sample, its size, and geometry. In addition, the mass that gives the furnace its thermal stability also gives it thermal inertia. If one is establishing the effect of pyrolysis temperature on the products, or optimizing the production of fragments of interest, one must allow equilibration time between temperature changes to establish stability at the new temperature.

B   HEATED FILAMENT PYROLYZERS

Isothermal furnaces achieve fast sample heating by keeping the pyrolysis instrument hot while injecting samples into it. Heated filament pyrolyzers take the opposite approach in that the sample is placed directly onto the cold heater, which is then rapidly heated to pyrolysis temperature. Commercially, two heating methods are used: resistance heating—in which a controlled current is passed through the heating filament—and inductive heating—in which the current is induced into the heating filament that is made of a ferromagnetic metal. Each type achieves very fast heating rates by using a small filament; consequently, the sample size must be limited to an amount compatible with the mass of the filament. This sample size is usually in the low to high microgram range, which is compatible with gas chromatograph column capacities, providing the sample is essentially all organic in nature. For larger samples or materials of low organic content, filaments may not be able to heat the sample efficiently, and a furnace style may be more appropriate.

C INDUCTIVELY HEATED FILAMENTS: THE CURIE-POINT TECHNIQUE

It is well known that electrical current may be induced into a wire made of ferromagnetic metal using a magnet, and this is the principle used to heat the filaments of a Curie-point pyrolysis system. If one continues to induce a current, the wire will begin to heat—and continue to do so—until it reaches a temperature at which it is no longer ferromagnetic. At this point, the metal becomes paramagnetic, and no further current may be induced in it. Consequently, the heating of the wire stops—the Curie-point of the specific metal alloy being used. This Curie-point temperature is different for each ferromagnetic element, and for different alloys of these metals. For example, a wire made entirely of iron will become paramagnetic and stop heating at a temperature of 770 °C, while an alloy of 40% nickel and 60% cobalt will reach 900 °C.

In a Curie-point pyrolyzer, an oscillating current is induced into the pyrolysis filament by a high-frequency coil. It is essential that this induction coil be powerful enough to permit heating the wire to its specific Curie-point temperature quickly. In such systems, the filament temperature is said to be self-limiting, since the final or pyrolysis temperature is selected by the composition of the wire itself, and not by some selection made in the electronics of the instrument. Properly powered, a Curie-point system can heat a filament to pyrolysis temperature in milliseconds. Provided that wires of the same alloy composition are used each time, the final temperature is well characterized and reproducible.

1 Design

Curie-point pyrolysis systems must be designed to permit easy insertion of the wire containing the sample into the high-frequency coil in a chamber that is swept with carrier gas to the analytical instrument. Two approaches may be taken: (1) the pyrolysis chamber, which is surrounded by the coil, is opened and the sample wire is dropped or placed inside; or (2) the sample wire is attached to a probe that is inserted through a septum into the chamber, which is surrounded by the coil (Figure 2.2). The coil chamber itself may be attached directly to the injection port of the gas chromatograph, or it may be part of a larger module that is isolated from the GC by a valve. In the

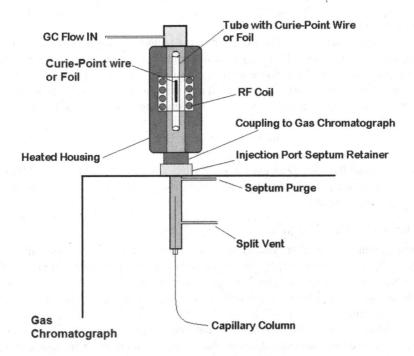

FIGURE 2.2 Curie-point pyrolyzer installed on gas chromatograph injection port.

latter design, the sample may be introduced and removed easily without interrupting the carrier gas flow of the gas chromatograph.

Because the Curie-point filament is heated inductively, no connections are made to the wire. This facilitates auto-sampling and permits loading the wires into glass tubes for sampling and insertion into the coil zone. Unlike the isothermal furnace, which is continuously on, the Curie-point wire is heated only briefly and is cold the rest of the time. This necessitates heating the pyrolysis chamber separately to prevent immediate condensation of the fragments made during pyrolysis. Therefore, Curie-point pyrolyzers have control for the parameters of the pyrolysis wire, and temperature selection for the interface chamber housing the wire.

2 Sample Introduction

To capitalize upon the very rapid heating rates possible using the Curie-point technique, the sample and wire should be kept to a low mass. The technique is best suited to the analysis of samples that may be coated onto the filament as a very thin layer. Soluble materials may be dissolved in an appropriate solvent, and the wire dipped into the solution. As the solvent dries, it leaves a thin deposit of the sample material, which will then heat rapidly and uniformly to pyrolysis temperature when the wire is heated. Paints, varnishes, and soluble polymers may be analyzed in this way quite easily.

Pyrolysis samples that are not soluble must be applied to the wires in some other fashion [4]. Finely ground samples may be deposited onto the wire from a suspension, which is then dried to leave a coating of particles on the wire. Another approach is to apply the sample as a melt, which then solidifies onto the wire. To attach materials that will not melt or form a suspension, some analysts flatten the wire, or create a trough in it. Some samples are held in place by bending or crimping the wire around the material, although this introduces concerns about the thermal continuity of a Curie-point material that has been distorted. To encapsulate a sample completely, some systems use a ferromagnetic foil [5] instead of a wire. The sample is placed in the center of a piece of metal foil, which is then wrapped around it and dropped into the high-frequency coil chamber, just as a standard wire would be.

3 Temperature Control

The control of pyrolysis temperature in a Curie-point instrument is achieved solely by the alloy of the ferromagnetic material used. Pyrolysis of a material at a variety of temperatures requires a selection of Curie-point wires of different alloy composition. These wires are generally purchased from the manufacturer of the instrument, and each manufacturer offers a range of alloys covering a wide temperature range. The analyst, therefore, has no access to temperature settings—repro-reproducible and accurate temperature control relies on the accuracy of the wire alloy, the power of the coil, and the placement of the wire into the system. One may be reasonably confident that the sample wire achieves the theoretical temperature if the instrument has enough power to induce current enough to heat the wire to its limit. Use of wires from one manufacture lot throughout an experiment will increase the likelihood of temperature reproducibility, as will the attention to sample loading and placement.

4 Advantages of Curie-Point Systems

The self-limiting temperature of an inductively heated wire and the rapidity with which it heats are major advantages of the Curie-point system. Using wires of the same manufacture and alloy, one may be confident that the sample is being heated to a specific and known temperature in a rapid and reproducible manner. Since there is no temperature control setting, there is no temperature calibration to perform.

Curie-point systems have some additional advantages from the standpoint of convenience. The sample wires may be coated with sample, then placed into a glass tube for storage so that one may prepare several samples at once and subsequently load them into the pyrolyzer. In addition, since the heating wire need not be connected to a source of power, insertion into the unit is simple and may be readily automated [6] by placing wires in glass tubes into a feeding magazine, or into a multi-position auto-sampler. Although the wires may be cleaned and reused, they may also be discarded after one use, eliminating concerns about carryover of sample from one analysis to another.

5   Disadvantages of Curie-Point Systems

Since the temperature of pyrolysis is a function of the Curie-point wire alloy composition, the wire, and consequently, the sample, may be heated to that temperature only. If it is desired to evaluate several different pyrolysis temperatures, or to study the behavior of a sample material at different temperatures, it is necessary to use a different Curie-point wire for each run. Therefore, it is not possible to optimize the pyrolysis temperature of a sample with a Curie-point system by placing the material into the instrument and increasing the temperature in a stepwise fashion, observing the pyrolysis products after each heating. The fact that the sample is in contact with a different metal at each different temperature produces concerns about the catalytic effect of the metal during heating for only very small samples, since in larger samples, energy is conducted through the sample material itself and pyrolysis takes place without contacting the metal surface. The analyst is, however, limited to the pyrolysis temperatures provided by the Curie-point wires or foils offered by the manufacturer. Curie-point materials are generally provided for a range of temperatures from 350–1000 °C, covered by 10–20 specific alloys. Investigations at temperatures between the Curie points of the materials offered are not possible.

Although most pulse pyrolysis analyses benefit from the very rapid temperature rise time experienced by a Curie-point filament, there are investigations when the analyst wishes to control the heating rate during pyrolysis. This has been attempted by modulating the power supplied to the coil, but for commercial instruments, slowing the heating rate is not possible. For experiments requiring the slow, linear heating of a sample material, a programmable furnace or resistively heated filament pyrolyzer, is required.

D   RESISTIVELY HEATED FILAMENTS

Like the Curie-point instruments, resistively heated filament pyrolyzers operate by taking a small sample from ambient to pyrolysis temperature in a very short time. However, the current supplied is connected directly to the filament, and not induced. This means that the filament need not be ferromagnetic, but that it must be physically connected to the temperature controller of the instrument. Filaments are generally made of materials of high electrical resistance and wide operating range like iron, platinum, and nichrome [7].

1   Design

As with Curie-point systems, the filament of a resistively heated pyrolyzer must be housed in a heated chamber that is interfaced to the analytical device. This interface chamber is generally connected directly to the injection port of a gas chromatograph, with column carrier gas flowing through it. The sample for pyrolysis is placed onto the pyrolysis filament, which is then inserted into the interface housing and sealed to ensure flow to the column (Figure 2.3). When current is supplied to the filament, it heats rapidly to pyrolysis temperatures, and the pyrolysate is quickly swept into the analytical instrument.

The pyrolysis filament may be shaped for convenience of sampling, and may be a flat strip, foil, wire, grooved strip, or coil. In the case of the coil, a small sample tube or boat is inserted into the filament so that the sample is not heated directly by the filament but is, in effect, inside a very small, rapidly heating furnace. The pyrolysis filament must be connected to a controller capable of supplying enough current to heat the filament rapidly, with some control or limit since the materials used for filaments are not self-limiting. The temperature of the filament may be monitored

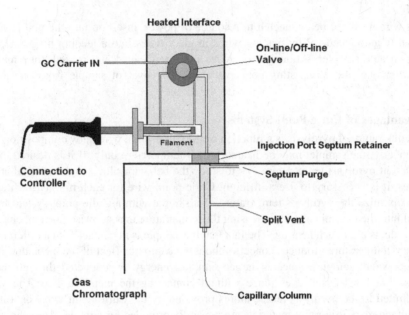

FIGURE 2.3   Resistively heated filament pyrolyzer installed on gas chromatograph injection port.

using the resistance of the material itself, or some external measure such as optical pyrometry [8] or a thermocouple [9].

The filament used can be quite small and is, therefore, capable of being inserted into instruments other than gas chromatograph—including the sample cell of infrared spectrometers and the ion source of a mass spectrometer [10].

2   Sample Introduction

Samples may be applied to resistively heated filaments in the same manner used for Curie-point wires. Soluble materials may be deposited from a solvent, which is then dried before pyrolysis. However, the solution is generally applied to the filament from a syringe, instead of dipping, since the filament is attached to a probe or housing. Insoluble materials may be melted in place to secure them before pyrolysis. Since the filament may be a flat ribbon or contain a grooved surface, placement of some solid materials may be simpler than when using a Curie-point wire.

Since the sample must be placed onto the filament before it is inserted into a heated chamber with carrier flow, placement of fibers and fine powders presents a problem. If these materials cannot be melted onto the filament, they may fall off or blow off in the carrier gas stream before pyrolysis. Consequently, these sample types are generally analyzed using a small quartz tube that is inserted into a coiled filament. The sample may be placed into the tube, held in position using plugs of quartz wool, weighed, and then inserted into the coiled element for pyrolysis. It must be remembered that the sample is now insulated by the wall of the quartz tube, so the temperature rise time and final temperature will not be the same as for a sample pyrolyzed directly on a filament. Use of a quartz tube, however, extends the use of a filament pyrolyzer to materials such as soils, ground rock samples, textiles, and small fragments of paint.

Viscous liquids such as heavy oils may be applied directly to the surface of a filament or may be pyrolyzed while suspended on the surface of a filler material, such as quartz wool, inside a quartz tube. Lighter liquids, especially anything easily vaporized, will probably be evaporated from the filament by the heat of the interface before pyrolysis, and are probably better studied using a furnace-type pyrolyzer.

3   Interfacing

Filament pyrolysis instruments may be designed small enough to insert directly into the analytical device, especially the injection port of a gas chromatograph or the ion source of a mass spectrometer. In these cases, one need only ensure that the filament is positioned so that the pyrolysate can enter the analytical portion of the instrument, and that the filament probe is sealed into the unit so that there are no leaks. For most gas chromatographs equipped with capillary inlets, however, considerable attention has been paid to the elimination of dead volume, and there is not enough room to accommodate a pyrolyzer. In these cases, some heated interface must be attached upstream of the injection port to house the pyrolyzer and ensure efficient transfer of the pyrolysate onto the column. This interface should have its own heater—independent of the pyrolysis temperature—to prevent condensation of pyrolysate compounds, and be of minimal volume. Flow from the gas chromatograph is brought up into the interface, past the filament, and then back into the injection port. In most cases, this means that the chromatograph is open to air when samples are being inserted, so the column oven is cooled before samples are introduced or removed. An alternative is to place a valve between the pyrolyzer and the injection port to isolate the chromatograph flow, permitting removal of the filament during a run [11].

4   Temperature Control

The temperature of a resistively heated filament is related to the current passing through it. Since resistively heated filaments are not self-limiting in the sense that Curie-point filaments are, exacting control of the filament current is essential for temperature accuracy and reproducibility. Early versions of resistively heated pyrolyzers relied on rather approximate control, and, given the difficulty of measuring and calibrating the final temperature, produced some mixed results. It must be remembered that the pyrolysis temperature of the filament depends upon the resistance of the filament, which is affected by both temperature and physical condition. A length of resistive wire could be expected to produce the same temperature each time it is heated using the same current, but relating that temperature to an instrument set-point—or calibrating the unit after replacing a broken filament—proved problematic.

Current versions of resistively heated pyrolyzers incorporate small computers to control and monitor the filament temperature. These computers may be used to control the voltage used, adjust changes in resistance as the filament heats, and compensate for differences when broken filaments are replaced. In addition, instruments have been designed to include photo-diodes [8], which are used by the computer to measure the actual temperature of the filament during a run. Other instruments include a small thermocouple welded directly to the filament for temperature readout, or use the computer to measure the resistance of the filament itself and make adjustments as needed during a program.

Since the temperature and the rate of heating the filament are completely variable, the instrument may control these parameters independently, as single or multiple steps. This gives the analyst the control to select any final pyrolysis temperature, and heat the sample to this temperature at any desired rate. Instruments are commercially available that heat as slowly as 0.01 °C per minute and as rapidly as 30,000 °C per second.

5   Advantages of Resistively Heated Filament Pyrolyzers

The central advantage of a resistively heated pyrolyzer is that the filament may be heated to any temperature over its usable range, at a variety of rates. This permits the examination of a sample material over a range of temperatures without the need to change filaments for each temperature. A sample may be placed onto the filament, heated to a set-point temperature and the products examined, then heated to higher temperatures in a stepwise fashion, without removing the sample or filament from the analytical device. This ability also allows pyrolysis at temperatures between the discrete values permitted by Curie-point filaments, and frequently at temperatures higher than those permitted by furnaces.

The ability to control the rate at which the filament—and the sample—is being heated extends the use of filament pyrolyzers in two ways. First, it permits the examination of materials and how they are affected by slow-heating, duplicating processes such as thermogravimetric analysis (TGA). Second, it permits the interface of spectroscopic techniques with constant scanning for three-dimensional, time-resolved thermal processing. A sample may be inserted directly into the ion source of a mass spectrometer, or placed in the light path of an FT-IR, and the products monitored real-time throughout the heating process [12].

6   Disadvantages of Resistively Heated Filament Pyrolyzers

The main disadvantage of a resistively heated pyrolyzer results from the fact that the filament must be physically connected to the controller. The temperature control of a resistively heated filament is based on the resistance of the entire filament loop, including the filament as well as its connecting wires. Anything that damages or alters the resistance of any part of the loop will influence the actual temperature produced by the controller.

An additional disadvantage may be produced by the fact that the filament must be housed in a heated zone. Introduction of some samples into a heated chamber before pyrolyzing them may produce volatilization or denaturation, altering the nature of the sample before it is degraded. However, instruments that permit insertion of the sample into a cold interface are available. The interface is then heated to a higher temperature to prevent condensation of the pyrolysate, and the filament is then automatically fired to the pyrolysis temperature [13].

IV   PYROLYSIS AT SLOW OR PROGRAMMED RATES

Although for most analytical pyrolysis techniques it is advisable to heat the sample to its final temperature as rapidly as possible, there are times when the opposite is required. To simulate thermal processes—such as TGA on a small scale—or to analyze the degradation products produced from a sample as they are being generated, slower, controlled heating is required. These slow-rate experiments demonstrate, for example, that poly(vinyl chloride) degrades in a two-step process, producing HCl at a relatively low temperature and aromatics at elevated temperatures.

While a sample material is slowly heated to its degradation temperature, the volatile products may be swept by a carrier gas into an analytical instrument or collected onto a trap for analysis as a composite. Alternatively, the sample may be pyrolyzed directly in the analytical instrument, while scans are being collected continuously. This produces a time-resolved picture of the production of specific products, as measured by the abundance of specific masses, or absorbance at specific wavelengths.

Slow pyrolysis is generally produced using either a programmable furnace or resistively heated filament pyrolyzer.

A   PROGRAMMABLE FURNACES

The design and interfacing of furnace pyrolyzers has been discussed. The same mass and thermal stability that make typical furnaces useful for isothermal applications may be used to provide reproducible heating profiles at slow rates for large samples. Several programmable furnace pyrolysis instruments are available commercially [14], with the most common application being the examination of fuel source samples [15]. These materials are generally ground rock from oil well core drillings or shale deposits, and most of the sample is inorganic and nonvolatile. Consequently, a rather large sample must be heated to extract the organics, generally performed at a rate of 50–100 °C/minute.

Programmable furnaces may be interfaced to a gas chromatograph, generally with some sort of intermediate trapping, or interfaced directly to spectroscopic techniques including mass spectrometry and FT-IR [16]. A furnace pyrolyzer [17] interfaced to a GC/MS may be adapted to perform slow-heating directly to the mass spectrometer by replacing the analytical column with a short piece of uncoated, narrow-bore fused silica. Although Curie-point pyrolyzers are generally not capable of providing slow heating rates, a hybrid Curie-point/furnace system [18] that can provide controlled heating rates is available.

B RESISTIVELY HEATED FILAMENTS AT SLOW RATES

Since the temperature and rate of heating of a resistive filament pyrolyzer are functions of the current, these instruments are frequently used to provide slow programs in addition to pulse-heating pyrolyses. The advantage of the resistive filament is that the sample may be placed directly onto the filament during heating, so there is no thermal gradient as with a furnace design. In practice, however, the sample is more likely placed into a quartz tube and heated using a coiled filament, creating a small quartz furnace inside of the resistive element.

As with the slowly heated furnace, a collection step is generally required when producing a sample for gas chromatography, since the pyrolysate may be produced over the course of several minutes. For some applications, collection may be done directly onto the gas chromatograph column, either at ambient or sub-ambient temperatures. For direct mass spectrometry or FT-IR spectrometry in a time-resolved fashion, the pyrolyzer is either inserted into an expansion chamber—which is flushed or leaked into the spectrometer—or the pyrolyzer is inserted directly into the instrument. Specially-designed probes permit operation at pyrolysis temperatures directly in the ion source of a mass spectrometer if it was originally equipped with a solids probe. For mass spectrometers without a sample probe inlet, such as a GC detector, the GC may be modified as with the furnace pyrolyzer using a short piece of fused silica to provide time-resolved pyrolysis-MS or EGA analysis. For FT-IR, a heated pyrolysis chamber may be flushed with carrier gas into a light pipe, or a cell used that positions the filament of the pyrolyzer directly below the light path for real-time pyrolytic analysis.

V OFF-LINE INTERFACING

For many experiments, including the slow rate pyrolyses discussed above, it may be advantageous or necessary to install the pyrolysis device away from the analytical instrument, with a collection or trapping system between. This may be accomplished in a variety of ways, both manually and automatically.

For manual isolation, the pyrolysis device may be flushed with a stream of carrier gas that is routed through a trap, either sorbent or thermal, using dry ice or liquid nitrogen. The pyrolysate is swept from the pyrolyzer and through the trap, where the carrier gas is vented, but the organics are frozen or adsorbed. The trap is then disconnected from the pyrolyzer and taken to the analytical instrument where the pyrolysate is either thermally revaporized or extracted with a solvent and injected. An early method of performing pyrolysis-IR was to allow the pyrolysate to condense onto a window, which was then inserted into the IR and scanned. These techniques also permitted the use of pyrolysis devices to heat samples in atmospheres incompatible with the analytical instrument, or at pressures or flows not permissible in the analyzer. For example, a sample may be heated in oxygen for combustion studies, with the reaction products collected, then transferred to a gas chromatograph [19]. Normal installation of a pyrolyzer would require the use of oxygen as the GC carrier, which is incompatible with column phases.

An analyst performing many off-line pyrolyses or reactant gas studies will probably select an automatic means of isolating the pyrolysis end of the experiment from the analytical. This is typically done by employing a valve in a heated oven to place the collection device alternately in line with the pyrolyzer, or with the analytical device. Commercial systems that incorporate this valve in a programmable controller to automate the process are available, as well as in manually operated modes. The pyrolysis chamber carrier gas is directed through the valve to the trap, or alternately to vent. During an experiment, the pyrolysis gas—inert or reactive, such as air—is routed through a sorbent or cold trap, where the pyrolysate is collected. During this time, the analytical device, generally a gas chromatograph, is swept with inert carrier as usual. When the valve is rotated, only the trap is inserted into the GC analytical flow, and the collected analytes are heated and transferred to the column for analysis. The pyrolyzer itself, and its flow, never touch the analytical instrument. This design permits heating the sample material slowly or for as long as desired, with transfer of the composite-collected pyrolysate for a single GC analysis. Systems have been designed as well with multiple traps for the stepwise collection of multiple fractions from the sample as it is being heated, with subsequent GC analysis of the material from each trap.

VI MULTI-STEP ANALYSES

Frequently, the samples analyzed in a laboratory by pyrolysis are complex materials that include a polymer matrix plus other volatile or semi-volatile organic compounds. Pyrolysis permits the analysis of the polymer, but in heating the sample, the other components may also be pyrolyzed, desorbed, or volatilized. These include plasticizers, antioxidants, dyes, fillers, and other intentional ingredients, as well as residual monomers, solvents and processing agents, or contaminants. It is frequently possible to liberate the volatile and semi-volatile compounds from the polymer thermally at temperatures too low to produce pyrolysis, so the matrix remains intact. This has led to the use of pyrolysis instruments to perform thermal "extractions" at lower temperatures, in addition to pyrolysis. The same sample may be heated sequentially to increasingly higher temperatures to deliver selected ranges of compounds for analysis. Sometimes two steps, a low temperature followed by pyrolysis is enough, but increasingly multiple step programs [20] are being employed to provide greater specificity. These techniques are attractive for several reasons, including the elimination of solvent extractions, and the ability to have the steps automated, starting the GC each time. They are especially useful in discriminating between monomer that is produced during pyrolysis from the residual monomer in the product, and in distinguishing products that may come from an additive instead of the matrix.

To be practical in these analyses, the pyrolyzer must have the ability to heat the sample material over a wide range of temperatures, and to operate at lower temperatures without pre-heating the sample or introducing cold-spots. Isothermal interfaces are usually a problem in that if they are hot enough to transfer all the pyrolysis products to the GC, they are probably too hot for the desorption steps, and volatiles will be lost while the sample is being inserted into the unit. What is generally needed is a programmable interface, or a separate heating zone for the desorption and pyrolysis steps.

Curie-point pyrolyzers are generally not used in this stepwise fashion, since they are limited to one temperature per sample because of the way heating is controlled. Microfurnaces, however, have been designed with a separate desorption zone [21] so that a sample may be manually lowered into a low temperature zone for a first run, retrieved, and then lowered into the pyrolysis zone for a second run. Filament pyrolyzers are now available with a low-mass, programmable interface zone along with computer controlled filament temperature. Samples may be placed into the cold interface that is programmed to heat rapidly for desorption, and then cooled between runs. Since the filament temperature is controlled by the computer, any number of subsequent runs at higher temperatures may be made automatically, heating the interface to an appropriate temperature before the filament is heated.

VII AUTO-SAMPLERS

In the interest of efficiency, auto-samplers have become increasingly important in analytical laboratories. Auto-sampling systems are now available for all three types of pyrolysis equipment – furnaces, Curie-point and resistively heated filament.

A FURNACE AUTOSAMPLERS

Most of the auto-sampling pyrolysis systems available use a sample magazine or carousel to introduce prepared samples into a common heater. In the furnace type [17], samples are placed into deactivated steel cups which are dropped sequentially into the furnace for pyrolysis, then ejected pneumatically after the analysis. Temperatures up to 800 °C are achievable, and the carousel has space for up to 48 sample cups.

B CURIE-POINT AUTO-SAMPLERS

One of the first approaches to pyrolysis auto-sampling involved the interface of multiple Curie-point coil interfaces to a common injection port [22]. This provided up to 12 analyses, each one with its own Curie-point wire, so the different samples could be pyrolyzed at different temperatures. This design has been replaced with the approach of dropping the individual samples into a common high frequency coil, which has increased the number of samples possible. In one design, the sample wires are held in glass tubes which are loaded from a carousel into the coil [23]. In another, the samples are folded into a foil of ferromagnetic metal and delivered magnetically from a stack of sample foils [18]. In either case, each sample is limited to one temperature, although different metals (and therefore, different temperatures) may be used for each sample. The stacked magazine style of sample introduction permits the analysis of up to 20 samples, while the tube carousel design provides for 24.

C RESISTIVE HEATING AUTOSAMPLERS

Two approaches to auto-sampling with resistively heated filaments are available, one in which multiple filaments are used and one in which a common filament is used. The former uses up to 14 different filament interfaces [24], each of which is connected sequentially to the gas chromatograph for pyrolysis. The computer permits individual control of each filament, so each sample may be analyzed using a different program. Like the Curie-point systems, the samples may be placed directly onto the heating filament.

In the other version [25], a common coil filament is used, into which samples placed into quartz tubes are dropped. The quartz tubes are delivered from a carousel, which can hold up to 47 samples. Different pyrolysis conditions may be programmed for each sample, and multiple runs may be performed on each tube. Between runs, the sample is taken off-line from the GC and purged to vent, then placed back on-line for the next analysis.

VIII SAMPLE HANDLING AND REPRODUCIBILITY

It is not enough to have a pyrolyzer for which the endpoint temperature and heating rate are well characterized to guarantee reproducible results. Sampling, sample handling, introduction, and transfer from the pyrolyzer into the analytical device must be performed with attention to all the inaccuracies that may be introduced. The effects of instrument design and interfacing have been discussed. The most common sources of error in sample manipulation will now be briefly described. The most important areas of concern are sample preparation, including size and shape, homogeneity, and contamination.

Using microsyringes, it is relatively easy to introduce a 1 μL or smaller sample into a gas chromatograph injection port. Preparing and inserting a solid sample that is only a few micrograms, on the other hand, presents some difficulties. This is particularly the case if the sample is an insoluble material of an inconsistent make-up throughout, such as plant material or layers of paint. Even analysts confident in their ability to prepare small slices of a sample are concerned whether that piece is representative of the whole.

A SAMPLE SIZE AND SHAPE

If a sample material is soluble, microliter-sized portions of the solution may be deposited onto the surface of the pyrolysis instrument using a syringe, which not only regulates the amount of sample material, but also causes the sample to be spread as a thin film that heats and pyrolyzes readily. One of the advantages of pyrolysis as an analytical technique, however, is its benefits in the analysis of difficult solids, which have limited solubility, such as polymers. In these cases, it is necessary to make samples of the same size and shape consistently for reproducible results. Since analytical pyrolyzers are designed to heat small samples rapidly, it is easy to overload the pyrolyzer with too much sample. This not only affects the rate at which the sample heats, related to the thermal gradient through the thickness of the material, but may also overload the analytical device, causing contamination and carryover into the next analysis. Generally, 10–50 μg of sample is desirable for direct pyrolysis-GC, and about twice as that for direct pyrolysis-FT-IR. A thin slice placed on its side is preferable to a cube or sphere and melting the sample into a film is sometimes helpful if the sample will cooperate. In any event, it is important to use as nearly the same size and shape sample each time to be sure that at a minimum, the sample material goes through the same heating process each time.

B HOMOGENEITY

Ensuring that a sample of solid material only a few micrograms large is homogeneous and, therefore, representative of the material from which it was taken presents a constant problem to analysts. The inability to obtain several samples of identical composition frequently casts suspicion on the technique of analytical pyrolysis, since the results obtained show poor reproducibility. This lack of reproducibility may be interpreted as the result of unexplained or unreliable reactions occurring in the sample during pyrolysis, when in fact it is an indication that the samples are different from each other.

Some samples have such large, obvious differences in composition from end to end that non-homogeneity is self-evident. Materials such as plant leaves, soils and rocks, scrapings, textiles, and laminated products will clearly provide different materials from different physical areas sampled. Since few analysts have the luxury of analyzing pure, homogeneous materials, these concerns must be addressed and the effects of non-homogeneous samples on the pyrolysis results established for each analysis.

Analysts sampling materials where homogeneity is a concern have devised several methods to deal with the problem. If possible, the sample material may be ground to a fine powder from which small portions are taken for analysis. There is sometimes a concern that the grinding process—or the heat produced during—may alter the sample so that the results are no longer representative. Many materials, however, have been ground successfully under cryogenic conditions, resulting in a powdered sample that has not been heated sufficiently to volatilize any of its constituents. Other analysts have chopped samples finely using a scalpel and then analyzed several of the small fragments together. Again, if the material or materials comprising the sample are soluble, dissolving them and working from a solvent provides an easy way to ensure homogeneity and sample size reproducibility.

Some analysts have elected to pyrolyze large samples when concerned that a small sample cannot be made to be representative. A sample of about 1 mg, if pyrolyzed using an instrument powerful enough to heat it effectively, may produce better reproducibility than a smaller sample which is less representative. In such cases, it is also essential to limit the amount of the pyrolysate entering the analytical instrument, generally by use of a splitter with a large split ratio, or by passing the pyrolysate in a carrier gas through a small sample loop attached to a valve that is interfaced to the analytical unit. It is also important to be mindful of the increased amounts of residue and char produced, requiring cleaning to prevent contamination from run to run.

Another way to deal with non-homogeneous samples is to analyze the individual constituents independently. Since analytical instruments are sensitive enough to respond to very small samples, it is sometimes possible to remove specific portions of an overall sample and investigate them separately. For example, specs of contamination in a polymer melt, discolorations, particles in papers or pulp, different strata in mineral samples, layers of paint, etc. may be studied independently as well as part of the whole sample material. It is important to remember that the individual constituents in a mixture will each pyrolyze in a way consistent with its molecular make-up, and that the entire sample will amount to the sum of its individual parts.

REFERENCES

1. S. Prati, D. Fuentes, G. Sciutto, and R. Mazzeo, *J. Anal. Appl. Pyrol.*, *105*: 327–334 (2014).
2. J. Zeaiter, F. Asisi, M. Lameh, D. Hamza, Y. Ismail, and A. Abbas, *Renewable Energy*, *123*: 44–51 (2018).
3. S. Tsuge and H. Matsubara, *J. Anal. Appl. Pyrol.*, *8*: 49–64 (1985).
4. A. Venema and J. Veurlink, *J. Anal. Appl. Pyrol.*, *7*, 3: 207–214 (1985).
5. N. Oguri and P. Kim, *Internat. Lab.*, *19*, 4: 59–62 (1989).
6. H.-R. Schulten, W. Fischer, and H. J. Walistab, *J. High Res. Chromatogr. Chromatogr. Commun.*, *10*: 467–469 (1987).
7. E. M. Anderson and I. Ericsson, *J. Anal. Appl. Pyrol.*, *3*: 13–34 (1981).
8. I. Ericsson, *Chromatographia, 6*: 353–358 (1973).
9. R. S. Lehrle, J. C. Robb, and J. E. Suggate, *Eur. Polym. J.*, *8*: 443–461 (1982).
10. J. B. Pausch, R. P. Lattimer, and H. L. C. Meuzelaar, *Rubber Chem. Technol.*, *56*, 5: 1031–1044 (1983).
11. T. P. Wampler and E. J. Levy, *Am. Biotechnol. Lab.*, *5*: 56–60 (1987).
12. J. W. Washall and T. P. Wampler, *Spectroscopy*, 6. 4: 38–42 (1990).
13. CDS Analytical, Oxford, PA, USA.
14. Wildcat Technologies, LLC, Humble, Texas.
15. B. Horsfield, *Geochim. Cosmochim. Acta*, *53*, 4: 891–901 (1989).
16. P. R. Solomon, M. A. Serb, R. M. Carangelo, R. Bassilakis, D. Gravel, M. Baillargeon, F. Baudais, and G. Vail, *Energy Fuels*, *4*, 3: 319–333 (1990).
17. Frontier Laboratories Ltd, Koriyama, Japan.
18. Japan Analytical Industry Co. Ltd. Tokyo, Japan.
19. T. P. Wampler and E. J. Levy, *J. Anal. Appl. Pyrol.*, *8*: 153–161 (1985).
20. K. Sam, *American Lab, 50*, 3 (April 2018).
21. S. Tsuge, H. Ohtani, C. Watanabe, and Y. Kawahara, *Am. Lab. 35*, 1: 32–37 (2003).
22. Fischer Labor-und Verfahrenstechnik, Meckenheim, Germany.
23. GSG Mess-und Analysengeräte, Bruchsal, Germany.
24. Pyrol AB, Lund, Sweden.
25. CDS Analytical, *Inc.*, Oxford, PA, USA.

3 Pyrolysis Mass Spectrometry: Instrumentation, Techniques, and Applications

C. J. Maddock[1], T. W. Ottley[1], and Thomas P. Wampler[2]
[1]Horizon Instruments, Ltd., Heathfield, East Sussex, England
[2]CDS Analytical, Inc., Oxford, Pennsylvania

I INTRODUCTION

A HISTORY

Pyrolysis-mass spectrometry has found many applications in microbiology, geochemistry, and soil and polymer sciences. However, its development has been historically hindered by the lack of competitively priced instrumentation that has restricted the number of laboratories able to use the technique. Furthermore, because of the nature and complexity of the data produced, a heavy reliance upon statistical techniques has been required for data analysis. This specific problem has benefited greatly from the development of powerful personal computers and the increased availability of suitable software, as well as the reduction in the price of mass spectrometers and the production of specific libraries for pyrolysis work.

Probably the first pyrolysis-mass spectrometer was described by Meuzelaar and Kistemaker [1]. This instrument was based on a Riber quadrupole mass spectrometer and used the Curie-point pyrolysis method first described by Giacobbo and Simon [2]. This development eventually produced two commercial instruments—the Extranuclear 5000 (Extranuclear Laboratories, Pittsburgh, PA), effectively a copy of the FOM machine, and the Pyromass 8–80 (VG Gas Analysis, Middlewich, England), based on the same principles but using a small magnetic mass spectrometer.

Both these systems were based on existing instrumentation used for conventional mass spectrometry. This made them prohibitively expensive and only a few were sold. A significant reduction in the cost of a pyrolysis-mass spectrometer could only be achieved through the development of a dedicated instrument. Subsequently, a truly dedicated and automated pyrolysis mass spectrometer, the PYMS-200X (Horizon Instruments, Sussex, England), was developed and found its principal use in microbiology. This instrument has been fully described elsewhere [3], together with typical applications [4–6]. Applications were limited because of the mass range offered by the instrument, 12–200 Da. Further development of the technique led to the production of the RAPyD-400 system, with a mass range of 12–400 Da, allowing more applications to be addressed.

Schulten [7] has coupled pyrolysis with field ionization mass spectrometry (Finnigan MAT 731) providing a soft ionization technique with the production of molecular ions after pyrolysis. This approach has been used to study complex natural materials such as coal and soil organic matter. Alternative techniques, adapting existing equipment, have also been employed, including replacing the solids probe of the mass spectrometer with a pyrolysis probe and modifying a GC/MS system to use the mass detector for pyrolysis-MS work.

B   INSTRUMENT DESIGN

1   Direct Insertion Probes

Many mass spectrometers—especially larger, research-oriented units (as opposed to GC detectors)—are equipped with a solids probe inlet port. This is a vacuum locking inlet that permits the insertion of a solid sample on a heatable probe directly into the ion source of the mass spectrometer. These probes are capable of heating a small sample of material to volatilize or desorb compounds, which are then ionized to produce a mass spectrum. Pyrolysis probes have been designed as direct replacements for these probes, but with a filament capable of reaching over 1000°C, so that the sample may be pyrolyzed directly into the ion source. A picture of such pyrolysis direct insertion probes is shown in Figure 3.1. The pyrolysis probes generally utilize a sample tube inserted into a resistively heated coil, so the heating rate may be fast or slow. With fast or pulse rates, a composite mass spectrum is produced, which may be quite complex, since it contains information about all of the compounds created during pyrolysis. When the sample is heated at slower rates, a thermal separation that can help isolate compounds and give evidence of which compounds are generated at a specific temperature are provided. This sort of experiment may be done at rates generally used for thermogravimetry, providing direct MS information regarding which compounds are evolved through a heating regime. Programmed heating also helps shed light on degradation mechanisms and in kinetic work, especially when the material degrades in several steps.

2   Atmospheric Pressure Interfaces

In addition to probe inlets operated at vacuum, interfaces that permit the analysis of solid samples at atmospheric pressure have been developed. These include the Direct Analysis in Real Time (DART [8]) interface and the Atmospheric Solids Analysis Probe (ASAP [9]) interface. These systems have been applied to the analysis of plant materials [10], polymers [11], drugs of abuse [12], and foodstuffs [13].

3   Modified GC/MS for Pyrolysis-MS

The use of a direct insertion probe for pyrolysis-MS requires that the mass spectrometer be equipped with an inlet for a solids probe. Most mass spectrometers used in analytical labs are configured as detectors for gas chromatographs, and are relatively simple and inexpensive, but are rarely equipped with a probe inlet. The only way for a sample compound to enter a mass spectrometric detector is via the capillary column inlet, configured to accept a piece of fused silica. Nevertheless, if the capillary column is removed and replaced with a piece of fused silica sufficiently restrictive to limit the flow into the mass spectrometer, pyrolysis-MS data may be

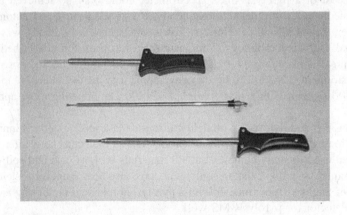

FIGURE 3.1   Direct insertion probes for mass spectrometers.

obtained. Sample loading is still controlled using the injection port splitter, so the sample is introduced into the mass spectrometer in a stream of helium at the same concentration that would be seen from a GC experiment. A 1-meter length of 0.1 mm fused silica is generally sufficient to transfer sample into the mass spectrometer at a flow rate capable of being handled by the vacuum pump. A diagram of such arrangement is shown in Figure 3.2.

The major drawback in trying to perform Py-MS using a modified GC system lies in the fact that the analytes must travel through the fused silica before they are ionized, so there is a short delay between the production of the compounds via pyrolysis and their ionization. Direct insertion systems, on the other hand, produce the pyrolysate right in the ion source. The consequence of this is that only stable molecules are seen, and early, reactive products may be missed. The same is the case for doing Py-GC/MS; so much useful information may still be obtained. Figure 3.3 shows a typical result obtained by pyrolyzing a piece of polystyrene directly to the MSD of an Agilent GC-MS. The run takes less than 0.5 minutes, and the inset shows the mass spectrum taken at the apex of the peak. If the heating rate is slowed, a time (or temperature) resolved analysis is produced. In Figure 3.4, a piece of polyurethane has been heated at 100°C/minute. A two-step evolution is evident, where the first peak consists mostly of toluene diisocyanate, while the second peak represents oxygenated compounds resulting from the degradation of the polyol portion of the polyurethane.

4 PyMS Instruments

Pyrolysis-mass spectrometer systems so far produced have contained common features of loading system, pyrolysis region, expansion chamber, and molecular beam before reaching the mass spectrometer ion source. The FOM Autopyms, the Pyromass 8–80, the Extranuclear 5000, and the PYMS-200X have all been detailed elsewhere [1,3,14]. A description of the RAPyD-400 will show how such dedicated instruments are configured.

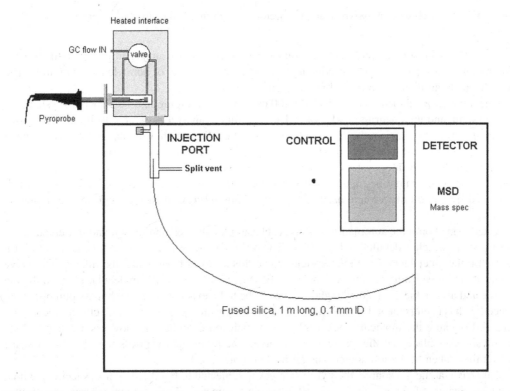

FIGURE 3.2 Modified GC for Pyrolysis/MSD.

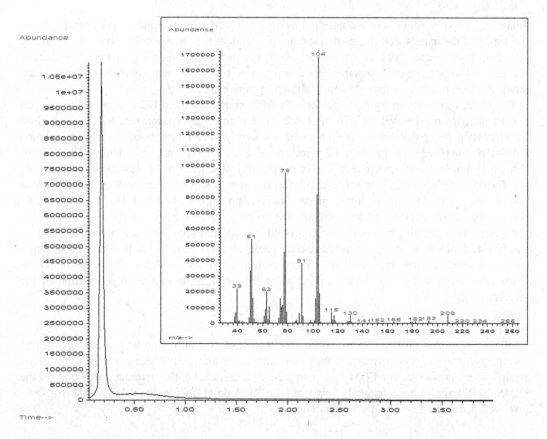

FIGURE 3.3   Pyrolysis of polystyrene at GC injector, through fused silica to mass detector.

The RAPyD-400 is a bench-top, quadrupole-based mass spectrometer utilizing the method of Curie-point heating for sample introduction. The system is capable of being loaded with up to 150 samples that can then be processed unattended.

The three basic elements of the RAPyD-400 can be seen in Figure 3.5—the vacuum system, the inlet system, and the quadrupole analyzer. The quadrupole analyzer—which itself consists of two separate parts—lies inside the main vacuum chamber. The high vacuum is attained using a turbomolecular pump that is backed by a dual-stage rotary pump mounted externally to the main system. The sample inlet system is connected to the ion source of the mass spectrometer via a heated molecular beam tube. Around the underside of the ion source is a copper cold finger, which is cooled by liquid nitrogen and used as a sample dump to prevent carryover from one sample to another.

Once loaded onto the automation system, each sample tube is taken in turn and presented to the inlet system, which is detailed in Figure 3.6. The o-ring on the open end of the tube is used to form a seal on the outer face of the inlet system. Once this seal has been made, the inlet system valve plunger can move from position A to position B, where it closes off the molecular beam to the ion source and allows the air inside the glass sample tube to be evacuated via the bypass pumping line. Once this has been achieved, the plunger moves from position B to position C, closing the bypass line and opening the molecular beam tube. An expansion chamber has now been created. Once pyrolysis takes place and the resultant gas evolves, the pyrolyzate expands into the vacant space and is then taken to the ion source via the heated molecular beam.

On arrival at the ion source, the pyrolyzate gas is subjected to bombardment by electrons with a nominal energy of 25 eV in classic EI mass spectrometry. The resultant charged molecules

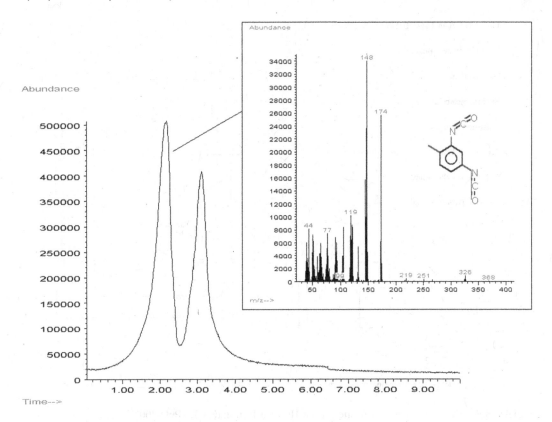

FIGURE 3.4 Polyurethane heated at 100°C/min.

produced are transmitted by the quadrupole and a mass spectrum of the pyrolysate is recorded by an electron multiplier operating in pulse-counting mode.

The pyrolysis-field ionization mass spectrometry system used by Schulten [15] produces pyrolysis directly in the ion source. A high electric field strength between the emitter and the cathode, typically 14 kV, produces soft ionization and creates the molecular ions with little secondary fragmentation. Heating rates of about 1°C/second permit the generation of thermograms through the heating range, in the same way that the collection of mass spectra from a detector produce a total ion chromatogram in CC/MS.

As in the example using the oven of the gas chromatograph as a heated transfer zone from the pyrolyzer to the mass spectrometer, any mass spectrometer that is compatible with a heated inlet could be used to make a direct pyrolysis-mass spec system. Voorhees [16] has taken this approach in creating a micro-fabricated pyrolyzer interfaced to a quadrupole ion trap mass spectrometer. Time of flight mass spectrometers (TOF MS), for example, are also easily adapted to a "direct" pyrolysis MS interfacing. Figure 3.7 shows the screen of a TOF analysis (Comstock MiniTOF, Oak Ridge, TN) of a sample of a styrene/butadiene copolymer. The inlet to the mass spectrometer was a piece of 0.1 mm fused silica, which was connected to a filament pyrolyzer with a heated interface. The lower window shows the composite peak produced when the copolymer was heated to 750°C for a few seconds. The upper windows show the averaged spectra for the whole experiment on the left and a selected spectrum on the right. The copolymer was 85% styrene, and the mass spectrum contains a large peak at mass 104 for styrene (as well as the other masses produced from styrene) and a peak at 39, a prominent mass in the spectrum of butadiene. Figure 3.8 shows a comparison of the spectra from two different styrene/butadiene copolymers, 28% styrene on the top and 85% styrene on the bottom, produced with this system. The ability to produce clearly

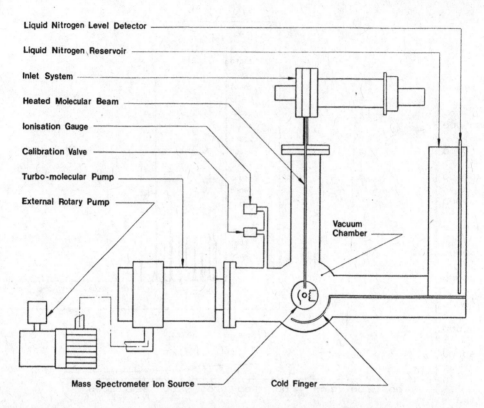

FIGURE 3.5   Vacuum system schematic for the Horizon Instruments RAPyD-400.

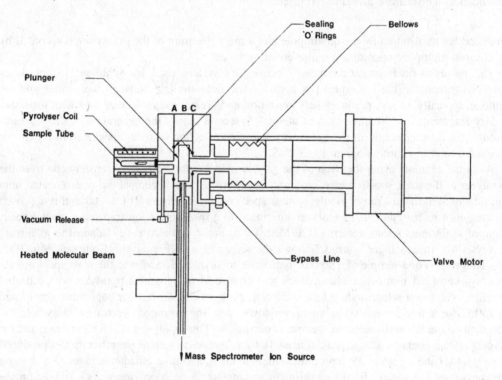

FIGURE 3.6   Inlet system for the Horizon Instruments RAPyD-400.

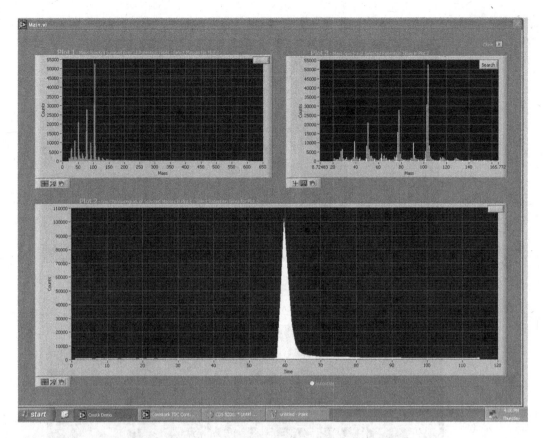

FIGURE 3.7 PyTOF analysis of styrene/butadiene copolymer at 750°C.

different spectra even from copolymers using the same monomers permits the creation of spectral libraries for the pyrolysates of various polymers. These libraries may be edited and searched using the same software used for searching the spectra of individual peaks in a chromatogram. The differences between the spectra generated from similar sample materials are the basis for many of the PyMS applications, some of which may necessitate more sophisticated computer analysis to derive the distinctions and associations required.

II DATA ANALYSIS

A MULTIVARIATE STATISTICS

Chemometrics is the application of numerical techniques to the identification of, or discrimination between, chemical substances. Expressed more simply, it is the attempt to make relatively simple quantitative measurements and use the data obtained in various ways to give the maximum amount of information regarding the differences between samples.

In truth, most analytical methods make use of the concept of chemometrics, at least to some extent, but some techniques are much more suitable than others. Obviously, all forms of spectroscopy can yield a large amount of numerical data, but mass spectroscopy is ideally suited for two reasons.

In a mass spectrum, ions are recorded over a certain mass range, but within that range (at low-to-medium resolution), only integer mass values are possible, the separation between peaks corresponding approximately to the mass of a neutron.

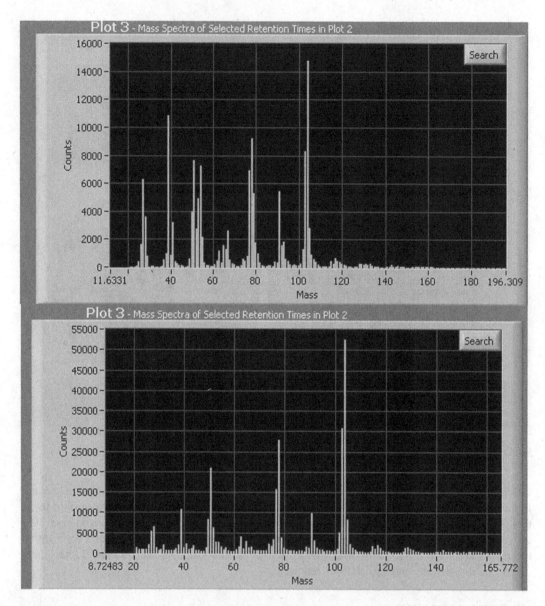

FIGURE 3.8   PyTOF spectra of two styrene/butadiene copolymers (top, 28% styrene, bottom, 85% styrene).

Second, it is possible to use single ion counting methods to record an actual numerical result (the number of ions) directly at each mass position. Thus, a two-dimensional data matrix is the immediate result.

Multivariate statistical methods, as part of the general chemometrics discipline, are particularly well-suited for the analysis of complex mass spectra. An excellent introduction to these techniques has been written by Manly [17].

One of the main reasons for using multivariate statistics is to reduce a large amount of very complex data down to a form where it can be readily understood. This usually means that any graphical output needs to be in two or three dimensions only. In effect, this means using principal components, canonical variates, and clustering analyses.

Principal components analysis is a well-established multi-variate statistical technique that can be used to identify correlations within large data sets and to reduce the number of dimensions required to display the variation within the data. A new set of axes and principal components (PCs) are constructed, each of which accounts for the maximum variation not accounted for by previous principal components. Thus, a plot of the first two PCs displays the best two-dimensional representation of the total variance within the data. With pyrolysis mass spectra, principal components analysis is used essentially as a data reduction technique prior to performing canonical variates analysis, although information obtained from principal components plots can be used to identify atypical samples and/or outliers within the data and as a test for reproducibility.

Canonical variates (CVs) analysis is the multivariate statistical technique that takes into account sample replication or any other a priori structure within the data and attempts to discriminate between the sample groups. CVs are derived in a similar manner to PCs except that their axes are constructed to maximize the ratio of between-group to within-group variance. Thus, a plot of the first two CVs displays the best two-dimensional representation of the sample group discrimination. Examination of the associated mass loadings/weightings allows chemical interpretation. Large mass loadings reflect masses significant to the discrimination in a particular direction. The application of graphical rotation facilitates the construction of factor spectra showing the contribution of masses to the discrimination of a specific sample group. Factor spectra are often representative of pure compounds or classes of components aiding chemical interpretation of the observed discrimination.

The use of hierarchical cluster analysis leading to the construction of dendrograms—and minimum spanning trees for data sets where the samples analyzed have a high degree of relatedness—is often the most useful way to display the total discrimination within the data because all the variance is displayed in two dimensions.

Although there are other multivariate techniques that can be applied to mass spectral data, the ones discussed here tend to be the most widely applied. In reality, there are more checks on validity that are made at different stages. This is to avoid any lack of confidence in the results obtained.

B Artificial Neural Networks

A related approach to multivariate statistics is the use of artificial neural networks (ANN), which are, by now, a well-known means of uncovering complex, nonlinear relationships in multivariate data. ANNs can be considered as collections of very simple computational units that can take a numerical input and transform it (usually via a weighted summation) into an output [18–20]. The relevant principle of supervised learning in ANNs is that they take numerical inputs (the training data) and transform them into desired (known, predetermined) outputs. The input and output nodes may be connected to the external world and to other nodes within the network. The way in which each node transforms its input depends on the so-called connection weights (or connection strengths), which are modifiable. The output of each node to another node or to the external world depends on both its weight-strength and on the weighted sum of all its inputs, which are then transformed by a (normally nonlinear) weighting function referred to as its activation function. For the present purposes, the great power of neural networks stems from the fact that it is possible to train them. Training is affected by continually presenting the networks with the known inputs and outputs and modifying the connection weights between the individual nodes—typically according to a back-propagation algorithm—until the output nodes of the network match the desired outputs to a stated degree of accuracy. The network, the effectiveness of whose training is usually determined in terms of the root mean square (RMS) error between the actual and the desired outputs averaged over the training set, may then be exposed to unknown inputs and will immediately output the best fit to the outputs. If the outputs from the previously unknown inputs are accurate, the trained ANN is said to have generalized.

The reason this method is so attractive to pyrolysis-mass spectroscopy (Py-MS) data is that it has been shown mathematically [21] that a neural network consisting of only one hidden layer, with an arbitrary large number of nodes, can learn any arbitrary (and, hence, nonlinear) mapping to an arbitrary degree of accuracy. ANNs are also considered to be robust to noisy data, like those that may be generated by Py-MS.

A neural network usually consists of three layers—representing inputs, output, and a hidden layer—that are used to make the connections (Figure 3.9). By training a neural network with known data, it is possible to obtain outputs that can accurately predict such things as polymer concentration mix.

ANNs can be trained using pyrolysis-mass spectra as the inputs and the known concentrations of target analytes as the outputs. For each input (one mass spectrum), there should normally be one output. The trained network can then be tested with the pyrolysis-mass spectra of "unknowns" to accurately predict the concentration or authenticity of the unknowns.

III APPLICATIONS

A MICROBIAL CHARACTERIZATION

The rapid accurate discrimination and characterization of microorganisms is the goal of many diagnosticians. Py-MS has proved to be a rapid and inexpensive epidemiological typing technique, applicable to a wide range of bacterial pathogens, and it can be used to identify the source of an infectious outbreak as well as give an assessment of the relation of bacterial isolates [5,22,23]. Work has been done on whole cells [24] and in the detection of bacillus spores [16].

B ORGANIC GEOCHEMISTRY

The characterization of sedimentary organic matter in terms of type and maturity is a prerequisite in the determination of the petroleum-generating potential of sediments. The nonextractable material—kerogen—consisting of Complex high-molecular-weight fragments, when subjected to Py-MS and multivariate analysis, gives excellent discrimination into the three universally recognized types [25]. Coals [26,27] have been studied extensively from the standpoint of product formation and structure. The organic material in soils has also been investigated by pyrolysis-MS, including contributions from plants to soil carbon [28], humic acids [15] and fatty acids [29].

C HIGH-VALUE PRODUCTS

Authentication and the detection of adulteration are serious problems within the citrus juice industry. Traditional multicomponent analysis methods are limited by the time required to perform the individual analyses and to construct the database required. Py-MS rapidly provides fingerprints of the original juice, which facilitates the use of multivariate pattern recognition procedures to detect potentially adultered samples and confirm authentication, as well as help in quality control. More recently, the combination of Py-MS and ANNs has been applied to problems of adulteration within the olive oil industry [30], shellfish [31], and milk [32].

D FORENSICS

In applications for the characterization of car, house, and fine art paints, Py-MS is valuable in obtaining quick and reproducible fingerprints that can subsequently be matched for identification purposes, whether it be by the forensic scientist or the polymer chemist [33].

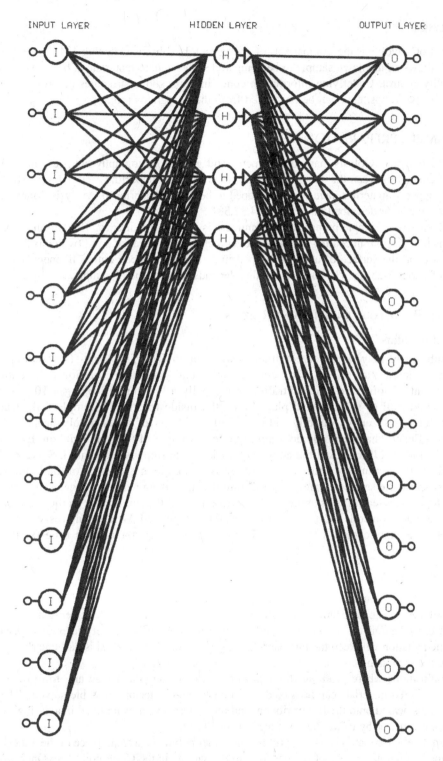

FIGURE 3.9 Typical artificial neural network structure.

E POLYMERS

Identification involving the comparison of pyrograms by library matching [34] or multivariate analysis allows an unknown sample to be compared with standard materials in both basic research and quality control. Underlying information concerning backbone structures based on monomer, dimmer, trimer ratios, etc., can be elucidated from the pyrograms [35,36].

IV SAMPLE PREPARATION

The techniques of pyrolysis-mass spectrometry and gas chromatography both rely upon a good physical contact between the sample to be analyzed and the supporting material. Over the years, various shapes of materials have been developed to hold samples prior to pyrolysis. Many of these methods have been described elsewhere [14,37,38].

In all aspects of sample preparation, cleanliness is of vital importance. The aim in preparing a sample is to obtain a thin coating of material over the inside surfaces of a foil (which is formed into a "V"), around the surface of a wire, crimped in a ribbon or placed onto a filament, with a dry weight of about 5-25 µg. The method for this depends on the type of sample.

A CURIE-POINT FOILS AND FILAMENT RIBBONS

1 Soluble Solids

Many substances may be dissolved either in water or an organic solvent. The technique is to dissolve sufficient material so that a conveniently small volume of solution will contain the right amount of solid. This concentration is normally a few mg/ml. Using a 10 µl syringe, extract about 3 µl of solution and place along the inside fold of the sample foil, if using a Curie-point pyrolyzer, or directly onto the foil or ribbon of a resistively heated filament pyrolyzer. Ensure that the liquid does not run over the edges of the foil or ribbon. Even with 5 µl, this is not too difficult, but the addition may be made stepwise if desired. Solvent may be dried by gently heating the Curie-point foil, or by using the dry function of the filament unit. Once dry, push the sample Curie-point foil into the tube using a depth gauge. The same depth gauge should be used for all samples. By ensuring that all the samples lie in the same position within the induction coil, another potential variable is removed. When using resistively heated filaments, the sample should be applied to the center of the ribbon each time for the same reason.

2 Insoluble Solids

The method here is similar to that preceding, except that the material must first be ground down to a fine powder, using for example a pestle and mortar, and then kept in suspension while being pipetted onto the foil or ribbon. A suitable suspension medium is either water or acetone, and ultrasonic agitation will keep the fine particles in suspension. This method is particularly attractive for samples such as glycogen.

In applications where profiling (fingerprinting) is more important than any form of quantification, powdered materials can be mixed into a paste with water or a suitable solvent. The paste can then be smeared onto the foil or ribbon, and the water/solvent evaporated in a drying oven for Curie-point foils, or by using the dry function for filaments.

Alternatively, for materials such as rubber and other polymers, a small piece of the material may be crimped with a heavy pair of forceps to hold the sample in the Curie-point foil. Once released, foils can be inserted into glass sample tubes with the same pair of forceps and positioned using a depth gauge. For resistively heated filaments, the sample may be placed onto the ribbon that is quickly heated to melt or fix the sample to the surface of the filament.

For samples that are difficult to grind or suspend and are not soluble—paint chips, fibers, minerals and so on—an alternative is to place the sample into a quartz tube that is surrounded by either a coil of ferromagnetic material for Curie-point units or resistive wire for filaments.

3 Bacteria and Yeasts

In the case of actively growing colonies on plates, all that is necessary is to pick up a small portion with a disposable loop or flamed wire and smear on both inside faces of the Curie-point foil or on top of the filament ribbon. Avoid picking up any of the substrate with the culture. With a little practice, it is possible to obtain the correct amount of material (20 µg) without difficulty. For some bacteria, particularly those in the Actinomycetes family, it is preferential to grow them on a filter placed on top of the medium. This will also prevent the media from being sampled with the colony.

In the preparation of the glass tubes, quartz tubes, and foils prior to sample loading, it is important that none are touched by hand. The method described by Sanglier et al. [5] has proved to work well for foils used inside glass tubes. In this method, pyrolysis foils and tubes are washed in acetone, then dried overnight at 27°C. A single foil is inserted by flamed forceps into each pyrolysis tube to protrude about 6 mm from the mouth. The tubes may then be stored at 80°C in a clean, dry oven or vacuum desiccator until needed. For each strain, small amounts of biomass (25 µg) are scraped from three different areas of the inoculated plate and smeared onto the protruding foil. The assembled foils are placed in an 80°C oven for 15 minutes to dry the biomass.

Liquid cultures can be handled in a way similar to suspended solids in that they can be pipetted onto foils or ribbons. Cultures presented on slopes can be treated as if they had been plated out.

4 Blood or Blood Cultures

These may be treated in the same way as soluble compounds but the volume of liquid applied to the foil should be much less (1-2 µl). Also, the samples should not be dried in warm air, instead they should be left in a vacuum desiccator for a short time.

5 Liquids

Many liquids requiring analysis will, in fact, be solutions, but often too dilute to use directly. In the case of whiskey, for example, the concentration must first be increased by a factor of ten before a sample can be taken. This can be done by blowing a jet of dry nitrogen over the surface of about 10 ml of liquid contained in a beaker, and evaporating down to 1 ml. This takes about 30 minutes. 3 µl of the concentrate may then be applied to the foil or ribbon and allowed to dry. Alternatively, the whole 10 ml aliquot can be evaporated to dryness and then reconstituted with 1 ml of the original sample. Either of these two methods should be used in preference to dispensing ten individual 3 µl aliquots on top of each other once the previous aliquot has dried.

In liquid samples such as fruit juice, olive oil, and milk, the concentration of dissolved solids is sufficient that no preconcentration is required.

B Curie-Point Ribbons

For some solid samples. it may be more convenient to use a Curie-point ribbon sample carrier. These ribbons take the place of wires in instruments specifically designed to take them.

Taking a ribbon with the folding lines facing upwards and a flat-ended pair of pliers, fold up the two sides until they are vertical. At this stage, it may be advisable to clean the ribbons before further use.

Place a sample of material in the ribbon and then fold down both sides to cover the sample, making a tent. Now fold the long end of the ribbon completely over the top of the tent and squeeze gently.

Most of these ribbons are made from pure iron so they need to be protected from water vapor. As supplied, they are normally packed under argon, which should prevent oxidation. After initial folding and cleaning, it may be preferable to store them in a pure solvent—hexane is recommended for this purpose.

C   CURIE-POINT WIRES

For solutions, suspensions and sticky materials (bacteria, resins, etc.), it is quite acceptable to use a straight ferromagnetic wire as the sample holder. For solid samples, a wire is prepared to firmly hold the sample in place.

Coating of a wire evenly with a solution or suspension is most desirable to enhance reproducibility. To this end, the use of wire coating units are of great help [1]. This can be achieved by rotating the wires slowly about their axes while applying the liquid medium. As the solvent evaporates, a uniform coating is deposited, which aids reproducibility. Up to six wires can be prepared at a time. Wires are held 10mm by the end with tweezers or pliers, and inserted into one of the holes in the guide plate at the front of the wire coater. The wire is then pushed into a small socket that will hold it firmly. This part of the wire can be cut off before analysis, so slight contamination is not important at this stage. With the rotation speed set to a mid value, solutions of approximately 3 µl are applied using a small syringe or pipette, one drop at a time, to the same position on each wire. After a short time, the coating will form uniformly as the solvent evaporates. If necessary, rotation speed can be varied to improve the uniformity of coating.

REFERENCES

1. H. L. C. Meuzelaar and P. G. Kistemaker, Techniques for fast and reproducible fingerprinting by pyrolysis mass spectrometry, *Anal. Chem.*, *45*: 587 (1973).
2. H. Giacobbo and W. Simon, *Pharm. Acta Helv.*, *39*: 162 (1964).
3. R. E. Aries, C. S. Gutteridge, and T. W. Ottley, Evaluation of a low cost, automated pyrolysis-mass spectrometer, *J. Anal. Appl. Pyrol.*, *9*: 81–98 (1986).
4. K. Orr, F. K. Gould, P. R. Sisson, N. F. Lightfoot, R. Freeman, and D. Burdess, Rapid interstrain Comparison by pyrolysis mass spectrometry in nosocomial infection with *Xanthomonus malthophilia*, *J. Hosp. Infect.*, *17*: 187–195 (1991).
5. J. J. Sanglier, D. Whitehead, G. S. Saddler, E. V. Ferguson, and M. Goodfellow, Pyrolysis mass Spectrometry as a method for the classification and selection of actinomycetes, *Gene*, *115*: 235–242 (1992).
6. R. Goodacre, D. B. Kell, and G. Bianchi, Olive oil quality control by using pyrolysis mass spectrometry and artificial neural networks, *Olivae*, *47* (1993).
7. H.-R. Schulten, *Int. J. Mass Spectrom. Ion Phys.*, *32:* 97 (1979).
8. IonSense, Saugus, MA, USA.
9. M&M Mass Spec Consulting, Harbeson, DE, USA.
10. R. B. Cody, A. J. Dane, B. Dawson-Andoh, E. O. Adedipe, and K. Nkansah, Rapid classification of White Oak and Red Oak by using pyrolysis direct analysis in real time (DART) and time-of-flight mass spectrometry, *J. Anal. Appl. Pyrol.*, *95*: 134–137 (2012).
11. S. Trimpin, K. Wijerathne, and C. N. McEwen, Rapid methods of polymer and polymer additives identification: Multi-sample solvent free MALDI, pyrolysis at atmospheric pressure and atmospheric solids analysis probe mass spectrometry, *Anal. Chim. Acta,* *654*, 1: 20–25 (2009).
12. R. Lian, Z. Wu, X. Lu, Y. Rao, H. Li, J. Li, R. Wang, C. Ni, and Y. Zhang, Rapid screening of abused drugs by direct analysis in real time (DART) coupled to time-of-flight mass spectrometry (TOF-MS) combined with ion mobility spectrometry (IMS), *Forensic Sci. Int.*, *279*: 268–280 (2017).
13. J. Hajslova, T. Cajka, and L. Vaclavik, Challenging applications offered by direct analysis in real time (DART) in food-quality and safety analysis, *Trends Anal. Chem.*, *30*, 2: 204–218 (2011).
14. W. J. Irwin, *Analytical Pyrolysis: A Comprehensive Guide*. Marcel Dekker, New York (1982).
15. H.-R. Schulten, in *Humic and Fulvic Acids: Isolation, Structure and Environmental Role* (J. Gaffney, N. Marley and S. Clark, eds.), ACS Symposium Series, 651 (1996).
16. C. D. Havey, F. Basile, C. Mowry, and K. J. Voorhees, *J. Anal. Appl. Pyrol.*, *72*: 55–62 (2004).
17. B. J. F. Manly, *Multi-variate Statistical Methods*, Chapman and Hall, London (1986).
18. P. D. Waiserman, *Neural Computing: Theory and Practice*, Von Nostrand-Reinhold, New York (1989).
19. D. E. Rumelhart, J. L. McClelland, and the PDP Research Group (Eds.), Parallel distribution processing, Experiments in the microstructure of cognitation, Vol. 2. Cambridge, MA: MIT Press (1986).
20. J. Hertz, A. Krogh, and R. G. Palmer, *Introduction to the Theory of Neural Computation*, Addison-Wesley, CA (1991).

21. K. Hosnik, M. Stinchcombe, and M. White, *Neural Net.*, *3*: 551–560(1990).
22. L. A. Shute, C. S. Gutteridge, J. R. Norris, and R. C. W. Berkeley, Curie-point pyrolysis mass spectrometry applied to characterization and dentification of selected *Bacillus* species, *J. Med. Microbiol.*, *130*: 343–355 (1984).
23. R. Freeman, F. K. Gould, R. Wilkinson, A. C. Ward, N. F. Light-foot, and P. R. Sisson, Rapid interstrain comparison by pyrolysis mass spectrometry of coagulase—Negative staphylococci from persistent CAPD peritonitis, *Epidemiol. Infect.*, *106*: 239–246 (1991).
24. P. Miketova, C. Abbas-Hawks, K. J. Voorhees, and T. L. Hadfield, *J. Anal. Appl. Pyrol.*, *67*: 109 (2003).
25. T. I. Eglinton, S. R. Larter, and J. J. Boon, Characterization of kerogens, coals and asphaltenes by quantitative pyrolysis mass spectrometry, *J. Anal. Appl. Pyrol.*, *20:* 25–45 (1991).
26. N. E. Vanderborgh, J. M. Williams, and H.-R. Schulten, *J. Anal. Appl. Pyrol.*, *8*: 271–290 (1985).
27. N. Simmleit, Y. Yun, H. L. C. Meuzelaar, and H.-R. Schulten, in *Advances in Coal Spectroscopy* (H. L. C. Meuzelaar, ed.), Plenum Publishing, New York (1992).
28. Y. Kuzyankov, P. Leinweber, D. Sapronov, and K. Eckhardt, *J. Plant Nutr. Soil Sci.*, *166*: 719–723 (2003).
29. G. Jandi, P. Leinweber, H.-R. Schulten, and K. Eusterhues, *Eur. J. Soil. Sci.*, *55*: 459 (2004).
30. R. Goodacre, D. B. Kell, and G. Bianchi, *Olivae*, *47*: 36–39 (1993).
31. M. Cardinal, C. Viallon, C. Thonat, and J.-L. Berdague, *Analusis*, *28*: 825–829 (2000).
32. R. Goodacre, *Appl. Spectrosc.*, *51*, 8: 1144–1153 (1997).
33. D. A. Hickman and I. Jane, Reproducibility of PyMS using three different pyrolysis systems, *Anahyst*, *104*: 334–347 (1979).
34. K. Qian, W. Killinger, and M. Casey, *Anal. Chem.*, *68*: 1019–1027 (1996).
35. C. J. Curry, Pyrolysis mass spectrometry studies of adhesives, *J. Anal. Appl. Pyrol.*, *11*: 213–225 (1987).
36. W. A. Westail, Temperature programmed pyrolysis mass spectrometry, *J. Anal. Appl. Pyrol.*, *11*: 3–14 (1987).
37. T. W. Ottley, *RAPyD-400 Pyrolysis Mass Spectrometer Instruction Manual*, Horizon Instruments, Ltd.
38. T. W. Ottley, *Curie-Point Pyrolyser Instruction Manual*, Horizon instruments, Ltd.

4 Microstructure of Polyolefins

Shin Tsuge[1] and Hajime Ohtani[2]
[1]Nagoya University, Nagoya, Japan
[2]Nagoya Institute of Technology, Nagoya, Japan

I INTRODUCTION

Polyolefins — such as polyethylenes (PEs), polypropylenes (PPs), and ethylene-propylene copolymers [P(E-co-P)s] — are among the most commonly utilized synthetic polymers. Despite having simple elemental compositions consisting of only carbon and hydrogen, their physical properties are known to be highly dependent on their microstructural features, such as short- and long-chain branchings, stereoregularities, chemical inversions during monomer enchainment, sequence distributions, etc. [1].

The structural characterization of polyolefins has most commonly been carried out using molecular spectroscopy techniques, such as Fourier transform infrared spectrometry (FTIR), [1]H-nuclear magnetic resonance spectrometry (NMR), and ^{13}C-NMR [1]. High-resolution pyrolysis-gas chromatography (Py-GC), which incorporates the pyrolysis-hydrogenation technique and high-resolution capillary column separation, also provides a simple but powerful technique to study the microstructures of polyolefins [2–4]. In this chapter, the instrumental aspects of high-resolution Py-hydrogenation-GC (Py-HGC) are first discussed briefly. In the remainder of the chapter, its applications to the rapid estimation of short-chain branching in low-density PEs, stereoregularity differences and chemical inversion of the monomer units in PPs, and the study of sequence distribution in P(E-co-P)s are demonstrated.

II INSTRUMENTATION FOR PYROLYSIS HYDROGENATION-GAS CHROMATOGRAPHY

When saturated polyolefins such as PE and PP are exposed to high temperatures under an inert atmosphere, they produce various hydrocarbon fragments that reflect the microstructures of the original polymers. These consist mainly of various α,ω-diolefins, α-olefins, and n-alkanes. If short-chain branches exist in the polymer chain, analysis of the resulting degradation products will be further complicated by the presence of additional diastereomeric, geometrical, and positional isomers. Thus, the number of possible isomers in such fragments increases as a function of the number of carbon atoms they contain, complicating their complete chromatographic separation even when a high-resolution capillary column is used.

Figure 4.1 presents a flow diagram for a typical Py-HGC system that was developed in the authors' laboratory [4–6]. In the inlet liner between the furnace-type pyrolyzer and the splitter, the hydrogenation catalyst Diasolid H (80–100 mesh), coated with 5 wt% of Pt, is packed. If necessary, a precut region containing the same packing material coated instead with an ordinary polydimethylsiloxane liquid phase can be placed in the upper region of the inlet. This would protect the activity of the catalyst and the high efficiency of the capillary column from tarry or less-volatile degradation products that could decrease the resolution. Both the precut and the catalyst in the inlet is maintained at 200°C. About 0.1-0.5 mg of the polymer sample is pyrolyzed, typically at 650°C under a 50 ml/min flow of the carrier gas hydrogen; the hydrogen also acts as the hydrogenation gas. A fused-silica or deactivated stainless-steel capillary column is used in

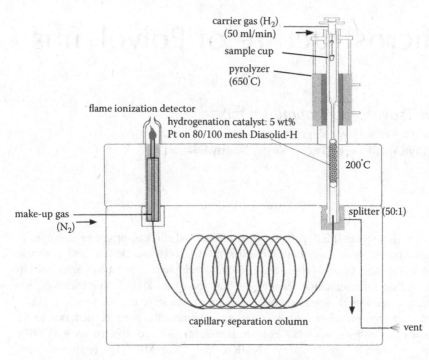

FIGURE 4.1 Schematic flow diagram of the high-resolution pyrolysis-hydrogenation gas chromatographic system.

temperature-programmed mode with typical temperature ranges extending from 40°C to 300°C or more. A flame ionization detector (FID) is used for peak detection in the pyrograms, and the peak assignment is mainly carried out using a directly coupled GC/mass spectrometer (MS) system.

This technique results in pyrograms that are not only drastically simplified, but also highly resolved, since α-olefin and α,ω-diolefin peaks with the same carbon number are combined into the associated alkane peak, and geometrical isomers are eliminated. This situation is illustrated in Figure 4.2, which shows the changes in the pyrogram of low-density PE (LDPE) after hydrogenation, and in Figure 4.3, which illustrates the typical C_{13} products of PP before and after hydrogenation.

III ANALYSIS OF POLYOLEFINS

A SHORT-CHAIN BRANCHING IN LDPE

Models of the probable branching structures of three practically utilized PEs are illustrated in Figure 4.4. LDPE — which are synthesized using a high-pressure method — are known to exhibit both short-chain branching (SCB) and long-chain branching (LCB). Based on the polymerization mechanism, the SCBs are mainly ethyl (C_2), butyl (C_4), and amyl (C_5) groups, while methyl (C_1), propyl (C_3), and groups larger than hexyl ($>C_6$) are minor components, if present at all. The type and concentration of SCBs in LDPE vary depending on the polymerization conditions and affect many properties of these polymers. Therefore, the characterization of SCBs plays a very important role in clarifying the structure-property relationships of LDPE.

Although ^{13}C-NMR has been used extensively in the quantitative analysis of SCB, this method requires relatively long measurement times (from hours to days) and relatively large sample sizes (10-100 mg) [7,8]. In contrast, Py-HGC requires a relatively small amount of sample (about 0.1 mg) and involves a simple and rapid (about 1 h) operation.

As shown in Figure 4.2, the pyrogram of LDPE before hydrogenation (A) consists mainly of a series of triplets corresponding to α,ω-diolefins, α-olefins, and n-alkanes with additional weak but

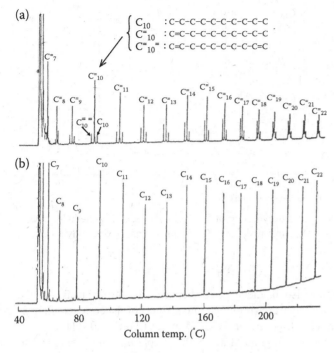

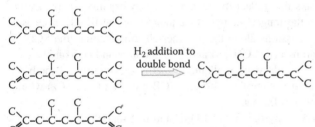

FIGURE 4.2 Typical high-resolution pyrograms of low-density PE before (A) and after (B) hydrogenation [5].

FIGURE 4.3 Hydrogenation of the C_{13} pyrolysis products of PP.

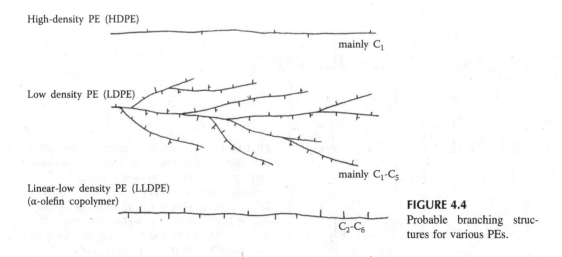

FIGURE 4.4 Probable branching structures for various PEs.

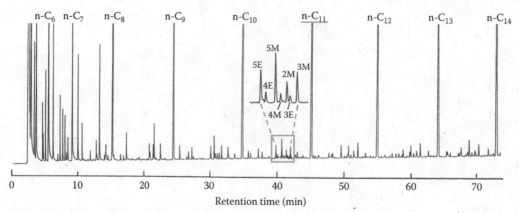

FIGURE 4.5 Typical high-resolution hydrogenated pyrogram of LDPE at 650°C. 2M, 2-methyldecane; 3M, 3-methyldecane; 4M, 4-methyldecane; 5M, 5-methyldecane; 3E, 3-ethylnonane; 4E, 4-ethylnonane; 5E, 5-ethylnonane [12].

highly complex isoalkane, isoalkene, and isoalkadiene peaks between the strong peaks of the triplets. After hydrogenation (B), these triplets were simplified into singlets of the corresponding *n*-alkanes and various isoalkanes that are characteristic of the SCB in the LDPE [5,9–11].

Figure 4.5 shows a typical high-resolution pyrogram of an LDPE sample measured using the in-line hydrogenation technique [12]. As shown in the expansion of the C_{11} component region of the pyrogram in Figure 4.5, the isoalkane peaks that reflect the SCB are clearly separated between the serial *n*-alkane peaks attributed mainly to the longer methylene sequences in the LDPE; such iso-alkanes would be formed via two thermal scissions along the polymer chain containing the SCB:

For example, in the case of a methyl branch ($R = CH_3$), α and ω, β and ω, γ and ω, δ and ω, or ε and ω scission followed by hydrogenation would yield *n*-alkane, 2-methyl-, 3-methyl-, 4-methyl-, or 5-methyl-isoalkanes, respectively. Similarly, the other possible SCB moieties (C_1 to C_6) would be expected to produce the corresponding isoalkanes.

Figure 4.6 shows the expansions of typical pyrograms of LDPE and five model copolymers for methyl, ethyl, butyl, amyl, and hexyl branches in the C_{11} fragment region [12]. Once the relative

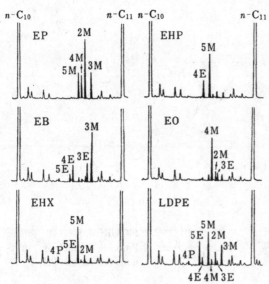

FIGURE 4.6 Expansions of the pyrograms of the hydrogenated reference copolymers ethylene-propylene (EP), ethylene-1-butene (EB), ethylene-1-hexene (EHX), ethylene-1-heptene (EHP), and ethylene-1-octene (EO), along with that of LDPE, in the C_{11} region. The labels 2M, 3M, 4M, 5M, 3E, 4E, and 5E are the same as those in Figure 4.5, and 4P is 4-propyloctane [12].

TABLE 4.1
Estimated SCB Contents in Low-Density Polyethylenes [12].

Sample	Short chain branches/1000 carbons[a]						
	Methyl	Ethyl	Butyl	Amyl	Hexyl	Longer[b]	Total[c]
LDPE-A	1.3 (0.4)	4.9 (6.4)	8.3 (7.4)	2.3 (2.6)	0.2	(2.8)	17.0 (19.9)
LDPE-B	1.5 (0.5)	7.2 (7.2)	11.2 (8.5)	2.2 (2.7)	0.3	(3.5)	22.4 (24.9)
LDPE-C	1.3 (0.5)	4.8 (5.4)	8.6 (6.4)	1.9 (2.2)	0.4	(2.4)	17.0 (17.3)
LDPE-D	1.0 (0.1)	2.1 (2.3)	4.5 (2.4)	0.6 (0.7)	0.3	(0.1)	8.5 (6.7)

Notes

a Data in parentheses were obtained by ^{13}C-NMR.

b Branches longer than C_7 were not considered in the case of Py-HGC.

c 13C-NMR values were determined from the propyl branch content between 0.2 and 0.5.

peak intensities characteristic of the SCBs were determined using the well-defined model polymers containing known amounts of the possible SCBs, the relative abundance of the SCBs in LDPE could be easily estimated by simulation of the observed isoalkane peaks in the LDPE pyrogram [12–15].

Table 4.1 summarizes the SCB contents obtained using this method for four LDPE samples, along with those found by ^{13}C-NMR [12]. Overall, the estimated contents of each type of short branch and of the total short branches were in fairly good agreement with those obtained by ^{13}C-NMR. Thus, high-resolution Py-GC combined with in-line hydrogenation was found to be an effective tool for the determination — only requiring a short analysis time and a small amount of sample — and detection of SCBs at concentrations as low as a few SCBs per 10,000 carbons [14]. In addition, the existence of paired and branched branches in the LDPE was suggested [12,16]. This technique was also successfully applied to study the SCB distribution as a function of the molecular weight for a given LDPE sample [17] and to determine the SCB content in poly(vinyl chloride) after the replacement of the chlorine atoms in the polymer chain with hydrogen atoms [6,14].

B MICROSTRUCTURES OF POLYPROPYLENES

Many physical solid-state and solution properties of PPs are strongly affected not only by the average molecular weight and the molecular weight distribution, but also by configurational characteristics such as the average stereoregularity, stereospecific sequence length, and the degree of chemical inversion of the monomer units along the polymer chains. Py-HGC has also been successfully applied to the conformational characterization of various PPs with different stereospecificities [18–21].

Before further discussion of the characterization technique, the relationship between the possible pyrolysis products of PP and the stereoregularity of the initial polymer is summarized in Table 4.2. The decomposition of the polymer sample into monomers and dimers results in the complete loss of information regarding its stereoregularity. Trimers also lack stereospecific information regarding the polymer chain, although enantiomers are observed. Tetramers are the smallest fragment that can provide information regarding the stereoregularity of the initial polymer, as they provide meso and racemo diastereoisomers. Hence, to study the stereoregularity of PP, the tetramer — or larger products in the observed programs — should be examined.

The possible pentamer components (C_{14} to C_{16}) that could be formed by the thermal degradation of PPs followed by hydrogenation are illustrated in Figure 4.7, to show the origin of the complicated mixture of diastereoisomers. This figure demonstrates that the C_{14}, C_{15}, and C_{16} triads

TABLE 4.2

Possible Pyrolysis Fragments of PP.

fragments	chemical structure(s) [asymmetric C:C*]	enantiomers	diastereoisomers

monomer	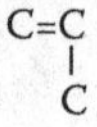 C=C
	  |
	  C
		X	X

dimer	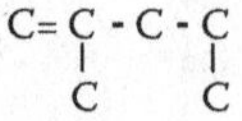 C= C - C - C
	   | |
	   C C
		X	X

trimer	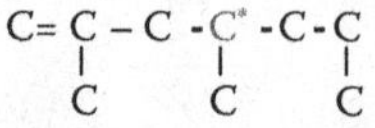 C= C – C - C* - C - C
	   | | |
	   C C C
		✓	X

tetramer	meso (m): 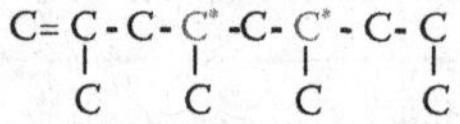 C= C - C - C* - C - C* - C - C racemo (r): C= C - C - C* - C - C* - C - C 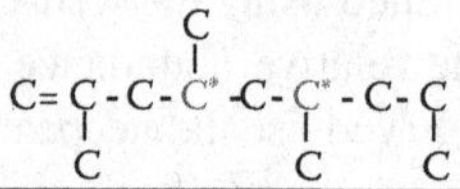	✓	☑

observed in the pyrogram in Figure 4.8 were actually composed of a triplet (mm, mr, rr), a quartet (mm, mr, rm, rr), and a triplet (mm, mr, rr), respectively, where m and r are the meso and racemo conformations, respectively [19].

The authors have studied PP samples with a variety of stereoregularities synthesized in the presence of different Ziegler-Natta (ZN) or metallocene (ML) catalysts, namely five isotactic-PP (ZN-$I_{1,2}$ and ML_{1-3}), two atactic-PP (ZN-A and ML-A), and five syndiotactic-PP (ZN-S and ML-S_{1-4}) polymers. Figure 4.8 shows the hydrogenation pyrograms of an isotactic-PP (ZN-I_2), atactic-

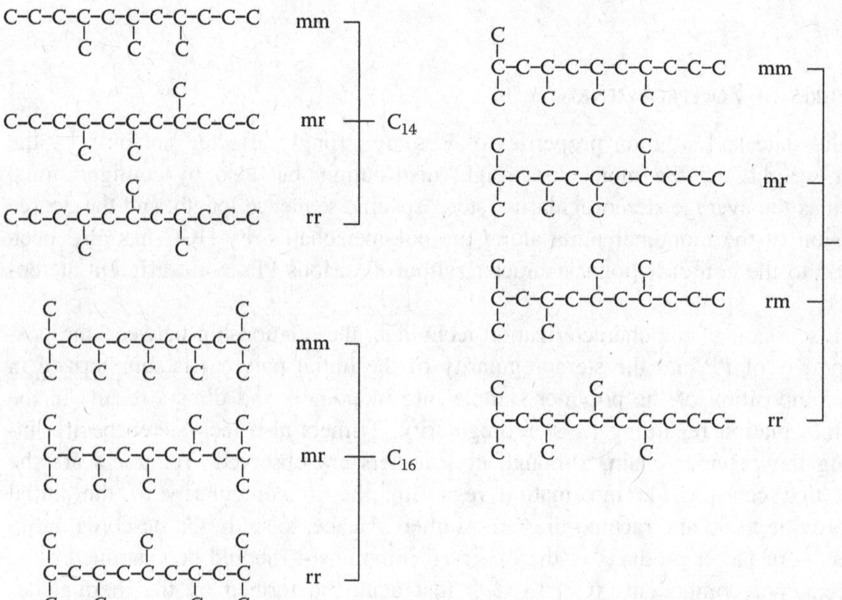

FIGURE 4.7 Possible diastereomeric isomers of the C_{14}, C_{15}, and C_{16} pentamer clusters. m and r are the meso and racemo configurations, respectively [19].

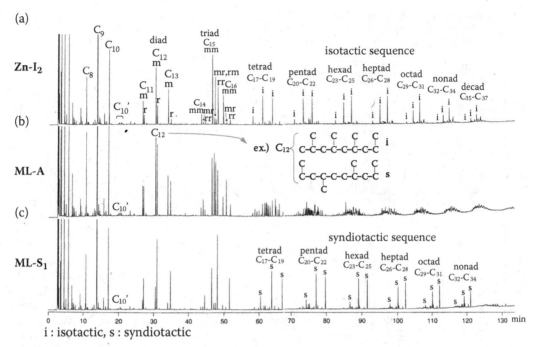

FIGURE 4.8 Typical high-resolution hydrogenated pyrograms of (A) isotactic PP (ZN-I₂), (B) atactic-PP (ML-A), and syndiotactic-PP (ML-S₁). i, s, and h represent isotactic, syndiotactic, and heterotactic products, respectively.

PP (ML-A), and syndiotactic-PP (ML-S₁) obtained at 650°C using a fused-silica capillary column (0.25 mm i.d. × 100 m) coated with polydimethylsiloxane [20,21]. In the pyrogram of isotactic-PP, a series of triplet peaks with long isotactic sequences were observed up to the decads (C_{35} to C_{37}), while the intensities of the peaks associated with syndiotactic and atactic sequences became negligibly small at higher carbon number regions. These data suggest that the isotactic-PP (ZN-I₂) contained fairly long isotactic sequences. Similarly, in the pyrogram of syndiotactic-PP(ML-S₁), a characteristic series of fairly intense triplets associated with long syndiotactic sequences was seen up to the nonads (C_{32} to C_{34}). On the other hand, in the pyrogram of atactic-PP (ML-A), complex atactic multiplets were observed up to the decads (C_{35} to C_{37}).

Using the relative intensities of the tetramer and pentamer peaks, the values %m and %r (r + m = 100 %), as well as %mm, %mr plus %rm, and %rr (mm + mr + rm + rr = 100 %), can be estimated. From the obtained values, the average tactic sequence lengths of the propylene units can be calculated using the following equation:

$$N_s = 2\frac{\%r}{\%mr}, \quad N_i = 2\frac{\%m}{\%mr}$$

where N_s and N_i are the average syndiotactic and isotactic sequence length, respectively [22].

Figure 4.9 shows two partial pyrograms for atactic-PP, ZN-A, and ML-A — which have nearly identical (50%) in %m and %r values. Expectedly, the meso and racemo peak pairs of the tetramer region showed almost comparable intensities for both atactic PP samples. However, it is interesting to note that the relative intensities of the heterotactic peaks (mr and rm) in the pentamer region were comparable to those of mm and rr for ML-A, while the heterotactic peak intensities for ZN-A were significantly weaker than those of ML-A. These data suggested that ML-A was a typical atactic PP with relatively smaller N_s and N_i values, while ZN-A had longer tactic sequences. This

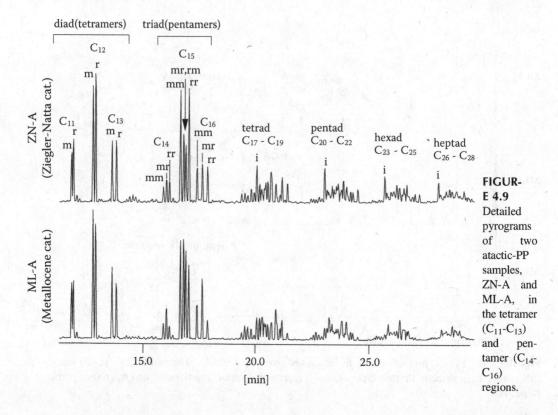

FIGURE 4.9 Detailed pyrograms of two atactic-PP samples, ZN-A and ML-A, in the tetramer (C_{11}-C_{13}) and pentamer (C_{14}-C_{16}) regions.

result was also supported by the data in Table 4.3, where the values of N_s and N_i obtained from the C_{13} tetramers and C_{16} pentamers in the Py-HGC data are shown to be in fairly good agreement with those obtained by ^{13}C-NMR [20,21].

In practical applications of PP materials, small differences in the stereoregularity of highly isotactic PPs are often critical. Although the pentad tacticity of PP is generally evaluated using ^{13}C-NMR, it can also be estimated more sensitively and rapidly using Py-HGC/MS. Figure 4.10 shows an expansion of the pyrogram of atactic PP (ML-A) and the ten possible diastereoisomeric structures of the C_{22} products, which reflect the pentad tacticity. Thus, the original pentad tacticity in the PP sample should be estimated from the relative peak intensities in the heptamer (C_{20} to C_{22}) region. Unfortunately, it was difficult to resolve all the peaks observed in this region in

TABLE 4.3

Average Syndiotactic Sequence Length (N_s) and Isotactic Sequence Length (N_i) Estimated Using Py-HGC and ^{13}C-NMR [21,22].

PP sample	Py-HGC		^{13}C-NMR	
	N_s	N_i	N_s	N_i
ML-I₁	1.7	5.0		
ML-I₃	5.2	31.9	2	65
ML- A	1.6	2.2	2	2
ML-S₁,ML-S₄	25.219.2	4.64.6	6548	13
ZN-I₂	5.7	39.8	2	198
ZN- A	2.8	2.9		
ZN-S	10.5	4.7	8	3

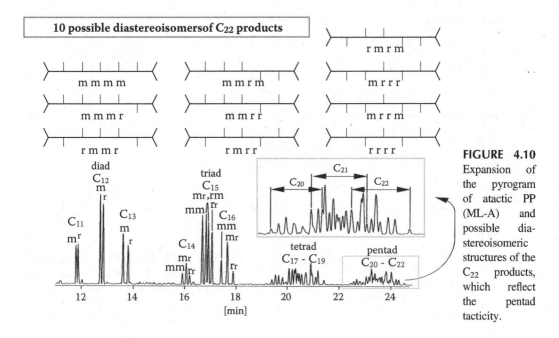

FIGURE 4.10 Expansion of the pyrogram of atactic PP (ML-A) and possible diastereoisomeric structures of the C$_{22}$ products, which reflect the pentad tacticity.

both the pyrogram obtained using FID and the total ion chromatogram obtained using MS because the peaks of products with different carbon numbers partially overlapped, as shown in Figure 4.10.

The C$_{22}$ products, therefore, were selectively observed in mass chromatograms using Py-HGC/MS at $m/z = 310$ corresponding to the molecular ion of C$_{22}$ products. Figure 4.11 shows the typical pyrograms of several PP samples in mass chromatogram mode at m/z 310. For the sample ML-A, ten clearly separated peaks were present in the mass chromatogram that corresponded to the number of possible C$_{22}$ products. The pentad tacticity of these ten peaks was assigned, as shown in Figure 4.11, based on the general elution tendency (products with the mesostructure elute faster) and the observed peak intensities for the PP samples with various stereospecificities.

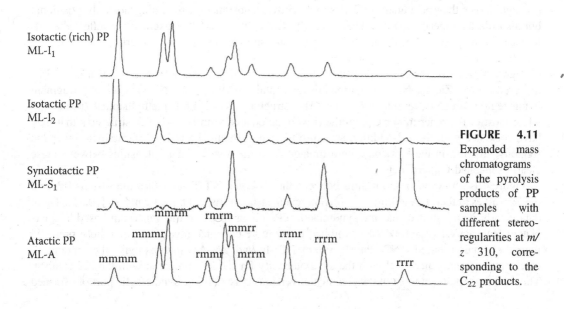

FIGURE 4.11 Expanded mass chromatograms of the pyrolysis products of PP samples with different stereoregularities at m/z 310, corresponding to the C$_{22}$ products.

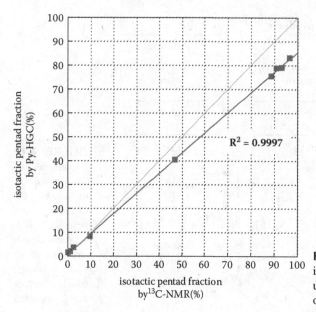

FIGURE 4.12 Relationship between the isotactic pentad fraction values estimated using Py-HGC/MS and ^{13}C-NMR. The diagonal line represents the ideal relationship.

Figure 4.12 illustrates the relationship between the relative peak intensities of the isotactic pentad (mmmm) fractions obtained using Py-HGC/MS and ^{13}C-NMR. Although the values obtained using Py-HGC/MS were slightly smaller probably due to stereoisomerization during pyrolysis [21], a good linear relationship was observed over the entire stereoregularity range. In particular, slight differences in the isotactic pentad in the highly isotactic region (mmmm fraction greater than 90%) could be evaluated with Py-HGC/MS using a trace amount of sample and a relatively short measurement period. These features make this technique suitable for the characterization of PP materials in practical applications.

So far, repeated head-to-tail (H-T) structures have been assumed in the discussion of the stereoregularity of PP. However, regioirregular (head-to-head (H-H) and tail-to-tail (T-T)) structures (chemical inversions) should coexist with the 1,3-addition structure along the polymer chain, with the extent of the regioirregularity depending on the polymerization conditions [1]. PP degradation products larger than pentamers ($>C_{11}$) should provide information regarding not only positional but also diastereomeric isomers. Therefore, the C_{10} components that exclusively reflect the differences in the positions of the methyl groups were carefully separated to determine the extent of the chemical inversion of the monomer units [19–21].

Figure 4.13 shows the expansions of the C_{10} region of the pyrograms of highly syndiotactic PPs prepared using a Zieglar-Natta and a metallocene catalyst (ML-S_1 and ZN-S). The main fragment in this region, which represented successive H-T structures, was 2,4,6-trimethylheptane (peak C_{10}), while most of the other minor C_{10} products were associated with H-H or T-T structures that may have coexisted with the 1,3-addition structure to some extent. The degree of chemical inversion (regioirregularity) in the PPs could be estimated from the relative peak intensities between these clearly separated C_{10} products.

Figure 4.14 shows the relationships between the observed %T-T values for the various PPs and the %r values in their C_{13} tetramers, which give a measure of its syndiotactic nature [20,21]. These data clearly demonstrated that the syndiotactic PPs prepared using the vanadium-based Zieglar-Natta (ZN) catalyst system (ZN-S) exhibited many more chemical inversions than those prepared with the titanium-based ZN catalyst system (ZN-Is and ZN-A). Interestingly, the amount of chemical inversions increased with the syndiotacticity for the PPs prepared with the ZN catalyst. This fact suggested that most of the chemical inversions existed in the syndiotactic portions formed

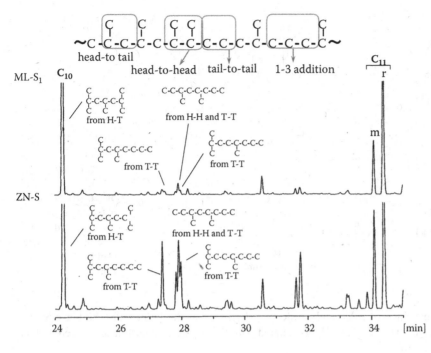

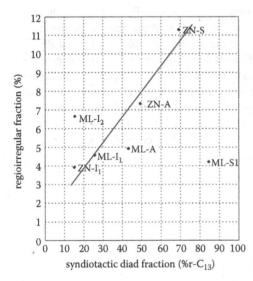

FIGURE 4.13 Expansions of the C_{10} region of the pyrograms of the highly syndiotactic PPs prepared using metallocene (ML) and Zieglar-Natta (ZN) type catalysts.

FIGURE 4.14 Relationships between the observed tail-to-tail% and syndiotacticity for various PPs.

during polymerization using the ZN-type catalyst, while catalytic control with isotactic-specific propagation yields mostly H-T linkages. On the other hand, the degree of chemical inversion was almost unchanged regardless of the tacticity for the metallocene catalyst system.

C SEQUENCE DISTRIBUTION IN ETHYLENE-PROPYLENE COPOLYMERS

As shown in Figure 4.15, the pyrograms of P(E-co-P)s generally consist of peaks associated with PE, PP, and the hybrid products of ethylene and propylene units [23]. Interestingly, the P(E-co-P) sample synthesized using a Ti catalyst showed much stronger serial n-alkane peaks than the sample

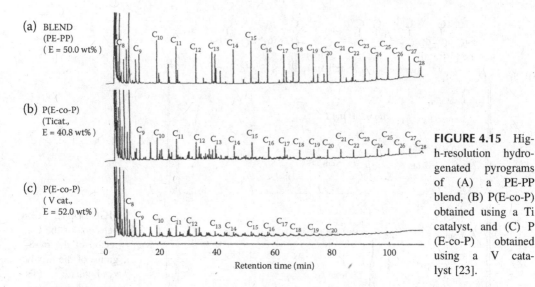

FIGURE 4.15 High-resolution hydrogenated pyrograms of (A) a PE-PP blend, (B) P(E-co-P) obtained using a Ti catalyst, and (C) P(E-co-P) obtained using a V catalyst [23].

synthesized in the presence of a V catalyst, despite the fact that the latter had a higher ethylene content than the former. This was closely associated with the sequence distribution of the ethylene units in the polymer chain.

To illustrate the differences among the ethylene sequences in the P(E-co-P)s, the relationships between the carbon number and the relative molar yields (N_c) of the serial n-alkane peaks in the pyrograms of the copolymers are shown in Figure 4.16, together with those for PE and a blend of PE and PP. The intensities of the n-alkane peaks in the pyrograms of PE are known to decrease as a semilogarithmic function versus the carbon number of the n-alkanes because random scission is the predominant thermal degradation mechanism of PE [24–27]. Expectedly, the physical blend of PE and PP showed a gentle slope almost equivalent to that of PE. On the other hand, the steeper slope corresponded to the shorter average sequence length of ethylene units in the P(E-co-P)s. Hence, this method should be effective in comparing the average sequence lengths in copolymer systems containing ethylene units.

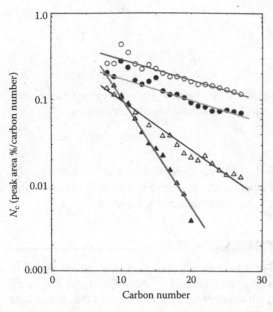

FIGURE 4.16 Relationships between the carbon number and molar yield (N_c) of the series of n-alkane peaks in the pyrograms of P(E-co-P)s and the related polymers at 650°C. ○, PE; ●, blend (PE/PP=50/50 wt%); △, P(E-co-P) (Ti catalyst, E = 40.8 wt%); ▲, P(E-co-P) (V catalyst, E = 52.0 wt%) [24].

REFERENCES

1. S. van der Ven, *Polypropylene and other Polyolefins; Polymerization and Characterization, Studies in Polymer Science 7*, Elsevier, Amsterdam, The Netherlands (1990).
2. J. van Schotten and J. K. Evenhuis, *Polymer, 6*: 343–360 (1965).
3. L. Michajlov, P. Zugenmaier, and H.-J. Cantow, *Polymer, 12*: 70 (1971).
4. S. Tsuge, *Trends Anal. Chem., 1*: 87 (1981).
5. Y. Sugiura and S. Tsuge, *Macromolecules, 12*: 512 (1979).
6. S. Mao, H. Ohtani, S. Tsuge, H. Niwa, and M. Nagata, *Polym. J., 31*: 79 (1999).
7. T. Usami and S. Takayama, *Macromolecules, 17*: 1756 (1984).
8. J. C. Randall, *J. Macromol. Sci. Rev. Macromol. Chem. Phys., C29*: 201 (1989).
9. M. Seeger and E. M. Barrall, II, *J. Polym. Sci. Polym. Chem. Ed., 13*: 1515 (1975).
10. D. H. Ahlstrom and S. A. Liebman, *J. Polym. Sci. Polym. Chem. Ed., 14*: 2478 (1976).
11. O. Mlejnek, *J. Chromatogr., 191*: 181 (1980).
12. H. Ohtani, S. Tsuge, and T. Usami, *Macromolecules, 17*: 2557 (1984).
13. Y. Sugimura, T. Usami, T. Nagaya, and S. Tsuge, *Macromolecules, 14*: 1787 (1981).
14. S. A. Liebman, D. H. Alstrom, W. H. Starnes, Jr., and F. C. Schilling, *J. Macromol. Sci. Chem., A17*: 935 (1982).
15. J. Tulisalo, J. Seppala, and K. Hastbacka, *Macromolecules, 18*: 1144 (1985).
16. M. A. Haney, D. W. Johnston, and B. H. Clampitt, *Macromolecules, 16*: 1775 (1983).
17. T. Usami, Y. Gotoh, S. Takayama, H. Ohtani, and S. Tsuge, *Macromolecules, 20*: 1557 (1987).
18. M. Seeger and H.-J. Cantow, *Makromol. Chem., 176*: 2059 (1975).
19. Y. Sugimura, T. Nagaya, S. Tsuge, T. Murata, and T. Takeda, *Macromolecules, 13*: 928 (1980).
20. S. Tsuge and H. Ohtani, *Analytical Pyrolysis, Technique and Applications* (K. J. Voorhees, ed.), Butterworth, London, p. 407 (1984).
21. H. Ohtani, S. Tsuge, T. Ogawa, and H.-G. Elias, *Macromolecules, 17*: 465 (1984).
22. B. D. Coleman and T. G. Fox, *J. Polym. Sci. Part A, 1*: 3183 (1963).
23. S. Tsuge, Y. Sugimura, and T. Nagaya, *J. Anal. Appl. Pyrol., 1*: 221 (1980).
24. M. Seeger, H.-J. Cantow, and S. Marti, *Z. Anal. Chem., 276*: 267 (1975).
25. M. Seeger and H.-J. Cantow, *Makromol. Chem., 176*: 1411 (1975).
26. M. Seeger and R. J. Gritter, *J. Polym. Sci. Polym. Chem. Ed., 15*: 1393 (1977).
27. M. Seeger and H.-J. Cantow, *Polym. Bull., 1*: 347 (1979).

5 Degradation Mechanisms of Condensation Polymers
Polyesters and Polyamides

Hajime Ohtani[1] and Shin Tsuge[2]
[1]Nagoya Institute of Technology, Nagoya, Japan
[2]Nagoya University, Nagoya, Japan

I INTRODUCTION

The characterization of polymers via analytical pyrolysis often necessitates a detailed understanding of the polymer degradation process during heating *in vacuo* or in an inert atmosphere. Elucidating the degradation pathways of condensation polymers with heteroatoms in their backbone chains is not an easy task, because their thermal degradation yields a number of complex polar compounds [1]. However, the development of highly specific pyrolysis devices and advanced gas chromatographs (GC) and mass spectrometers (MS) has made it feasible to deduce the complex thermal degradation behavior of condensation polymers using analytical pyrolysis methods such as pyrolysis-GC (Py-GC) and pyrolysis-MS (Py-MS). In this chapter, the thermal degradation of two typical classes of condensation polymers—polyesters and polyamides—are discussed, with a focus on the mechanisms of pyrolysate formation under flash pyrolysis conditions, in which the polymer sample is rapidly exposed to a temperature of 500 °C or higher.

II POLYESTERS

Poly(alkylene terephthalate)s—such as poly(ethylene terephthalate) (PET) and poly(butylene terephthalate) (PBT)—are widely used in the fiber and thermoplastic fields. A homologous series of poly(alkylene terephthalate)s, namely, PET, poly(trimethylene terephthalate) (PTT), PBT, poly (pentamethylene terephthalate) (PPT), and poly(hexamethylene terephthalate) (PHT) have been studied using Py-GC [2,3]. Figure 5.1 shows the high-resolution pyrograms of (a) PET, (b) PTT, (c) PBT, (d) PPT, and (e) PHT [3]. The assignments of the characteristic peaks and their relative yields are listed in Table 5.1. Analogous patterns are observed in these pyrograms. The fragments containing two terephthalic acid units (g and h) with high molecular weights (up to 494) provide information regarding relatively long sequences of the polymer chain.

The degradation of these polyesters has been proposed to be initiated by random scission of ester linkages through a six-membered cyclic transition state to give two fragments with an alkenyl and a carboxyl end group, respectively [4–6]:

$$--- PhCOOCH_2CH_2 --- \longrightarrow \left[---Ph-C \begin{matrix} O---H \\ \downarrow \\ O---CH_2 \end{matrix} CH--- \right] \longrightarrow ---PhCOOH + CH_2{=}CH ---$$

Furthermore, at the elevated temperatures used for flash pyrolysis, the acid-form terminals are often reduced to more stable phenyl terminals via the elimination of carbon dioxide. Thus, the

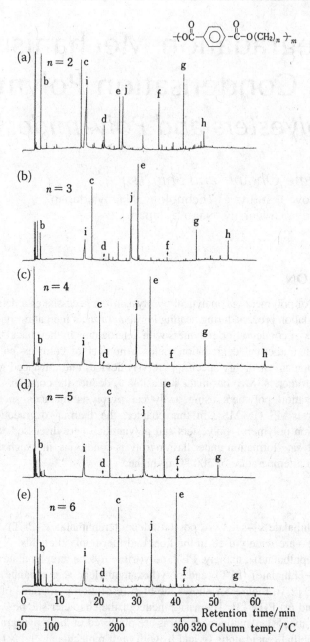

FIGURE 5.1 Pyrograms of (a) PET, (b) PTT, (c) PBT, (d) PPT, and (e) PHT at 590°C [3]. Peak notations correspond to those in Table 5.1.

main pyrolysis products is comprised of compounds with alkenyl-, phenyl-, and/or carboxyl end groups, depending on the polymer structure. However, dibasic acids—such as terephthalic acid, which would also be expected to form—were not detected. This was attributed to their high polarity. Clearly separated peaks corresponding to fragments with an acid end group were observed (i and j). Additionally, the low-boiling-point α and ω-dienes most likely originated from PBT, PPT, and PHT through the scission of two neighboring ester linkages via the cyclic transition state [7]. Based on the results described above, the common degradation pathways of poly(alkylene terephthalate) are summarized in Figure 5.2 [3].

TABLE 5.1

Assignment and Relative Intensities of the Characteristic Peaks in the Pyrograms of Terephthalate Polyesters (Peak Notations as in Figure 5.1) [3].

Peak notations	Structure[a]	Relative peak intensity,[b] %				
		PET	PTT	PBT	PPT	PHT
Small products[c]		19.0	9.6	37.6	34.8	38.6
b	(benzene ring)	5.1	2.3	1.5	1.7	2.7
c	$\text{-C(=O)-O-(CH}_2)_{n-2}\text{-CH=CH}_2$	5.9	8.5	6.7	7.5	7.0
i	-C(=O)-O-H	32.0	14.2	10.4	13.3	15.1
d	(biphenyl)	1.8	>0.3	>0.3	>0.3	>0.3
e	$\text{CH}_2\text{=CH-(CH}_2)_{n-2}\text{-O-C(=O)-}\langle\text{ring}\rangle\text{-C(=O)-O-(CH}_2)_{n-2}\text{-CH=CH}_2$	3.7	21.7	13.2	17.6	11.3
j	$\text{H-O-C(=O)-}\langle\text{ring}\rangle\text{-C(=O)-O-(CH}_2)_{n-2}\text{-CH=CH}_2$	9.2	26.0	21.7	21.0	14.2
f	$\langle\text{ring}\rangle\text{-C(=O)-O-(CH}_2)_n\text{-O-C(=O)-}\langle\text{ring}\rangle$	3.4	0.6	0.7	0.6	>0.5
g	$\langle\text{ring}\rangle\text{-C(=O)-O-(CH}_2)_n\text{-O-C(=O)-}\langle\text{ring}\rangle\text{-C(=O)-O-(CH}_2)_{n-2}\text{-CH=CH}_2$	5.7	4.5	4.1	2.2	0.9
h	$\text{CH}_2\text{=CH-(CH}_2)_{n-2}\text{-O-C(=O)-}\langle\text{ring}\rangle\text{-C(=O)-O-(CH}_2)_n\text{-O-C(=O)-}\langle\text{ring}\rangle\text{-C(=O)-O-(CH}_2)_{n-2}\text{-CH=CH}_2$	1.7	5.6	3.1	—	—

a. $n=2$ (PET), 3 (PTT), 4 (PBT), 5 (PPT) and 6 (PHT).
b. Relative peak area (%) among all peaks appearing on the pyrograms.
c. Total intensities of the peaks with shorter retention times than that of benzene.

The same basic degradation pathways were reported for PET and/or PBT in other Py-GC [8,9] and Py-MS [10–16] studies. Additionally, tetrahydrofuran was observed as a major product of PBT [9,13], and an additional pathway involving the transfer of a hydrogen atom and CO–O bond scission to yield a hydroxyl end group was also suggested by the Py-electron ionization (EI)-MS measurements [11,12,14]:

$$\text{----O-C(=O)-}\langle\text{ring}\rangle\text{-C(=O)-O-(CH}_2)_n\text{-O-C(=O)-}\langle\text{ring}\rangle\text{-C(=O)-O-(CH}_2)_n\text{-O-C(=O)-}\langle\text{ring}\rangle\text{-C(=O)-O----}$$

$$\longrightarrow \text{----O-C(=O)-}\langle\text{ring}\rangle\text{-C(=O)-O-CH-CH}_2(\text{CH}_2)_x\text{-O-C(=O)-}\langle\text{ring}\rangle\text{-C}\cdot \;+$$

$$\text{HO-(CH}_2)_n\text{-O-C(=O)-}\langle\text{ring}\rangle\text{-C(=O)-O----}$$

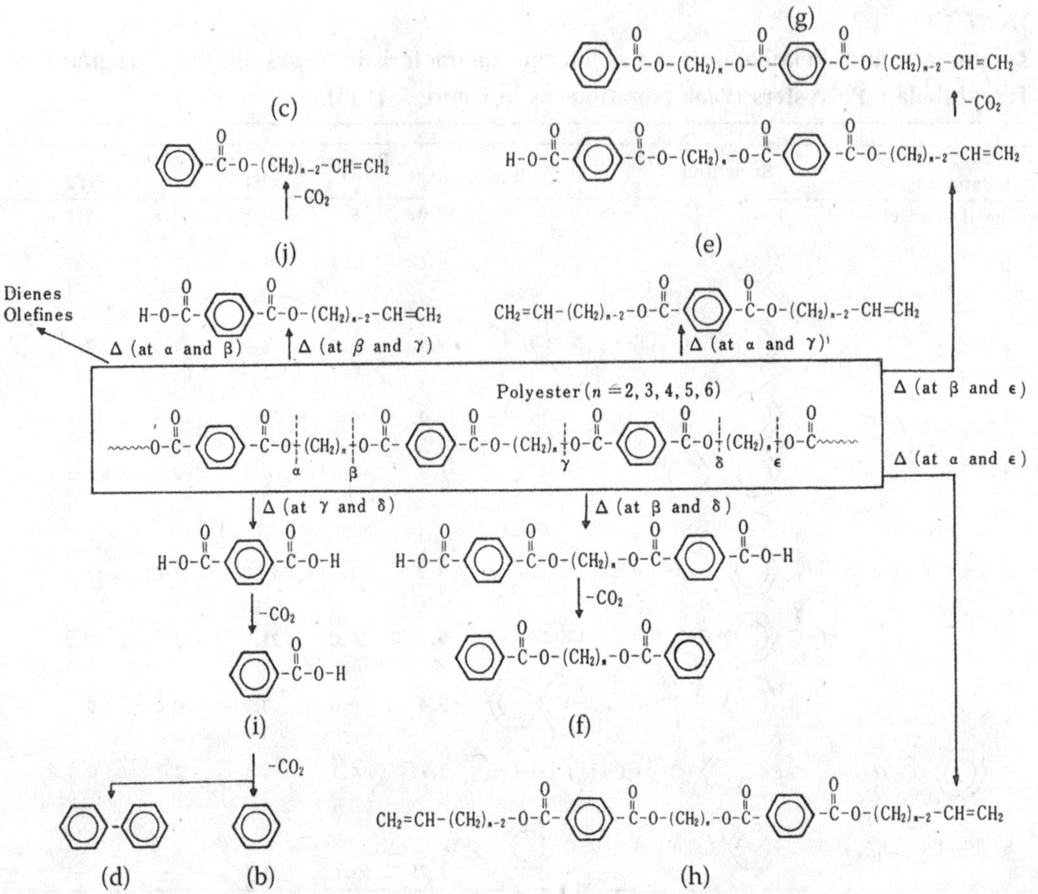

FIGURE 5.2 Thermal degradation mechanisms of terephthalate polyesters based on their pyrograms at 590°C [3]. △: Thermal cleavage. The labels b–j correspond to the peaks in Figure 5.1.

A series of poly(alkylene phthalate)s [17] and some unsaturated polyesters [18,19] were also investigated using Py-GC. The major pyrolysis products of these polyesters were the corresponding diols and anhydrides, such as phthalic anhydride and maleic anhydride, as well as products formed through the six-membered cyclic transition state. Moreover, various cyclic ethers and cyclic esters were observed in considerable quantities. A general scheme depicting the degradation pathways of poly(alkylene phthalate)s is shown in Figure 5.3 [17].

The thermal degradation of aliphatic polyesters has mainly been investigated using Py-MS [20–27], and its dominant degradation products were found to be the cyclic monomer (lactone), cyclic oligomers, and linear oligomers with ketene, hydroxyl, carboxyl, or alkenyl end groups. The relative yields of these products strongly depended on the chain length and the branching structures between neighboring ester groups in the polymer chain, as well as the degradation conditions used. For example, Py-chemical ionization (CI)-MS analysis of polylactones with various chain lengths indicated that poly(β-propiolactone) was the only polylactone to mainly undergo degradation via the six-membered ring transition state to form linear products with a carboxyl and a vinyl end group; the other polylactones predominantly decomposed via intramolecular transesterification reactions to form cyclic oligomers [25].

The thermal degradation of poly(butylene succinate-*co*-butylene adipate) was studied using Py-GC, with a focus on its biodegradability [28]. The pyrolysates observed in the pyrogram obtained at 500 °C included linear ester products with carboxylic, olefinic, and alkyl end groups,

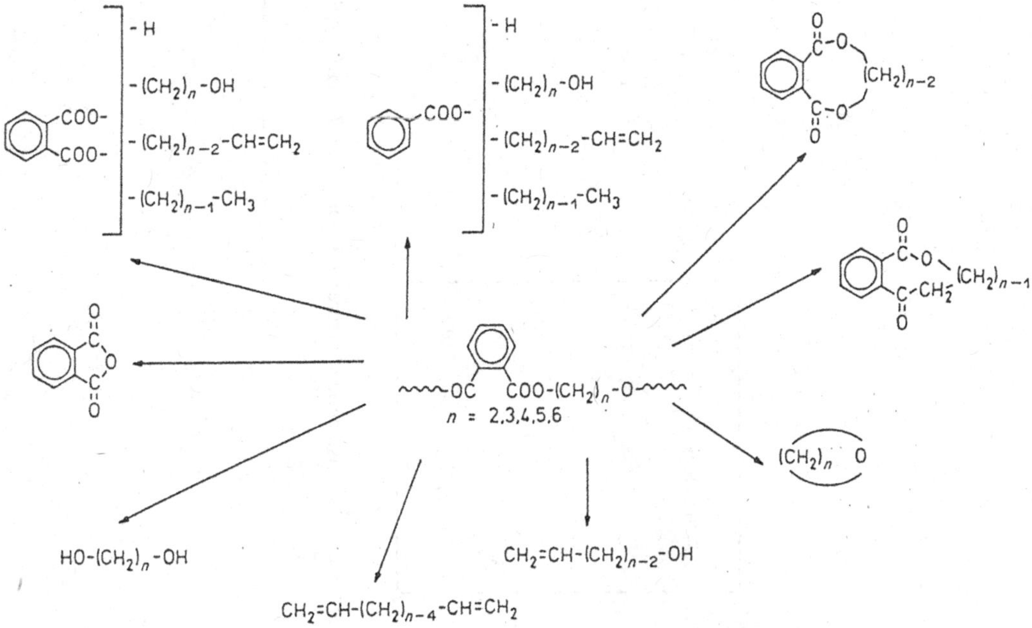

FIGURE 5.3   General scheme of the thermal degradation of poly(alkylene phthalate) [17].

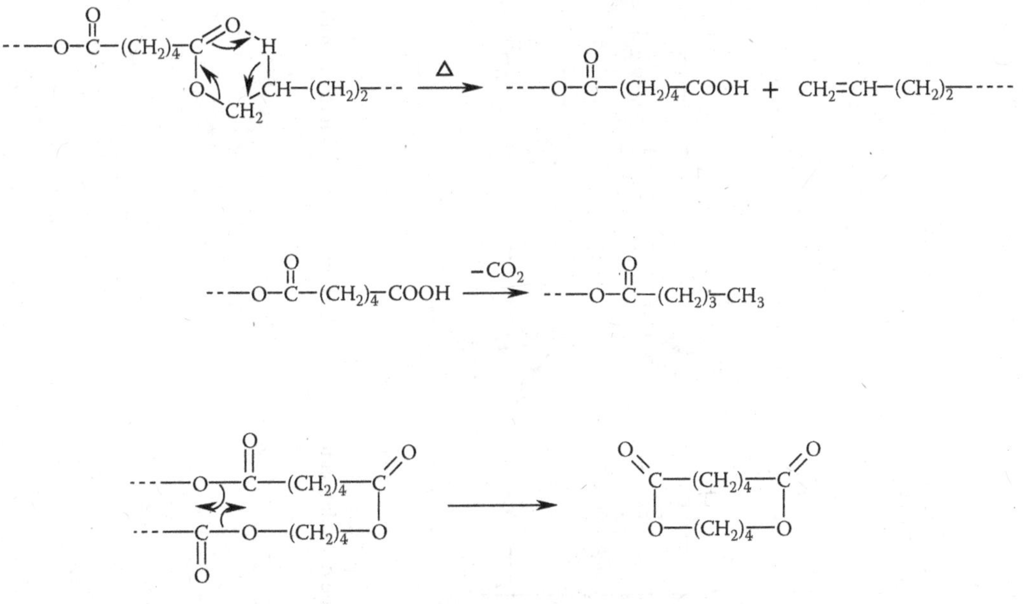

FIGURE 5.4   Major degradation pathways for poly(butylene succinate-*co*-butylene adipate) [28].

various cyclic monomers and oligomers, cyclopentanone derived from the adipic acid unit, and tetrahydrofuran derived from the butanediol unit. Figure 5.4 shows the formation pathways for the linear and cyclic ester products. Interestingly, the relative yields of the alkyl ester products were found to be a good indicator of the degree of biodegradation [28].

The major pyrolysis product observed in the Py-GC of poly(ε-caprolactone) (PCL) was ε-caprolactone monomer; minor amounts of cyclic and linear oligomers were also detected [29,30].

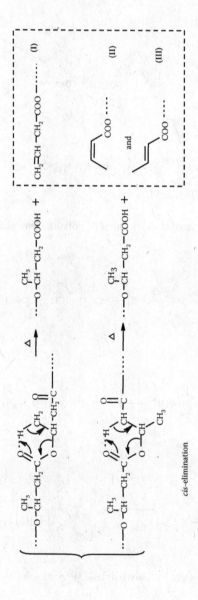

FIGURE 5.5 Pathways for the degradation for poly(3-hydroxybutylate), which result in the formation of three types of isomeric structures. (H. Sato et al., *J. Anal. Appl. Pyrolysis*, 74: 193 (2005)).

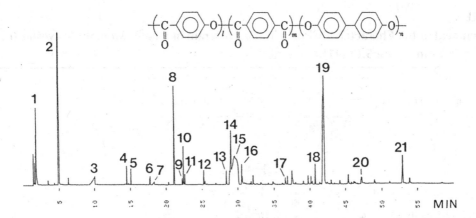

FIGURE 5.6 Pyrogram of a fully aromatic liquid crystalline polyester consisting of p-hydroxybenzoic acid, biphenol, and terephthalic acid at 650°C (PHB/PB/TA = 2/1/1) [47]. Peak numbers correspond to those in Table 5.2.

The distribution of these products varied considerably depending on the microstructure of the polymer, such as its end groups. The thermal degradation of poly(lactic acid) (PLA) has also been widely studied using Py-GC and Py-MS [29,31–39]. The main pyrolysis products were D-, L-, and meso-lactides; a large number of oligomeric products were also produced in small amounts. The pyrolysates of PCL and PLA should be formed in a manner similar to that shown in Figure 5.4.

Poly(3-hydroxybutyrate) is a representative bacterial polyester. Monomeric (butenoic) and oligomeric (mainly dimeric and trimeric) acids were observed in its pyrograms; presumably, these products are predominantly formed via the six-membered transition state [29,32,40–45]. Due to the presence of a methyl side chain, each oligomer can produce three types of isomeric structures, as shown in Figure 5.5. Among these, the trans type inner olefinic products usually predominated.

Fully aromatic polyesters based on *p*-hydroxybenzoic acid (PHB) have been recognized as high-performance liquid crystalline polymers (LCPs), and are especially useful in fields where very high thermal stability is required. The thermal degradation mechanisms of typical LCPs prepared from PHB, biphenol (BP), and terephthalic acid (TA) were studied using Py-GC/MS [46,47]. Figure 5.6 shows a typical pyrogram of an LCP sample (PHB/BP/TA = 2/1/1) obtained at 650 °C; the peak assignments are summarized in Table 5.2 [47]. General pathways for the formation of the characteristic products (Figure 5.7) were formulated by examining LCPs with different comonomer ratios or containing deuterated terephthalate units [47] based on the following observations:

1. Phenol is almost exclusively formed from the PHB moiety.
2. Benzene is mainly formed from the TA units.
3. Larger products, such as biphenol, *p*-hydroxyphenyl benzoate, and 4,4′-biphenyldibenzoate are mainly derived from the TA or BP moieties.
4. Biphenyl is mainly formed from the PHB and TA moieties via recombination, rather than directly from the BP moiety.
5. Phenyl benzoate is mainly formed by recombination reactions between phenoxy and benzoyl radicals from the PHB and TA moieties, respectively.

These observations suggest that:

1. The C–O bonds between a carbonyl carbon and a phenolic oxygen are preferentially cleaved over those between an aromatic ring and a phenolic oxygen.

TABLE 5.2

Assignment of the Characteristic Peaks in the Pyrogram of a Fully Aromatic Polyester (Peak Numbers as in Figure 5.6) [47].

peak no.	products	structure
1	benzene	
2	phenol	
3	benzoic acid	
4	biphenyl	
5	diphenyl ether	
6	benzofuran	
7	p-hydroxybenzoic acid	
8	phenylbenzoate	
9	m-phenylphenol	
10	p-phenylphenol	
11	o-hydroxybenzofuran	
12	xanthone	
13	p-hydroxybenzophenone	
14	phenyl(p-hydroxybenzoate)	
15	biphenol	
16	4,4'-dihydroxybenzophenone	
17	biphenylbenzoate	
18	acetylbenzoate	
19	p-hydroxybiphenylbenzoate	
20	p-hydroxybiphenyl(p'-hydroxybenzoate)	
21	4,4'-biphenyldibenzoate	

2. The C–C bonds between an aromatic ring and a carbonyl carbon are preferentially cleaved over other types of C–C bonds.

3. The C–C bonds between aromatic rings are not easily cleaved.

Moreover, inter- and intramolecular ester exchange reactions resulting in the rearrangement of the polyester sequences at the pyrolysis stage were also demonstrated to take place for aliphatic [26] and aliphatic-aromatic polyesters [48] using Py-MS.

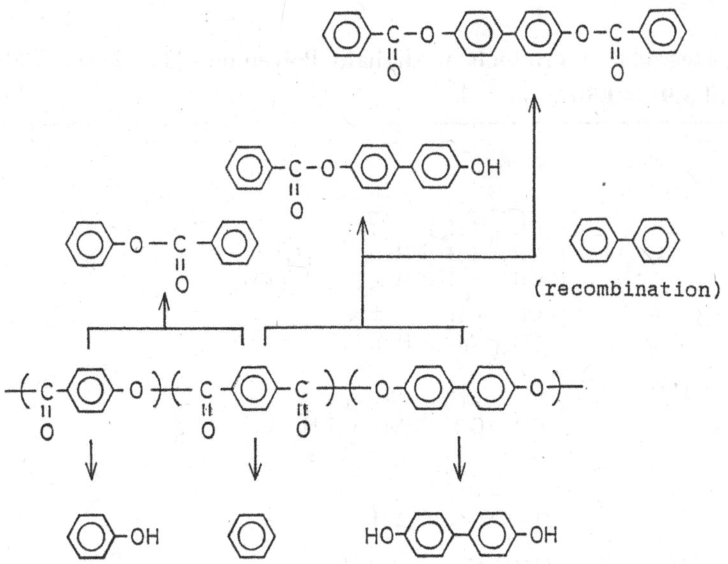

FIGURE 5.7 Thermal degradation mechanisms of the fully aromatic polyester [47].

III POLYAMIDE

Synthetic polyamides, or nylons, are widely exploited in fibers, moldings, and films. Py-GC was used to investigate the thermal degradation of a series of aliphatic polyamides [49]. Both lactam and diamine-dicarboxylic acid type nylon samples were pyrolyzed at 550 °C in a furnace pyrolyzer under flowing nitrogen as the carrier gas, and the resulting degradation products were continuously separated by a capillary separation column. Table 5.3 summarizes the various classes of characteristic products observed in the resulting pyrograms [49,50]. The degradation products were found to depend strongly on the number of methylene groups in the polymer chain units.

Figure 5.8 shows pyrograms of (a) nylon 6 and (b) nylon 6/6 obtained using a glass capillary separation column. The pyrolysis products in (a) consist of ε-caprolactam [4,10,49,51,52] and small amounts of hydrocarbons (HC) and nitriles (MN and MN(A)). Polylactams consisting of relatively short methylene chains, such as nylon 4 and 6, tend to regenerate the associated monomeric lactam upon heating.

As shown in Figure 5.6b, the most abundant product of the thermal degradation of nylon 6/6 was cyclopentanone (CP) [49,50,53–57], which is characteristic of adipic acid-based polyamides. Py-EI and CI-tandem MS have demonstrated that the formation of CP from nylon 6/6 occurs via a C–H hydrogen transfer reaction to nitrogen to give the primary products bearing amine and ketoamide end groups as follows [57]:

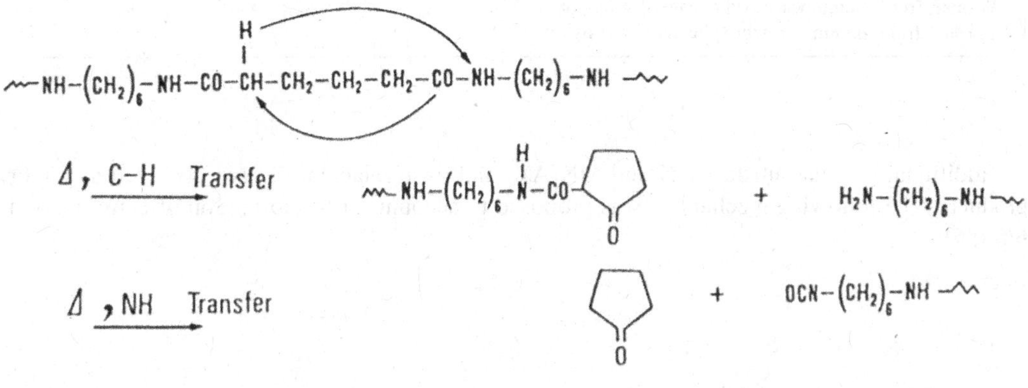

TABLE 5.3

Characteristic Degradation Products of Aliphatic Polyamides [31,32] (Peak Notations as in Figures 5.8 and 5.9) [49,50].

Class of compounds	Abbreviation	Structure
Hydrocarbons	HC	$CH_3-(CH_2)_m-CH_3$ $CH_2=CH-(CH_2)_{m-1}-CH_3$ $CH_2=CH-(CH_2)_{m-2}-CH=CH_2$
Mononitriles	MN	$CH_3-(CH_2)_m-C\equiv N$ $CH_2=CH-(CH_2)_{m-1}-C\equiv N$
Amines	AM	$CH_3-(CH_2)_m-NH_2$ $CH_2=CH-(CH_2)_{m-1}-NH_2$
Lactams	L	$O=C\underset{\diagdown}{\overset{(CH_2)_m}{\diagup}}NH$
Dinitriles	DN	$N\equiv C-(CH_2)_m-C\equiv N$
Cyclopentanone	CP	$\begin{matrix} CH_2-CH_2 \\ \vert \qquad\quad \diagdown \\ \qquad\qquad C=O \\ CH_2-CH_2 \diagup \end{matrix}$
Hydrocarbons containing one amide group	HC(A)	$CH_3-(CH_2)_m-\overset{O}{\overset{\Vert}{C}}-\overset{H}{\overset{\vert}{N}}-(CH_2)_n-CH_3$ $CH_2=CH-(CH_2)_{m-1}-\overset{O}{\overset{\Vert}{C}}-\overset{H}{\overset{\vert}{N}}-(CH_2)_n-CH_3$ $CH_2=CH-(CH_2)_{m-1}-\overset{O}{\overset{\Vert}{C}}-\overset{H}{\overset{\vert}{N}}-(CH_2)_{n-1}-CH=CH_2$
Mononitriles containing one amide group	MN(A)*	$CH_3-(CH_2)_m-\overset{O}{\overset{\Vert}{C}}-\overset{H}{\overset{\vert}{N}}-(CH_2)_n-C\equiv N$ $CH_2=CH-(CH_2)_{m-1}-\overset{O}{\overset{\Vert}{C}}-\overset{H}{\overset{\vert}{N}}-(CH_2)_n-C\equiv N$
Mononitriles containing one amide group	MN(A)**	$CH_3-(CH_2)_m-\overset{H}{\overset{\vert}{N}}-\overset{O}{\overset{\Vert}{C}}-(CH_2)_n-C\equiv N$ $CH_2=CH-(CH_2)_{m-1}-\overset{H}{\overset{\vert}{N}}-\overset{O}{\overset{\Vert}{C}}-(CH_2)_n-C\equiv N$

* Formed from ω-aminocarboxylic acid-type nylons.
** Formed from diamine-dicarboxylic acid-type nylons.

Additionally, some nitriles (DN and MN(A)′) and ε-caprolactam (L) are observed as minor products. The following mechanism was proposed to account for the formation of L from nylon 6/6 [56]:

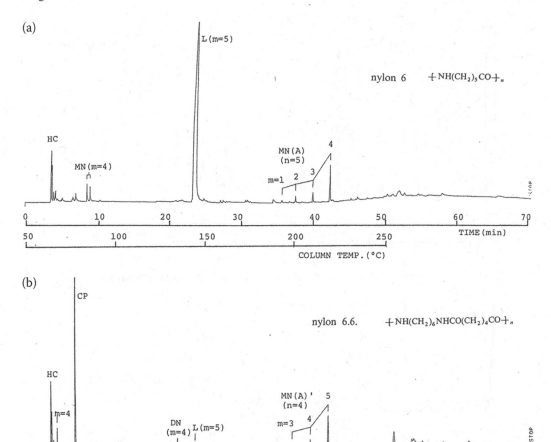

FIGURE 5.8 Pyrograms of nylons at 550°C observed by a glass capillary column: (a) Nylon 6 and (b) nylon 6/6 [49]. Peak notations correspond to those in Table 5.3.

Furthermore, peaks of amines arising from the cleavage of the amide bond were clearly observed in pyrograms of nylon 6/6 obtained using a fused-silica capillary separation column [50,58], although they are missing in Figure 5.8b.

Figure 5.9 shows the pyrograms of (a) nylon 11, (b) nylon 6/10, and (c) nylon 12/6 obtained using a fused-silica capillary column [50,58]. Fairly strong peaks corresponding to mononitriles (MNs) are observed in (a), while that of the associated lactam (L) is very small. This suggested that

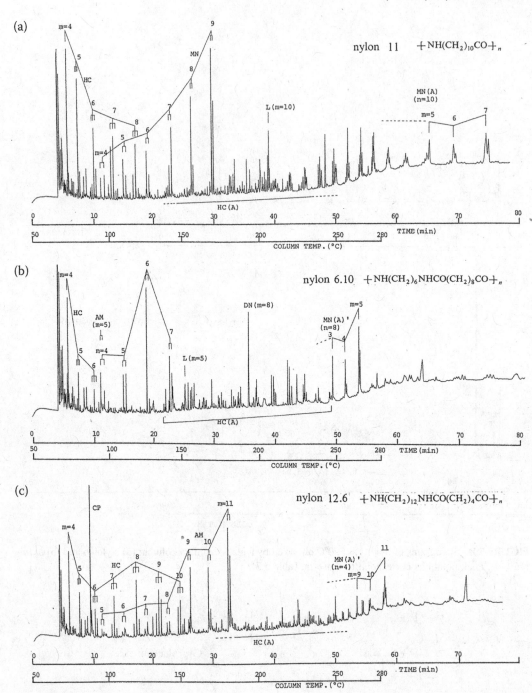

FIGURE 5.9 Pyrograms of nylons obtained at 550°C using a fused-silica capillary column: (a) Nylon 11, (b) nylon 6/10, and (c) nylon 12/6 [50,58]. Peak notations correspond to those in Table 5.2.

the thermal degradation of nylon 11 involved mainly homolytic cleavage of the CH_2–NH bond to form a double bond and an amide or a nitrile group through the six-membered cyclic transition state (*cis* elimination) [49,59]. The most intense olefinic MN peaks, whose carbon number corresponds to the number of successive methylene groups in the polymer chain plus 1, were formed through the following *cis* eliminations followed by the dehydration reaction [49]:

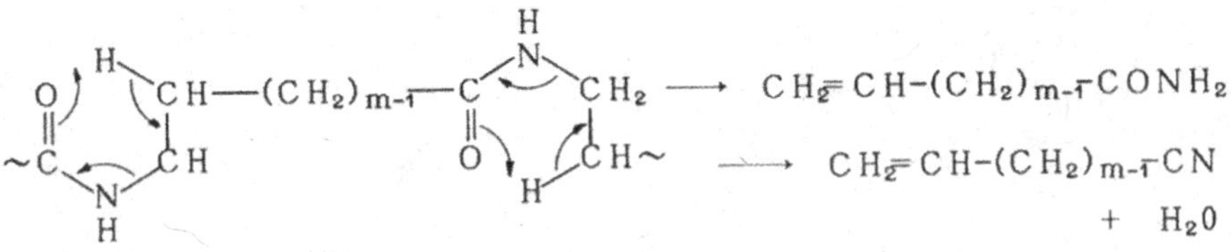

The minor peaks of the saturated MNs were always accompanied by those of the olefinic MNs. In addition, the smaller MNs and the associated hydrocarbons (HCs) were formed through further C–C bond cleavage. The main degradation pathway was basically the same as that of the polylactams, even for polymers with relatively longer methylene chains such as nylon 8, 11, and 12 [49].

One of the most characteristic products in Figure 5.9b is cebaconitrile (DN ($m = 8$)). In most diamine-dicarboxylic acid type polyamide chains—with the exception of adipic acid-based polyamides—large quantities of dinitriles were formed via *cis* elimination reactions at the neighboring amide groups across a dicarboxylic acid unit [49]. A series of mononitriles was also observed in the pyrograms. Among these, the second-longest unsaturated mononitrile exhibited the most intense peak. This fact suggests that the *cis* elimination reaction might take place at the opposite side of the amide bond as follows [60]:

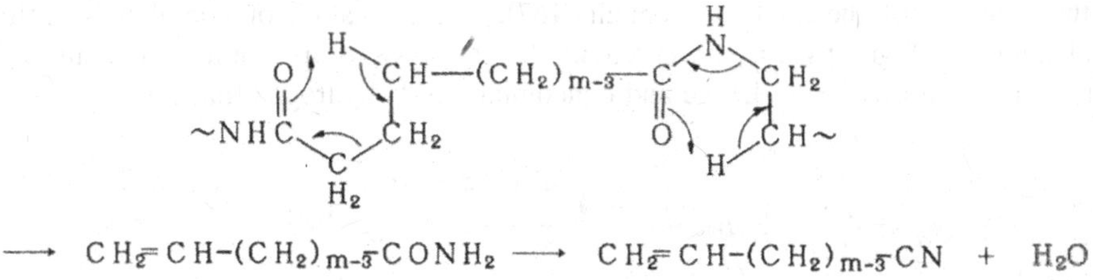

Similar to nylon 6/6, the most characteristic peak for nylon 12/6 was CP, which was formed from the adipic acid moiety. In addition, a series of HC triplets and amine (AM) doublets [50,58] with carbon numbers up to C_{12} were formed in considerable amounts through C–C and amide bond cleavage.

Thermal degradation of various aliphatic polyamides was also studied using Py-EIMS [10,59,61], Py-field ionization (FI), and field desorption (FD)-MS [52,60,62,63]. The Py-EIMS studies demonstrated that the polylactams mainly decomposed to cyclic oligomers, whereas the diamine-dicarboxylic acid type polyamides favored decomposition via the *cis* elimination reaction and cleavage of the amide bond—which also occurred in polylactams with a large number of methylene groups, such as nylon 12. On the other hand, in the Py-FIMS analysis of various diamine-dicarboxylic acid type polyamides, the main pyrolysis products were proto-nated dinitriles and protonated nitriles, as well as oligomers of up to 1000 Da, except in the case of polyamides containing adipic acid subunits, for which protonated amines and diamine were observed in large amounts [60]. A comprehensive review of the thermal degradation of aliphatic polyamides (nylons) has been published [64].

Thermal decomposition of aliphatic-aromatic polyamides has mainly been investigated using Py-MS. An early study found that the thermal degradation of the copolyamides formed from aliphatic and aromatic amino acids occurred almost exclusively via bond scission in the aliphatic moieties [65]. Py-EIMS results demonstrated that the thermal degradation of copolyamides of *p*-aminobenzoic acid and some aliphatic amino acids yielded poly-*p*-aminobenzoic acid via the elimination of lactams:

$$\text{~HN-}\bigcirc\text{-C(O)-NH-CH}_2\text{(C}_3\text{H}_6\text{)-C(O)N(H)-}\bigcirc\text{-CO~} \xrightarrow{360\ ^\circ C,\ 30\ min}$$

$$\text{~HN-}\bigcirc\text{-CO-NH-}\bigcirc\text{-CO~} + \text{ (succinimide ring) } H_2C\text{-CH}_2, HN\text{-CO}$$

The other favored reactions were the cleavage of the CO–NH bond to form *o*-aminophenyl end groups, *cis* elimination to form amide and vinyl end groups, and cleavage of the CH$_2$–CO bond to form isocyanate end groups [66]:

$$\sim CH_2\text{-}\{CO\text{-}NH\text{-}Ph\sim \xrightarrow{\Delta} \sim CH_3 + O=C=N\text{-}Ph\sim - e$$

Thermal decomposition processes of various aliphatic-aromatic polyamides have also been investigated using Py-GC-MS and Py-MS in both EI and CI modes [67–69]. The thermal decomposition of aromatic-diamine and aliphatic-dicarboxylic acid polyamides was strongly influenced by the structure of the aliphatic subunits [67]. The formation of compounds with succinimide and amine end groups was observed in the pyrolysis of polyamides containing succinic subunits via intramolecular exchange and concomitant N-H hydrogen transfer:

$$\text{~NH-Ph-N(H)-CO-CH}_2\text{-CH}_2\text{-CO-N(H)-Ph-NH~} \xrightarrow{\Delta,\ N\text{-}H\ Transfer} H_2N\text{-Ph-NH}_2$$

$$H_2N\text{-Ph-N}\langle CO\text{-}CH_2 | CO\text{-}CH_2 \rangle \qquad \langle H_2C\text{-}OC | H_2C\text{-}OC \rangle N\text{-Ph-N}\langle CO\text{-}CH_2 | CO\text{-}CH_2 \rangle$$

In the pyrolysis of polyamides containing adipic subunits, the primary products were compounds with amine and keto amide end groups. In addition, the *N*-methyl-substituted poly-amide decomposed via a hydrogen transfer process from the methyl group to the nitrogen atom, accompanied by the formation of compounds with amine and 2,5-piperidine end groups [68]:

$$\text{~N(CH}_3)\text{-Ar-N(CH}_2\text{-H)}\langle CH_2\text{-}CH_2, CO \rangle \quad \text{N(CH}_3)\text{-Ar-N(CH}_3)\langle CO \rangle \xrightarrow{\Delta,\ \alpha\ C\text{-}H\ Transfer}$$

$$H\text{-N(CH}_3)\text{-(-Ar-N(CH}_3)\text{-CO-CH}_2\text{-CH}_2\text{-CO-N-)}_n\text{-Ar-N}\langle CO\text{-}CH_2, CH_2, CH_2\text{-}CO \rangle$$

On the other hand, the primary thermal decomposition of polyamides of aliphatic-diamine and aromatic-dicarboxylic acid proceeded via the *cis* elimination process with the formation of the products containing amide and olefin end groups [69]. Nitrile end groups were also found to be formed through dehydration of the amide groups formed in the primary process.

TABLE 5.4

Tentatively Assigned Pyrolyzates of Aromatic Polyamides by Pyrolysis Field Ionization Mass Spectrometry [74].

m/z	Thermal degradation products	Relative abundance	
		Kevlar	Nomex
78	C_6H_6	0.5	5
93	$H_2N\!-\!C_6H_5$	7	10
103	$C_6H_5\!-\!CN$	15	35
108	$H_2N\!-\!C_6H_4\!-\!NH_2$	29	55
117	$NC\!-\!C_6H_4\!-\!CH_3$	0.7	3
118	$NC\!-\!C_6H_4\!-\!NH_2$	12	3
122	$C_6H_5\!-\!COOH$	1	13
128	$NC\!-\!C_6H_4\!-\!CN$	1	5
134	$H_2N\!-\!C_6H_4\!-\!N = C = O$	7	2
147	$NC\!-\!C_6H_4\!-\!COOH$	0.5	5
154	$C_6H_5\!-\!C_6H_5$	0.3	1
166	$HOOC\!-\!C_6H_4\!-\!COOH$	1	12
169	$C_6H_5\!-\!C_6H_4\!-\!NH_2$	1	1.5
179	$C_6H_5\!-\!C_6H_4\!-\!CN$	1	4
194	$H_2N\!-\!C_6H_4\!-\!C_6H_4\!-\!CN$	6	6
197	$C_6H_5\!-\!NH\!-\!CO\!-\!C_6H_5$	11	2
212	$H_2N\!-\!C_6H_4\!-\!NH\!-\!CO\!-\!C_6H_5$	100	70
222	$C_6H_5\!-\!NH\!-\!CO\!-\!C_6H_4\!-\!CN$	10	1
237	$H_2N\!-\!C_6H_4\!-\!NH\!-\!CO\!-\!C_6H_4\!-\!CN$	74	31
238	$O = C = N\!-\!C_6H_4\!-\!NH\!-\!CO\!-\!C_6H_5$	23	10
241	$C_6H_5\!-\!NH\!-\!CO\!-\!C_6H_4\!-\!COOH$	1	5
256	$H_2N\!-\!C_6H_4\!-\!NH\!-\!CO\!-\!C_6H_4\!-\!COOH$	4	100
282	$O = C = N\!-\!C_6H_4\!-\!NH\!-\!CO\!-\!C_6H_4\!-\!COOH$	2	5
316	$C_6H_5\!-\!CO\!-\!NH\!-\!C_6H_4\!-\!NH\!-\!CO\!-\!C_6H_5$ $C_6H_5\!-\!NH\!-\!CO\!-\!C_6H_5\!-\!CO\!-\!NH\!-\!C_6H_5$	55	14
331	$H_2N\!-\!C_6H_4\!-\!NH\!-\!CO\!-\!C_6H_4\!-\!CO\!-\!NH\!-\!C_6H_5$	21	25
341	$C_6H_5\!-\!CO\!-\!NH\!-\!C_6H_4\!-\!NH\!-\!CO\!-\!C_6H_4\!-\!CN$	37	12
346	$H_2N\!-\!C_6H_4\!-\!NH\!-\!CO\!-\!C_6H_4\!-\!CO\!-\!NH\!-\!C_6H_4\!-\!NH_2$	18	66
360	$C_6H_5\!-\!CO\!-\!NH\!-\!C_6H_4\!-\!NH\!-\!CO\!-\!C_6H_4\!-\!COOH$	2	19
385	monomer + 147	1	3
435	monomer + 197	2	0
450	monomer + 212	11	7
475	monomer + 237	1.5	1.5
476	monomer + 238	4	0.5

Various investigations of the thermal degradation of wholly aromatic polyamides (aramids) such as poly(1,3-phenylene isophthalamide) (Nomex) and poly(1,4-phenylene terephthalamide) (Kevlar) have also been reported [52,70–74].

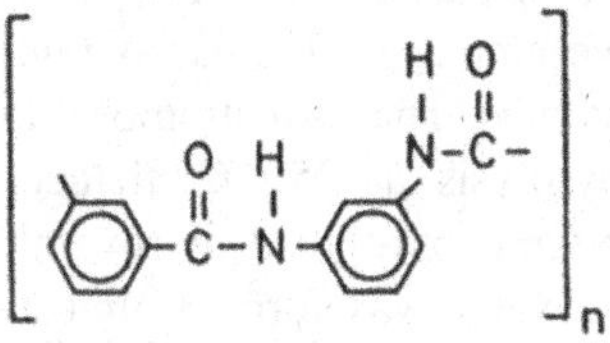

Nomex

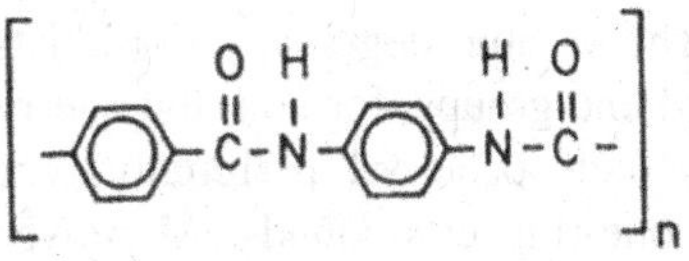

Kevlar

TABLE 5.5

Possible Intermediate Products of Aromatic Polyamides via Hydrolytic and Homolytic Decomposition [74].

Type of decomposition	Possible intermediate products
(a) Hydrolytic decomposition	
(b) Homolytic decomposition	

In an early Py-FIMS study [52], benzonitrile and other pyrolysates with amine and/or nitrile end groups were detected as the main pyrolysis products of Kevlar at 600 °C. In another paper dealing with the pyrolysis of Nomex at 550 °C, the primary low-boiling volatiles identified by GC/MS were CO, CO_2, H_2O, and benzonitrile [70]. Considerable amounts of benzene, methane, toluene, 1,3-tolunitrile, etc., were also observed. The presence of at least 17 additional degradation products in the condensable products was detected using high-performance liquid chromatography (HPLC); the two major components were 1,3-dicyanobenzene and 3-cyanobenzoic acid. These observations supported a mechanism involving the cleavage of an aromatic-NH bond followed by the loss of H_2O to form aromatic nitriles. Cleavage of the CO–NH bond, hydrolysis, and decarboxylation explained the other major products.

In another study, the pyrolysis products of a Nomex-type aramid and its chloro-derivative at 450° and 550 °C were identified using GC-Fourier transform infrared spectrometry (FTIR) and GC/MS [71]. The volatile degradation products were reported to have amine, nitrile, carboxylic acid, and phenyl end groups, for both the one-ring and two-ring compounds. The formation of the two-ring compounds occurred preferentially at 450 °C, whereas pyrolysis at 550 °C yielded predominantly one-ring compounds. Moreover, Nomex and Kevlar were pyrolyzed at several temperatures between 300° and 700 °C [72]. At lower temperatures, water was formed almost exclusively as the volatile degradation product, accompanied by traces of carbon dioxide. Hydrolysis products were formed with increasing temperature, followed by the formation of nitriles and products containing a phenyl end group. Toluene and biphenyl derivatives, HCN, and hydrocarbons were observed at high temperatures. The results reported in these two papers [71,72] suggested that homolytic cleavage involving all the nonaromatic ring bonds took place during the degradation of these aramids, along with hydrolytic reactions. At lower temperatures, the hydrolytic mechanism was dominant, whereas at higher temperatures, homolytic reactions became increasingly important. Furthermore, for Nomex, the cleavage of the N–H bond has been reported to be both thermodynamically and kinetically favored over the scission of other bonds in the polymer main chain at the beginning of thermal degradation [73].

The thermal degradation of Nomex and Kevlar was also studied using Py-FIMS and Py-GC [74]. In Py-FIMS, the polymer samples were pyrolyzed in the ion source by heating from 50° to 750 °C at a rate of 1.2 °C/s. The relative abundances of the thermal degradation products are listed in Table 5.4. Possible intermediates in the formation of the observed compounds through hydrolytic and homolytic decomposition processes are summarized in Table 5.5. Kevlar and Nomex produced similar signals overall, but strong differences were observed in the relative abundances of several of the signals. Higher abundances of degradation products containing carboxylic acid groups were generally observed for Nomex, whereas Kevlar yielded greater amounts of fragments resulting from the cleavage of the carbonyl/phenyl bond. In general, nitriles were preferentially formed over carboxylic acids at higher temperatures. As the subsequent decarboxylation reactions of the carboxylic acids are favored at higher temperatures, the amount of carboxylic acids produced from Nomex decreased with increasing pyrolysis temperature, and Kevlar produced even smaller quantities of carboxylic acids than Nomex because of the higher thermal stability of the former.

On the other hand, the pyrograms observed in flash Py-GC at 720 °C consisted almost entirely of fragments formed via homolytic degradation reactions, although many were identical to those observed using Py-FIMS. In addition, the differences between Nomex and Kevlar using Py-GC were much smaller than those in Py-FIMS. Moreover, the formation of secondary products, such as biphenyl derivatives, was much lower in Py-GC than in Py-FIMS. The differences between the results obtained using Py-GC and by Py-FIMS could be attributed to the differences in the final pyrolysis temperature and the heating rate.

REFERENCES

1. G. Montaudo and C. Puglisi, Thermal degradation of condensation polymers, *Comprehensive Polymer Science, 1st Supplement* (S. L. Aggarwal and S. Russo, eds.), Pergamon Press, Oxford, UK, pp. 227–251 (1992).
2. Y. Sugimura and S. Tsuge, *J. Chromatagr. Sci.*, *17*, 269 (1979).
3. H. Ohtani, T. Kimura, and S. Tsuge, *Anal. Sci.*, *2*, 179 (1986).
4. S. A. Liebman and E. J. Levy, eds., *Pyrolysis and GC in Polymer Analysis*, Chromatogr. Sci., Series, Vol. 29, Marcel Dekker, New York (1985); J. H. Flynn and R. E. Florin, *Degradation and Pyrolysis Mechanism*, pp. 179–186; D. H. Ahlstrom, *Microstructure of Synthetic Polymers*, pp. 256–269; R. Saferstein, *Forensic Aspect Pyrolysis*, pp. 350–353.
5. N. Grassie and G. Scott, *Polymer Degradation and Stabilization*, Cambridge University Press, Cambridge, UK, pp. 33–41 (1985).
6. I. C. McNeil, Thermal degradation, in *Comprehensive Polym. Sci., Vol. 6, Polymer Reactions* (G. C. Eastmond, A. Ledwith, S. Russo, and P. Sigwald, eds.), Pergamon Press, Oxford, UK, pp. 490–495 (1989).
7. V. Passalacqua, F. Pilati, V. Zamboni, B. Fortunato, and P. Manaresi, *Polymer*, *17*, 1044 (1976).
8. M. E. Bednas, M. Day, K. Ho, R. Sander, and D. M. Wiles, *J. Appl. Polym. Sci.*, *26*, 277 (1981).
9. C. T. Vijayakumar and J. K. Fink, *Thermochim. Acta*, *59*, 51 (1982).
10. A. Zeman, *Angew. Makromol. Chem.*, *31*, 1 (1973).
11. I. Luederwald and H. Urrutia, *Makromol. Chem.*, *177*, 2079 (1976).

12. I. Luederwald and H. Urrutia, Direct pyrolysis of aromatic and aliphatic polyesters in the mass spectrometer (C. E. B. Jones and C. A. Cramers, eds.), *Analytical Pyrolysis. Proceedings of the 3rd International Symposium on Analytical Pyrolysis, Amsterdam, 1976*, Elsevier, Amsterdam, The Netherlands, pp. 139–148 (1977).
13. R. M. Rum, *J. Polym. Sci. Polym. Chem. Ed.*, *17*, 203 (1979).
14. R. E. Adams, *J. Polym. Sci. Polym. Chem. Ed.*, *20*, 119 (1982).
15. D. C. Conway and R. Marak, *J. Polym. Sci. Polym. Chem. Ed.*, *20*, 1765 (1982).
16. I. Luederwald, *Pure Appl. Chem.*, *54*, 255 (1982).
17. C. T. Vijayakumar, J. K. Fink, and K. Lederer, *Eur. Polym. J.*, *23*, 861 (1987).
18. G. H. Irzl, C. T. Vijayakumar, J. K. Fink, and K. Lederer, *Polym. Degrad. Stab.*, *16*, 53 (1986).
19. C. T. Vijayakumar and K. Lederer, *Makromol. Chem.*, *189*, 2559 (1988).
20. I. Luederwald and H. Urrutia, *Makromol. Chem.*, *177*, 2093 (1976).
21. I. Luederwald, *Makromol. Chem.*, *178*, 2603 (1977).
22. H. R. Kricheldorf and I. Luederwald, *Makromol. Chem.*, *179*, 421 (1978).
23. E. Jacobi, I. Luederwald, and R. C. Schultz, *Makromol. Chem.*, *179*, 429 (1978).
24. M. Doerr, I. Luederwald, and H.-R. Schulten, *Fresenius Z. Anal. Chem.*, *318*, 339 (1984).
25. D. Garazzo, M. Gluffrida, and G. Montaudo, *Macromolecules*, *19*, 1643 (1986).
26. B. Plage and H.-R. Schulten, *J. Anal. Appl. Pyrol.*, *15*, 197 (1989).
27. B. Plage and H.-R. Schulten, *Macromolecules*, *23*, 2649 (1990).
28. H. Sato, M. Furuhashi, D. Yang, H. Ohtani, S. Tsuge, M. Okada, K. Tsunoda, and K. Aoi, *Polym. Degrad. Stab.*, *73*, 327 (2001).
29. Y. Aoyagi, K. Yamashita, and Y. Doi, *Polym. Degrad. Stab.*, *76*, 53 (2002).
30. H. Abe, N. Takahashi, K. J. Kim, M. Mochizuki, and Y. Doi, *Biomacromolecules*, *5*, 1480 (2004).
31. F.-D. Kopinke, M. Remmler, K. Mackenzie, M. Moeder, and O. Wachen, *Polym. Degrad. Stab.*, *53*, 329 (1996).
32. F.-D. Kopinke and K. Mackenzie, *J. Anal. Appl. Pyrolysis*, *40-41*, 43 (1997).
33. F. Khabbaz, S. Karksson, and A.-C. Albertsson, *J. Appl. Polym. Sci.*, *78*, 2369 (2000).
34. Y. Fan, H. Nishida, S. Hoshihara, Y. Shirai, Y. Tokiwa, and T. Endo. *Polym. Degrad. Stab.*, *79*, 547 (2003).
35. Y. Fan, H. Nishida, Y. Shirai, and T. Endo. *Polym. Degrad. Stab.*, *80*, 503 (2003).
36. H. Nishida, T. Mori, S. Hoshihara, Y. Fan, Y. Shirai, Y. Tokiwa, and T. Endo. *Polym. Degrad. Stab.*, *81*, 515 (2003).
37. Y. Fan, H. Nishida, Y. Shirai, Y. Tokiwa, and T. Endo. *Polym. Degrad. Stab.*, *86*, 197 (2004).
38. H. Abe, N. Takahashi, K. J. Kim, M. Mochizuki, and Y. Doi, *Biomacromolecules*, *5*, 1606 (2004).
39. Y. Fan, H. Nishida, T. Mori, Y. Shirai, and Y. Doi, *Polymer*, *45*, 1197 (2004).
40. B. E. Watt, S. L. Morgan, and A. Fox, *J. Anal. Appl. Pyrolysis*, *19*, 237 (1991).
41. R. S. Lehrle and R. J. Williams, *Macromolecules*, *27*, 3782 (1994).
42. R. S. Lehrle, R. J. Williams, C. French, and T. Hammond, *Macromolecules*, *27*, 3782 (1994).
43. F.-D. Kopinke, M. Remmler, and K. Mackenzie, *Polym. Degrad. Stab.*, *52*, 25 (1996).
44. S.-D. Li, J.-D., He, P. H. Yu, and M. K. Cheung, *J. Anal. Appl. Pyrolysis*, *19*, 237 (1991).
45. A. Gonzalez, L. Irusta, M. J. Fernandez-Berridi, M. Iriarte, and J. J. Iruin, *Polym. Degrad. Stab.*, *87*, 347 (2005).
46. B. Crossland, G. J. Knight, and W. W. Wright, *Br. Polym. J.*, *18*, 371 (1986).
47. K. Sueoka, M. Nagata, H. Ohtani, N. Nagai, and S. Tsuge, *J. Polym. Sci. Part A*, *29*, 1903 (1991).
48. M. Giuffrida, P. Marvigna, G. Montaudo, and E. Chiellini, *J. Polym. Sci. Part A*, *24*, 1643 (1986).
49. H. Ohtani, T. Nagaya, Y. Sugimura, and S. Tsuge, *J. Anal. Appl. Pyrol.*, *4*, 117 (1982).
50. S. Tsuge, *Chromatogr. Forum*, *1*, 44 (1986).
51. H. Senoo, S. Tsuge, and T. Takeuchi, *J. Polym. Sci.*, *9*, 315 (1971).
52. H.-J. Duessel, H. Rosen, and O. Hummel, *Makromol. Chem.*, *177*, 2434 (1976).
53. L. J. Peebles, Jr. and M. W. Huffma, *J. Polym. Sci. Part A*, *9*, 1807 (1971).
54. F. Wiloth, *Makromol. Chem.*, *144*, 263 (1971).
55. C. David, Thermal degradation of polymers, in *Comprehensive Chemical Kinetics, Vol. 14, Degradation of Polymers* (C. H. Bamford and C. F. H. Tipper, eds.), Elsevier, Amsterdam, The Netherlands, pp. 104–121, 130-153 (1975).
56. D. M. MacKerron and R. P. Gordon, *Polym. Degrad. Stab.*, *12*, 277 (1985).
57. A. Ballisteri, D. Garozzo, M. Giuffrida, and G. Montaudo, *Macromolecules*, *20*, 2991 (1987).
58. S. Tsuge, H. Ohtani, H. Matsubara, and M. Ohsawa, *J. Anal. Appl. Pyrol.*, *17*, 181 (1987).
59. I. Luederwald and F. Merz, *Angew. Makromol. Chem.*, *74*, 165 (1978).

60. H.-R. Schulten and B. Plage, *J. Polym. Sci, Part A*, *26*, 2381 (1988).
61. I. Luederwald, F. Merz, and M. Rothe, *Angew. Makromol. Chem.*, *67*, 193 (1978).
62. U. Bahr, I. Luederwald, R. Mueller, and H.-R. Schulten, *Angew. Makromol. Chem.*, *120*:163 (1984).
63. B. Plage and H.-R. Schulten, *J. Appl. Polym. Sci*, *38*:123 (1989).
64. S. V. Levchik, E. D. Well, and M. Lewin, *Polym. Int.*, *48*, 532 (1999).
65. H. R. Kricheldorf and E. Leppert, *Makromol. Chem.*, *175*, 1731 (1974).
66. I. Luederwald and H. R. Kricheldorf, *Angew. Makromol. Chem.*, *56*, 173 (1976).
67. A. Ballisteri, D. Garozzo, M. Giuffrida, P. Maravigna, and G. Montaudo, *Macromolecules*, *19*, 2963 (1983).
68. A. Ballisteri, D. Garozzo, G. Montaudo, and M. Giuffrida, *J. Polym. Sci. Part A*, *25*, 2531 (1987).
69. A. Ballisteri, D. Garozzo, P. Maravigna, G. Montaudo, and M. Giuffrida, *J. Polym. Sci. Part A*, *25*, 1049 (1987).
70. D. A. Chatfield, I. N. Einhorn, R. W. Michelson, and J. H. Futrell, *J. Polym, Sci. Polym. Chem. Ed.*, *17*, 1367 (1979).
71. Y. P. Khanna, E. M. Pearce, J. S. Smith, D. T. Burkitt, H. Njuguna, D. M. Hindenlang, and B. D. Forman, *J. Polym. Sci. Polym. Chem. Ed.*, *19*, 2817 (1981).
72. J. R. Brown and A. J. Power, *Polym. Degrad. Stab.*, *4*, 179 (1989).
73. A. L. Bhuiyan, *Eur. Polym. J.*, *19*, 195 (1983).
74. H.-R. Schulten, B. Plage, H. Ohtani, and S. Tsuge, *Angew. Makromol. Chem.*, *155*, 1 (1987).

6 The Application of Analytical Pyrolysis to the Study of Cultural Materials

Nathalie Balcar

Center for Research and Restoration of the Museums of France – C2RMF, Paris, France

I INTRODUCTION

The challenge for conservation scientists is to meet all the questions from conservators and curators about the materiality of a work of art, the stratigraphy of layers, materials used, state of preservation, etc., with only a single and tiny sample of matter (often smaller than 1 mm²). For the identification of the organic part, various chromatographic techniques are effective but first require materials to be soluble (e.g., high-performance liquid chromatographic [HPLC]) or volatile (e.g., gas chromatography [GC]), which may not be the case for instance with boiled oil used in old paintings or some synthetic polymers like plastics. Secondly, a preliminary chemical treatment is necessary and sometimes specific to the (organic) media, such as derivatization with N,O-bis (trimethylsilyl) trifluoroacetamide (BSTFA) with and without 1% trimethylchlorosilane (BSTFA-TCM) for gums [1], with *m*-(trifluoromethyl)phenyltrimethylammonium hydroxide TFTM (TMTFTH) for oils [2], or with ethyl chloroformate (ECF) for proteins (animal glue) [3]. This implies having hypotheses and luck, or successively applying all the treatments onto the same sample [4], which could increase the risk of material loss, or pollution. Moreover, it seems that the presence of some pigments could affect the treatment [5].

For all these reasons, the analytical pyrolysis, and mainly the pyrolysis-gas chromatography-mass spectrometry (Py-GC/MS) configuration, is considered in conservation laboratories at the international level (United States of America, France, Netherlands, Spain, etc.) as a powerful technique for the identification of binding media, varnish, and adhesive made with natural and synthetic polymers.

Several projects can be cited to illustrate the increasing use of Py-GC/MS in the field of cultural heritage, and related publications demonstrate how today, this technique is simple, can be rapidly easy-to-use, and powerful to enhance knowledge about craftsmen's and artists' techniques from antiquity up to contemporary art.

The use of Py-GCMS—thanks to the improvements in MS detectors—allows the identification of the main component of the material, and also some of its additives [6], that can be essential for curative or preventive conservation strategies.

Another attractive feature besides the identification of the material is the possibility to get information about its state of preservation. This will result in the loss of some components—for example, plasticizer in vinyl paints (PVAc) [7]—or, at the opposite, in the formation of degradation product coming from the material [8], both of which may be present at very low concentration.

The main advantage of Py-GC/MS is the absence of preparation of the sample—just put it in the sample vessel, pyrolyze, and wait for data. It is at the moment of interpretation of the results that the technique can seem complex, but this was the feeling of a new user in the 90s. Today, a multitude of books [9–13] and papers [14–17] have been published that makes the novice quickly become experienced users and, after some years, an expert.

There is a multitude of ways to pyrolyze a sample: at high or low temperature, with a fast or slow heating rate [18], for a long or short time, and, finally, with or without reacting agents [19]. By controlling all these parameters, it is possible to have the perfect fingerprint with a higher level of information and good signal intensity. But, selecting the perfect parameters means that you already "know" what you are analyzing—which is rarely the case in the field of heritage. For instance, a shiny paint could be achieved with linseed oil but also with a nitrocellulose lacquer; a handbag can appear leather but be made of textile coated with a PVC or PUR film, etc. To overcome this, a characteristic fingerprint can be obtained for a wide and diverse range of material with the same pyrolysis parameters—what can be called a screening method (see experimental detail). This technique has been developed in our laboratory in the early 2000s, and it was the beginning of the use of analytical pyrolysis with a CDS Pyroprobe 1200 connected to a chromatogram with FID detection. The final temperature of the filament selected was 650 °C applied for 10 s with a heating rate of 10 °C/msec. As it was possible with these parameters to identify most of the natural and synthetic polymers encountered in museum artifacts, this screening method was logically transposed to the Py-GC/MS equipment. MS detection is a necessity in the field of heritage, assuming that craftsmen/artists can prepare their paintings, varnish, etc. by mixing several ingredients; also because objects could be restored with a material different from the original. The samples studied are often complex mixtures.

This screening method has also been successfully tested with the TMAH [20,21], the first used reactive agent in cultural heritage laboratories to perform derivatization (methylation) that allows better separation and also enhances the method sensitivity for most of the natural organic polymers that contain polar functional groups. Most of the time, TMAH is added to the sample in the sample vessel just before pyrolysis, methylation occurs *in situ* simultaneously with the pyrolysis, avoiding loss of sample and time-consuming sample preparation. Some inconvenience has been reported with TMAH [22]. Other agents have also been tested and approved [23]: butylation with tetra-butylammonium hydroxide (TBAH) [24] and silylation with HMDS (hexamethyldisilazane) [25]; BSTFA alone or with TMCS as catalyst [26].

For most synthetic polymers, a conventional Py-GC/MS analysis—that is, without a reactive agent—is enough for the identification of the material. Acrylics, alkyds, poly-urethanes, polyamides, and epoxies can be identified with a high degree of confidence as each produces typical fragments. For some of these polymers, different monomers can be used for their formulation. The identification of these monomers can be satisfied only with the use of TMAH—this is the case of the polyols involved in alkyds resins or in polyurethane paints. This information is of great importance to better understand why two similar materials age differently, as it is often because one of the precursors is different. This information is crucial to establish the best preventive conservation, or select the best copolymer to meet specific needs in a curative project.

All these descriptions could suggest that the Py-GC/MS tool is very simple and can deal with any situation. It would be a lie to assert, but in the field of cultural heritage due to the uniqueness of the sample, the many possibilities of materials used—alone or in complex mixtures—and their various states of conservation, one can be affirmed that it is a very powerful technique. Of course, some disadvantages may be listed; for natural products, a reference library must be constructed even with detection by MS; some samples could be better identified with derivatization than without; pyrolysis may cause side reactions complicating the interpretation of data [27], etc. These can be overcome with a thorough and systematic practice of this tool; Py-GC/MS as a technique requires experience.

II INSTRUMENTAL CONSIDERATION

Different types of pyrolysis equipment are available on the market. Undoubtedly, there will be pros and cons of each system but this opinion is based on the user's experience. For 16 years, we used CDS Pyroprobe systems; the old model 1200 was bought in 2003, and it still operates. The newer model, 5200, has new functions allowing a more ergonomic work environment and, above all, offers new possibilities to explore the sample. It is a multifunctional pyrolyzer that gives access to additional information on materials of artworks. Among these new options, the multi-step pyrolysis is probably the one that can be applied routinely even in replacement of the single pyrolysis in the field of culture heritage with its unique sample.

Indeed, it often appears interesting to submit a sample to several 'pyrolysis' temperatures (multi-step) to obtain a more detailed composition. The lowest range (100–400 °C)—which corresponds to a more thermal desorption than to pyrolysis—will allow the detection of low molecular weight components such as additives in modern materials or oligomers from the degradation process, which can decompose partially or completely at high temperature and, therefore, can be difficult to detect (in the background noise) or invisible in a single shot pyrolysis normally sets above 500 °C. The highest range (400–900 °C) will fragment the remaining macromolecules like the synthetic polymers and the natural polymers such as wax esters (or fatty acids esters), polysaccharides, terpenes, etc.

This multi-step procedure can be particularly adapted for the study of the mixtures of compounds like the ones made by craftsmen or artists. Each ingredient is giving a more intense signal (greatest yield) at a specific temperature, making data interpretation easier [28].

The design of the equipment allows the multi-step to run in an automated way—that means the sample stays in the device during the series of increasing temperatures, but it is cooled near the room temperature between each pyrolysis.

The multi-step procedure can appear time-consuming but in some cases, it's the only one that could give an answer to multiple questions: Is this ready-made varnish based on natural or synthetic resin? Are there antioxidants added (like Irganox at low ppm levels)? Is the varnish cross-linked?

Another interesting option of the newer CDS Pyroprobes in the field of art is the trap that allows the collection of analytes from thermal desorption and slow rate pyrolysis.

The thermal desorption is used for the study of volatile organic compounds (VOCs). The off-gassing of VOCs is a very important current concern especially for indoor air pollution, but this also worries museums in terms of preventive conservation. Indeed in the collections, objects can be made of materials that emit harmful volatiles—such as acetic acid from cellulose acetate—so these must be displayed away from items made of wood, or metal sensitive to acids. The material used for the storage (foams, boxes, films, textile, etc.) and the exhibition (support, showcase, glues, paints, etc.) must also be free from any release of volatiles that could interact with artworks. It is the same for any products that can be used in a curative conservation treatment. With the trap option, it is possible to study the off-gassing of all these materials and help collection managers, restorers, and curators of the exhibition to make a good choice.

With the trap, we can also collect the analytes using slow rate pyrolysis that can be likened to artificial aging by heat. The results obtained can provide information on the natural aging process of the material.

III ANALYSIS OF MATERIALS

The goal of this chapter is to give future users of Py-GC/MS the keys to practice quickly, effectively, and confidently this technique, keeping them from the big doubts due to lack of knowledge faced with the practice of a new analytical tool (some specific case studies will be given to demonstrate how this technique could give a solution to what may appear to be an unsolvable problem).

Considering that this chapter is an update of the two previously published in 1995 and 2006, we will continue to organize the data according to the class of materials. We decided to present some results described in the literature focused on archeology and art field, and also some coming from our own experience of the use of Py-GC/MS in our everyday activity for conservation issues and research on materials.

A USUAL NATURAL MATERIALS: OILS, WAX, RESINS, GUMS, AND PROTEINS

Many natural macromolecules and polymers from vegetable and animal sources have been—and are—used in paint layers, varnishes, adhesives, and fillers for the creation of a piece of art as well as for conservation treatments.

Just for the record, the natural products are mainly formed by polar molecules such as proteins, lipids, carbohydrates, diterpenoids, and triterpenoid resins. All of them are what may be called 'good candidates' for analytical pyrolysis. Indeed, with the screening method already cited, the fragments resulting from pyrolysis are very different, and marker compounds can be listed for each category; they can, therefore, be unambiguously identified. One may note that carrying out the analysis of the sample with or without derivatization leads, in most cases, to the same conclusion concerning the material present inside. The use of a reactive agent—either TMAH, HMDS, or BSTFA—can improve the detection (better resolution and peak shape), but in some cases, it makes the identification more simple and/or increases the level of interpretation by giving significant information to determine the source of the material—as for instance, to distinguish an Arabic gum from a tragacanth gum [26]. Such precision can be essential to understand the different states of conservation for similar material and is always valuable for the history of art.

As many publications mention the mechanisms of pyrolysate formation for each product group, it will not be detailed in this chapter. From studies on natural materials—either as references material or present in samples from works of art—it was established that resulting fragments are fatty acids for oil [29,30], fatty acids and hydrocarbons for animal and vegetal wax [31–33], terpene structures for vegetal resins (dammar, mastic, colophony) [34–36], diketopiperazines structures for animal glue (gelatin, casein, egg) [17,37,38] and furfuryl structures for carbohydrates [19,39,40].

An important scientific literature is available on the analysis by Py-GC/MS of the major natural materials used in archeology and art, and list the main markers compounds allowing the material identification that are often the more intense peaks of the pyrogram.

Sometimes, unidentified fragments are systematically observed in addition to marker compounds, keep in mind that it's very important to add them to your own mass spectra library as one day they could give new keys for the interpretation of the data.

In the light of our experience, natural resins are probably the most widely used in the field of cultural heritage as they can be mixed with wax to make fillers on marble sculptures, used as adhesives for ceramics in old conservation treatments, varnishes for paintings and furniture, and even as additives in a binder of oil paints.

The use of pyrolysis at 650 °C (screening method, see detail in Table 6.1) with TMAH always enabled us to detect pure or mixed resins from fresh or historical samples and to distinguish the plant from which it was extracted. A recent work on ingredients used in European lacquers illustrates that the signal can be optimized by applying another temperature [41]. In this case, it's a lower temperature—480 °C with TMAH—that enhances the signal strength, even if this can appear less evident for resins such Congo and Manila copals. Also, when faced with a sample that is probably made of plant resins—given the age and the technique of the artwork, the function of the material sampled, and the results of the solvent tests made by the curators in the case of a conservation treatment, etc.—it could be interesting to use this lower temperature. Once again, it's your experience that will help decide.

TABLE 6.1

Experimental Details of the Screening Method

Samples were analyzed on a Shimadzu QP2010 gas chromatograph/mass spectrometer using a CDS 5200 fitted with a transfer line, valve oven, and accessory (pyrolysis chamber) set at 280 °C and purged with He at 22 ml/min.

Samples were placed into quartz tubes fitted with quartz wool, and for some materials, four microliters of 25% tetramethylammonium hydroxide (TMAH) in water were introduced for derivatization.

Samples were pyrolyzed using the following temperature program: 280 °C for 1 s, then ramped at 10 °C/mS to 650 °C and held isothermally for 10 s.

The split injector was at 290 °C (80:1 ratio for pure "resin" and 50:1 for resin + pigments and/or fillers) and the MS transfer line was set to 310 °C. A CP-Sil 8CB-MS capillary column (30 m × 0.25 mm × 0.25 μm) was used with He at 36 cm/s.

The GC oven temperature program was 5 min at 35 °C, then 10 °C/min to 320 °C, and 5.75 min isothermal. The solvent delay was 3 min with TMAH and no delay without TMAH. The mass spectrometer was scanned from m/z 30–650.

GCMS unit is a Shimadzu QP2010 with a capillary column CP-Sil 8CB-MS 30 m, diam 0.25 mm, thickness 025 μm.

In the face of the choice of temperature dilemma, one might question the advantage of the Py-GC/MS over a GC/MS analysis with prior treatment. Even if there are strong presumptions on the presence of natural ingredient(s) in the sample, there is always a doubt about its state. Often, they undergo cross-linking reactions by longtime exposure (in the field of heritage, it could be several hundred years) to light, air, and—in some cases—heat when the materials are boiled for the preparation of oil in some techniques. This polymerization can partially or totally affect the organic fraction of the sample and in this case, the ingredient(s) can be detected only with an analysis by Py-GC/MS.

This case has been encountered for the study of a sample taken at the time of the restoration of the Hall of Mirrors in the Palace of Versailles, France. This is an adhesive used to maintain the paintings on the ceiling; it comes in solid form and a very hard material, with an ochre color.

This adhesive, named "maroufle" in French, is often made with the residues of the oil paint collected in the base of pots. So, it was decided to use GC/MS with a prior BSTFA derivatization—one of the procedures to study oil painting. In fact, we never succeeded in solubilizing the sample, even after three weeks in a mixture of dichloromethane/methanol (95/5) with several ultrasonic steps. We tried on a second sample a Py-GC/MS analysis at 650 °C with TMAH and we detected all the methylated diacids (C8 to C10) and fatty acids (C14 to C18) that revealed the presence of an oil (Figure 6.1). The latter is, in fact, so polymerized that it becomes insoluble, and only thermal degradation allows the break down of the oily network formed.

Technically, it is often easy to take with a scalpel a sample of an artwork, without damaging it. We can choose the edges of the paint protected by the frame or those of an accident in the varnish layer. It is even easier when it comes to restoration products or accidental deposits—both must be ultimately removed. It becomes complicated in the degradation products that will most often also be eliminated because they form either a tiny layer or several small dispersed spots on the surface. In addition, they often appear white on objects and translucent when sampled. Then, it is very difficult to put them in the sampling tube for later extraction to introduce them into the quartz tube or other kind of device such as a stainless steel cup.

It is, therefore, necessary to find a support that transfers the degradation products from the surface to the quartz tube, which can be applied to the artwork surface without damaging it and whose fragments resulting from pyrolysis do not introduce confusion in the interpretation of the data. The best candidate is the sterile cotton swab—the ones used to take saliva in forensic science. This technique has been applied several times for the study of white blooming on the surface of unvarnished modern oil paint. In the pyrogram, the first area contains many peaks coming from the

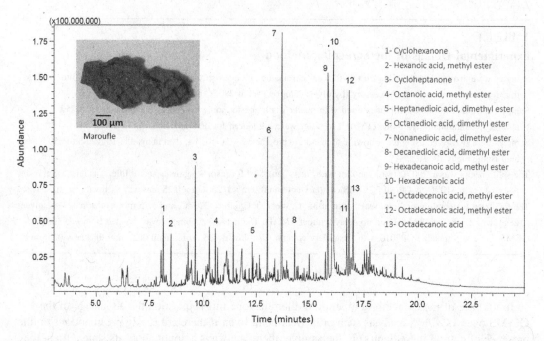

FIGURE 6.1 Pyrogram of the 'maroufle' sample. Series of diacids and acids methyl esters indicate that the binder is oil.

cotton; even if it is intense compared to the usual background, the peaks corresponding to the degradation products can be clearly detected and identified. So far, and for all the cases, the blooming results from accumulation on the surface of the same fatty acids (C16, C18) than those detected during an analysis of oil (Figure 6.2).

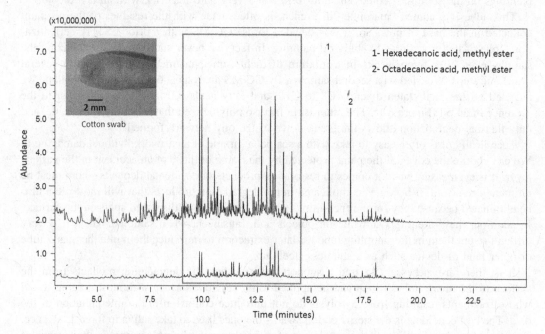

FIGURE 6.2 Pyrograms of a cotton swab applied on a painting with blooming (top) and for comparison the pyrogram of a cotton swab (bottom).

For some paintings, the blooming was so extensive that several samples can be taken always with a swab and analyzed with GCMS after silylation (BSTFA). The conclusions were the same but brought additional information—this lipid fraction is made of either only fatty acids or a mixture of fatty acids and glyceride structures (mono and di). This can be essential for a better understanding of the degradation mechanism of the oil film but the use of Py-GC/MS is enough to conclude that the blooming is once again due to lipid migration. It has saved time in the preparation of the sample and could be a plus if, one day, the white deposit is no longer lipid, but another product eventually cross-linked.

B AMBER AND BITUMEN

Among natural materials, amber and bitumen are probably those where simple citing immediately raises questions about their geographical origin and feeling of mistrust, since various mixtures of very different materials and even synthetic materials of the former [42] have been used to design look-alike materials. If their appearance makes them very different from each other, they do, however, have several points in common: they are classified in the category of fossilized natural substances—cross-linked and/or partially soluble materials. For conservation scientists, the pyrolysis prior to the analysis with gas chromatography appears to be an easy and efficient way to identify them.

Amber is a fossil resin formed from vegetal deposits. It is used to produce jewels, ornaments, and artifacts. Amber from the Baltic area is probably the most prominent of all, but this resin can come from several other countries in Europe (Germany, Romania, Italy) and around the world—Canada, Brazil, USA, Burma (Myanmar), Lebanon, etc. Characterization of the chemical composition of the amber would provide information regarding its geographical origin and reveal the trade routes between different civilizations.

Several techniques such as RMN, FT-IR, and Raman have been used to identify ambers but Py-GC/MS elucidate in detail the composition of the highly cross-linked part of this resin and, therefore, offer a relevant way to discriminate them by their geographical and botanical origin [43] and a better understanding of which materials can be used to make amber look-alikes [44].

As resin from vegetal sources contains molecules with polar functionalities, the use of a reactive agent is essential to improve their separation by GC. Both HMDS [45] and TMAH [46] could be used, each with their known and specific advantages and disadvantages. As amazing as it may seem, I never have to study objects made of amber. I know that succinic acid distinguishes Baltic amber from others, but I cannot give detailed results on the marker compounds that discriminate non-Baltic ambers. Luckily, during the last two decades, the trend is to use Py-GC/MS to study museum amber objects, their state of preservation, and the conservation treatments that have been applied; plentiful and related scientific literature is available [47–49].

Bitumen is a naturally-occurring product made of a mixture of low (solvent fraction) and high (resin fraction) molecular weight hydrocarbons. After biodegradation and/or evaporation of the volatile hydrocarbons, the bitumen becomes a black solid.

Bitumen comes from open-pit sources from several countries—Egypt (Dead Sea area), France, Germany, and Trinidad—and has been made from petroleum (artificial bitumen) after the middle of the eighteenth century.

Bitumen has various properties that make it useful as water-repellent, decorative coatings, adhesives, for medical practices (disinfectant and insecticide), lighting, etc. Even if in museum collections this material is most often related to ancient and old periods, some contemporary artists continue to use it, most often in paintings [50].

In the art field, the main question is : is it bitumen or not? for a better knowledge of the artists' and craftsmen's techniques and/or for a conservation issue. Its use is often suspected for black material observed on objects like mummies, furniture, or for the tinted varnishes for paintings from the nineteenth century [51] and sometimes for darkened varnish with an unusual aging ("craquelure anglaise") and insolubility attributed to bitumen [52]. In human science, getting detailed information about the geographical sources of the bitumen can help trace trade routes and bring information about old geopolitical and cultural contexts.

The first questions about bitumen came to me in 2004 for the study of several balm-soaked linen coming from human and animal mummies. The black materials were hard, mat, or shiny and have proven partly soluble after some tests in strong solvents like dichloromethane. The examples and the results described in the thesis of G. Languri's "Molecular studies of Asphalts, Mummy and Kassel earth pigments" [52] led us to apply Py-CG/MS to our samples. We decided to utilize our screening method—a pyrolysis temperature at 650°C, whereas the Curie point pyrolyzer used by G. Languri heated the sample at 770 °C. The chromatographic and mass spectrometry parameters were quite similar.

The fingerprint of the asphalt is very typical with the distribution of a series of peaks along an unresolved envelope (Figure 6.3). The main peaks result from a series of straight chains of alkenes (from C6 to C31) and alkanes (from C7 to C34). Other specific compounds are present and to trace them, it is necessary to use a specific m/z profile. The other families of compounds allowing the identification of asphalt include the C-ring monoaromatic steroids (m/z 253), the steroid hydrocarbons (m/z 217 + 231 + 245 + 253 + 267), the hopanes (m/z 191), the hopanoids (m/z 191), the alkylbenzenes (m/z 77; 91; 105; 119), the alkyl naphthalenes (m/z 128; 142; 156; 170), the alkylbenzothiophenes (m/z 148; 162; 176; 190), and the alkylbenzothiophenes (m/z 198; 212; 226; 240) [52].

Of the ten and so specimens of embalming black materials I analyzed, only two were found to contain asphalt, the others were mixtures of oil or wax and mastic or pine resins (Figure 6.4). The same ingredients were detected for the black material surrounding the white encrustation on the edges of an Iranian table from the nineteenth century made with the Khatam-kari technic. It was also this recipe added of pigments that have been used for the varnish (original or recent), applied to give a brown tone on a French painting from the seventeenth century.

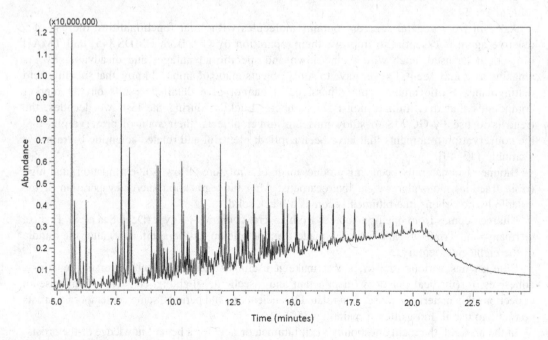

FIGURE 6.3 Representative pyrogram of bitumen sample.

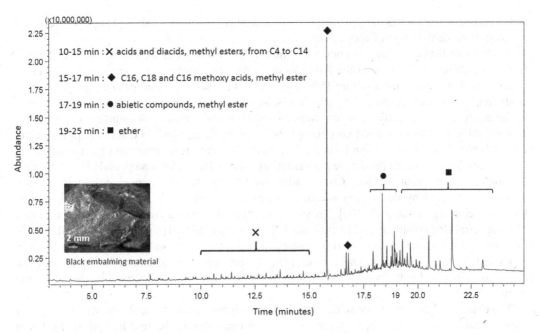

FIGURE 6.4   Pyrogram of the embalming material from a mummy of a ram: a mixture of beeswax ($\times$, ◆, ■) and diterpene resin (●).

C LACQUERS

Lacquerware is a decorative technique applied on a wide variety of objects: furniture, household items like screens, jewelry boxes, vases, and weapons [53]. The technique was developed in the orient (China and Japan) a thousand years ago, and the material used to make this coating is the sap collected from trees of the family *Anacardiaceae* (can be termed as anarcard resins) that mostly grew in Southeast Asia.

Lacquered objects came to Europe in the sixteenth century and immediately, this oriental lacquerware was appreciated and sought after. It was very difficult to get the Asian sap to answer to the demand, so many European craftsmen developed their own techniques of lacquer with familiar materials—western resins like copals, mastic, sandarac, and shellac often mixed with oils and dyes or colorant.

The drying process of the Asian lacquer (several days at room temperature and over 70% relative humidity) leads to a very sturdy and insoluble network lacquer film. For 20 years, Py-GC/MS has been used with success to discriminate the vegetal origins of lacquers [54,55] and to distinguish Asian from European lacquers on museum objects [56]. Several analytical conditions have been implemented with success but the use of a single shot at 550 °C with a methylation with TMAH appears to be the best setting capable of identifying in detail the components of lacquers from both parts of the world [41].

Three kinds of trees—*Rhusvernicifera* (Japan, China, and Korea), *Rhus succedanea* (Vietnam and Taiwan), and *Melanorrhoeausitata* (Thailand and Myanmar)—produce the saps involved in Asian lacquers. The main components of these different saps are urushiol, laccol, and thistsiol (catechol derivatives), respectively. These are blended with other materials water, glycoproteins, carbohydrates, and enzymes [57].

The studies of reference samples allowed to list the specific pyrolysis products that can be grouped in a series of catechols, phenyl catechols, phenols, phenyl phenol, alkylbenzenes, and hydrocarbons are produced [54,55]. For each of the three sap, there are predominant series with specific characteristics related to the number of carbon (maximum side chain length) and abundance. The studies of aged samples reported other compounds like acid catechols [58] and products from dehydration of alkylphenols [59].

To this long list of marker compounds for the anacard resins, we must add the markers coming from the additives mentioned in the Asian lacquers recipes and coming from drying oil, natural resins, proteins, starch, and colorant [58]. Moreover, the Asian and European lacquers are generally made of several ground and lacquer layers, and it is not feasible to sample layers one by one, so the analysis is frequently made on multilayered then multicomponent samples. For all these reasons, the use of extracted ion chromatogram is frequently required to detect the peaks of the urushiol, laccol, and thistsiol pyrolysis product markers in data from museum object samples.

The identification of all these potential materials using their marker compounds is a challenge, to which the scientists at the Getty Conservation Institute and the conservators at the J. Paul Getty Museum responded by developing a tool in a project named RAdICAL (Recent Advances in Characterizing Asian Lacquer) [60]. To sum it up, this tool uses a freeware named Automated Mass Spectral Deconvolution and Identification System (AMDIS) [61] to build a RAdICAL library of Asian and European lacquers marker compounds, to identify the peaks and export the result to an Excel RAdICAL workbook. Then, the users verify the presence or absence of materials with the aid of notes included in the workbook, coming from the knowledge of experts in lacquers over the world [58,62].

The example chosen to illustrate the technical investigation on an Asian lacquer by Py-GC/MS comes from our study of samples taken on a screen made during the first half of the twentieth century in France by Jean Dunand, who is considered as a French Art Déco artist working with Asian lacquer. Dunand learned the oriental lacquer techniques with the Japanese lacquer artist Seizo Sugawara and then developed several decorative techniques like the one used on the screen studied, which consists of eggshell fragments in lacquer [63]. The samples coming from the lacquered frame can be divided into two: the first one was used for a stratigraphic study that revealed a beige ground covered with three layers that have the same features under different illuminations and a final layer—different from the previous—that appears darker. The analysis of the second fragment (after elimination of the ground) by Py-GC/MS with TMAH and the screening method indicate the presence of oil, pine resin, and Asian lacquer (Figure 6.5). Considering the observation made on the cross-section, the Asian lacquer is likely to be present only in the final surface layer. This example demonstrates the value of supplementing Py-GC/MS analyses with a stratigraphy whenever possible.

D MODERN PAINTS

The twentieth century offers to the artists a wide variety of materials that give them the opportunities to experiment with new visual and color effects, to develop new means to express their message, and to work quickly and safely when the solvent of the paint is water, etc. Sometimes, the artists—for economic and/or sociologic reasons—use materials designed for industrial and everyday life applications. The artists have, in some way, a limitless toolbox that results in a nightmare for the conservation scientists. Numerous artworks produced during the modern and contemporary times are wholly or partly made with synthetic materials. As it is impossible to detail each of them here, only some focus will be discussed hereafter.

In the category of materials intended for artists' paints, acrylics are probably the most widely used of all the new mediums. They have been introduced on the market in the middle of the twentieth century; it was solvent-borne paints that had been quickly replaced with water-borne (dispersion) systems. Several paint manufacturers have developed these kinds of paints all over the world: Lefranc-Bourgeois and Sennelier (France), Lukas and Scmincke (Germany), Maimeri (Italy), Talens (Netherland), Lascaux (Switzerland), and Golden and Liquitex (USA).

The first acrylics Magna from Bocour introduced in 1947 were made with a poly(butyl methacrylate) p(BMA) dissolved in a solvent. This polymer—like other methacrylate polymers, ethyl

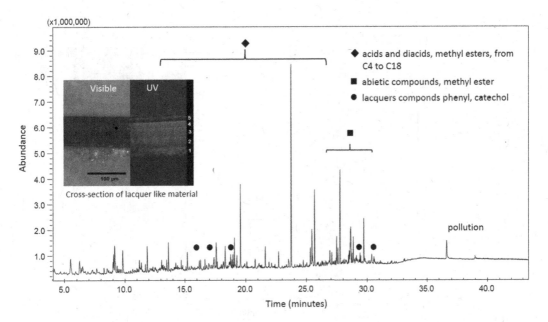

FIGURE 6.5 Pyrogram of the upper layers of the cross-section. A mixture of oil, diterpene resin, and lacquer that is probably only in the layer 5.

methacrylate (EMA), and methyl methacrylate (MMA)—unzips during the pyrolysis without loss of hydrogen, avoiding the formation of unsaturated linear or cyclic compounds. The pyrogram of the methacrylate is the simplest we can observe with just one peak corresponding to the monomer involve in the polymer [64,65].

Of the acrylate family, butyl acrylate, ethyl acrylate, and methyl acrylate give a little more complex fingerprint. Several oligomer structures are formed but the major peaks are always the trimer (three units of monomers). Dimer and monomer give peaks with an intensity around half as big. The tetra and pentamer can sometimes be detected.

The results from our study of acrylic artist paints coming from 12 countries and bought between 1985 and 2018 show that acrylic dispersion paints are mostly made with copolymers of acrylate and methacrylate. The oldest copolymer used to formulate artists' acrylics is an ethyl acrylate-methylmethacrylate (EA-MMA). Its pyrolysis leads mainly to the formation of the monomer. Several dimers and trimers are also formed, and in these two categories, the more intense peak is due to the acrylate part of the copolymer. Acrylate-methacrylate dimers and trimers give generally less intense peaks (Figure 6.6). The EA-MMA has replaced the BA-MMA copolymer at the end of the 1980s and the analyses of tubes bought in 2018 confirm that the BA-MMA is still used for some brands [66].

Even if for modern paint our knowledge about their natural aging is limited to a few decades compared to centuries-old oil paint, it is interesting to note that the fingerprint for acrylics seems to be quite similar for an old and a fresh paint. The analyses of samples in 2018 coming from a painting made in 1970 with Liquitex acrylic paint—with information given by the artist Vincent Bioules—show the expected pyrogram of acrylate-methacrylate copolymers—monomers, dimers, and trimers—of esters involved.

In some cases—and since the mid-1990s—some acrylic dispersion paints are bound with different terpolymers. One of them is a styrene-2-ethylhexyl acrylate-butyl methacrylate (Sty-2EHA-BMA) and fortunately, the complexity of the polymer does not lead to a much more complicated signal to interpret. The pyrogram has the monomer and dimer of the styrene, the monomer of the

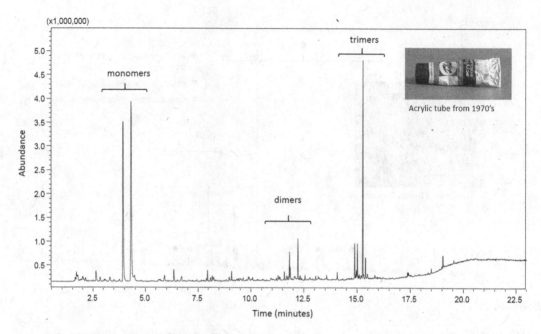

FIGURE 6.6 Pyrogram of an acrylic copolymer, an ethyl acrylate-methylmethacrylate (EA-MMA), that unzipped in monomers, dimers, and trimers.

methacrylate, and three compounds for the 2EHA:ethylhexyl acrylate monomer, 2-ethyl hexane, 2-ethyl hexanol. Another terpolymer including the 2EHA has been identified in a study made in the 2010s and in this case, it was added to the BA and MMA monomer (BA-2EHA-MMA).

As already mentioned, an artist can use paints other than those labeled 'artists or fine quality', and the binder detected can vary. Maybe a nitrocellulose paint, initially used for automobiles, can be considered as the oldest of modern paints. But in this case, the screening method does not give a characteristic fingerprint. It is often combined with a pre-analysis by FTIR technique, then the conclusions lead to the identification of nitrocellulose paints.

Vinyl paints (polyvinyl acetate binder) are household paints for interior and exterior uses. They represent a significant part of the paint used in modern and contemporary art. They first were formulated with an external plasticizer—the most widespread being the dibutyl phthalate—and then, in the middle of the 1960s, with an internal plasticizer. It is, in fact, a copolymer of PVAc and the vinyl ester of versatic acid (neodecanoic acid) VeoVa^{TM} that is still used today [67]. It is easy to distinguish these two forms of vinyl paints as you can see in Figure 6.7.

As for acrylics, the pyrolysis of aged vinyl paints displays a similar pattern to unaged indoor and outdoor artworks. Our study of samples coming from *The Tower* painted by Keith Haring in 1987 on a building at the Necker Hospital in Paris allows us to demonstrate, without any doubt, that the paint used is vinyl made with the PVAc-VeoVa copolymer (Figure 6.8).

An alkyd paint is another kind of household paint used by artists; the most famous who can be cited are Picasso [68] Le Corbusier, Picabia [69], and Pollock [70]. The paint is also called oil-modified alkyd paint because to plasticize the original alkyd named Glyptal that was too brittle, oils have been added. Alkyd paints can be detected with a single Py-GC/MS analysis but sometimes, interpretation can be confusing. The use of TMAH identifies the polyol involved as glycerol or pentaerythritol and gives a good detection for all the peaks with acidic function, like the phthalic esters (isophthalic, terephthalic, and anhydride) and the oily fraction (Figure 6.9). Compared to acrylic, there are a couple of vinyl and alkyd paints labeled for artists. The most known are Flashe

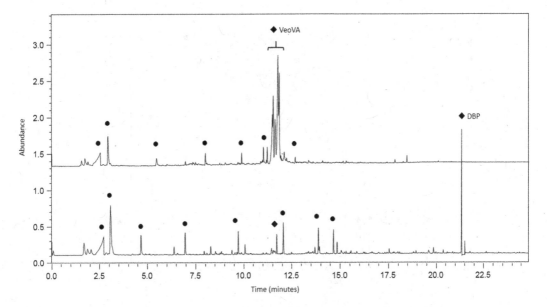

FIGURE 6.7 Comparison of the pyrograms of a PVAc (●) plasticized with dibutyl phthalate (DBP, bottom) and with vinyl versatate (copolymer PVAc-VeoVA, top).

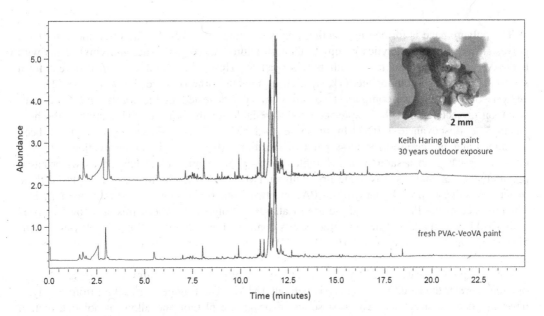

FIGURE 6.8 Comparison of the pyrograms of a fresh PVAc-VeoVA paint (bottom) and a 30 years old of outdoor exposure (top).

from Lefranc & Bourgeois and Griffin from Winsor & Newton. My choice was to explain these kinds of paints from the household market since many studies done were on artwork made with household paints.

The polyurethane and epoxy paints are the other 'non-artist' paints most frequently encountered on artworks. Their chemistry is very complicated since to prepare each, different monomers can be used.

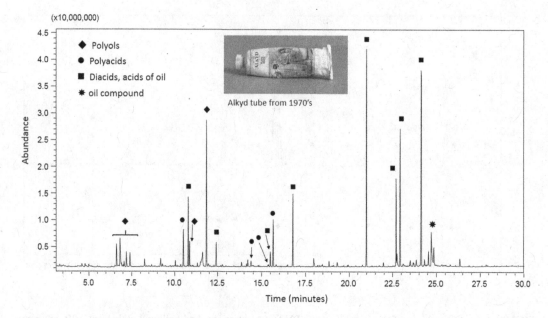

FIGURE 6.9 Pyrogram of an alkyd paint analyzed with TMAH. Polyols(♦), polyacids(●) and diacids, acids of oil(■) are typical markers.

The polyurethane is formed by reacting polyisocyanates and polyols, and often the latter have polyester, polyether, or acrylics groups in their structure. Alkyds, silicone, and vinyl resins with hydroxyl groups can also react with polyisocyanate. This paint could exist in different form systems: one or two components (1K or 2K), in solvent-borne or water-borne forms [71]. The analysis of different two-component solvent-borne paints tested in the frame of several conservation projects exhibit the marker compound reliable for this kind of PU coatings—the hexamethylene diisocyanate (HDI). In response to legislation, water-based systems have been developed and can be used for spray paint formulation. Always within a conservation project, a conservator gave me a sample of a Liquitex spray paint (white color) labeled as water-based polyurethane. Its analysis was made first without TMAH and the chromatogram displayed three major peaks due to phthalic anhydride (PA) and two isomers of isophorone diisocyanate (IPDI). The presence of the PA suggested the use of alkyds as polyols. To check this, a second analysis with TMAH was made and this hypothesis was confirmed with the detection of methylated short diacids (C8 and C9), fatty acids (C16, C18:1, C18), and other non-identified compounds observed in the analysis of alkyds resin references (Figure 6.10).

For the epoxy, the use of TMAH does not seem to provide a more informative signal—in my opinion, I prefer the result of the analysis without TMAH. The fragments resulting from pyrolysis have all phenolic structures. The first series of peaks are phenol and alkyl phenol like methyl phenol, dimethyl, methyl-ethyl isopropenyl phenol, etc. In the second part of the pyrogram, the biggest peak is due to bisphenol A (m/z = 228, 213, 119) that is involved in the most common epoxy resin. With TMAH, a big peak of a dimethylated Bisphenol A (m/z = 256, 241, 133) appears but other phenols are not methylated. I do not understand why I prefer the result without a reacting agent to verify. Before and after the bisphenol A, several small peaks with similar mass spectra feature can be observed (e.g. m/z = 242, 227, 119, m/z = 252, 237, 119: m/z = 266, 251, 119). There are likely alkyl-bisphenol, but as we cannot identify them, we decided to name them epoxy compounds a to g (Figure 6.11). It is very important to keep the mass spectra of unidentified molecules that are detected recursively in the library; they can be elucidated later or not, but all

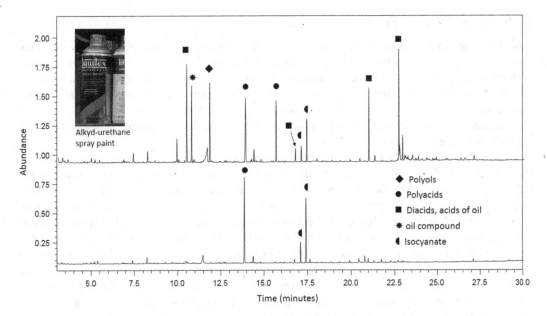

FIGURE 6.10   Analysis of alkyd urethane with TMAH (top) produces more markers than without (bottom). Identification is then more definite.

participate in the definition of a list of markers that make the interpretation more accurate. A total of 16 marker compounds to identify with a degree of confidence can be listed.

All the conclusions given above come from analyses of paints, but the same results were obtained for varnishes and adhesives prepared with these synthetic polymers and used either by artists or during old conservation treatments. This demonstrates that the addition of filler, mostly in mineral form like barium sulfate or calcium carbonate, and mineral or organic pigments do not modify the fragmentation pathway of the synthetic resin.

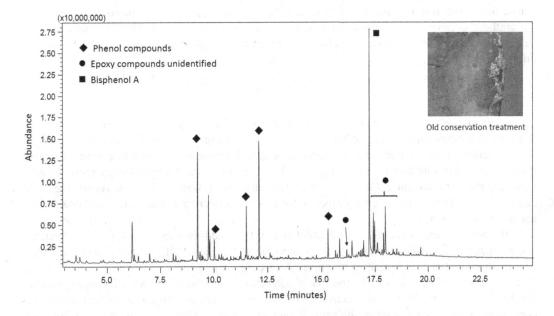

FIGURE 6.11   Pyrogram of an epoxy resin without TMAH, 16 compounds can be list as epoxy markers.

E SYNTHETIC ORGANIC PIGMENTS

First developed to replace natural organic dye, synthetic organic dyes have been transformed into synthetic organic pigments by precipitation with metallic salts. The expansion of synthetic organic pigments began in the twentieth century and are used in various applications including inks, paints, and the coloration of plastics. They have, for the newer products, interesting properties like lightfastness, heat stability, and excellent dispersion property. They exist in a limitless color tone and hue and in a specific range of tint—which are not available with natural pigments—and the daylight fluorescent pigment.

Due to their high tint strength, synthetic organic pigments represent a very small percentage in the formulation of a paint, so their identification by infrared FTIR analysis or X-ray powder diffraction frequently fails due to the presence of a higher concentration of mineral fillers. Raman can be considered as a powerful tool for their identification if you have the laser with the right excitation wavelength. The Py-GC/MS technique can also be used to identify some of these pigments (see below the groups listed), with the added benefit of informing about the nature of the paint binder in a single analysis—two important facts from only one sample from an artwork.

The synthetic organic pigments are often classified according to their chemical structure, and the first level of this classification is often between azo or non-azo pigments [72]. Sub-groups of the azo category are—by limiting to those identifiable by Py-GCMS—beta-naphthol, naphthol AS, disazo condensation, arylide, diarylide, and disazopyrazolone. For the non-azo groups, diketo-pyrrodopyrrole, phthalocyanines, Isoindolinone, perylene, and alizarin crimson can be listed [73]. All these groups correspond to yellow, orange, red, blue, violet, and green colors.

The molecular structures of these pigments are alike and with substituted poly-aromatic rings that lead to the formation of the corresponding mono and di aromatic structures, but also to aniline, benzofuran, and many unidentified products. Moreover, in each of these 11 groups, there can be up to two and eight different pigments. It is, therefore, impossible to detail here the pyrolysis products of all and you must refer to the results of research works published in articles and theses [74–77].

Several issues may justify the interest of undertaking the identification of synthetic organic pigments involved in colored layers. For preventive conservation, it can help to manage the exposure to light during an exhibition according to the lightfastness of the colors applied. In the context of curative conservation, it can guide the curator in the selection of a similar material to realize a mockup that will be used to evaluate the impact of products or protocols. Lastly, as the synthetic organic pigment has been invented at different periods, their identification can help date painted items in collections, gather objects from everyday life, and for those with no information available.

F ADDITIVES

In the formulation of modern materials (from 1950 to the present day) such as paints, adhesives, and varnishes based on synthetic polymers, there are components—additives —that are present at low (sometimes very low) concentrations but have a very important function to ensure either the initial state or the long lifetime of the product. These additives have very different purposes and include surfactants that allow the dispersion of the polymers in water, fillers that reduce the cost, anti-oxidant, and anti-UV that give protection for aging. There are also the wetting and coalescing agents, the defoamer, etc.

In the field of cultural heritage, the identification of these additives can be helpful to understand some repeating degradations of modern materials and to adapt the preventive or curative conservation treatment to minimize its impact on all the components.

These diverse functions of additives implicate that they belong to very different chemical families. So, during a Py-GC/MS analysis with optimal parameters for the identification of one component, such as the binder or the organic pigments, these additives can be more or less—or absolutely not—detected. Moreover, they can be considered as small molecules compared to the polymer.

The use of the multi-step mode could be an easy way to identify some of these additives as soon as they give an intense signal at a pyrolysis temperature different from that suitable for the detection of the binder.

Running the analysis in the SIM (Selected Ion Monitoring) mode could be another way to proceed with the detection of some additives but it is not a lasting solution, since additives change over the years to address performance, for economic, and toxicology needs. So, selecting ions for additives used during one period will be inappropriate for an earlier paint as for the one marketed within six months.

The detection of additives is more random than systematic, and according to my experience, it is due to the sample size, formula, age, etc. The plasticizers added to PVAc and some acrylics—both homopolymers—are probably the only additives that give a significant peak when analyzing a sample from an object.

For the identification of surfactants, we tried an extraction with water on micro-samples of reference paints and we were able to detect surfactants of type triton (Figure 6.12) [78]. This process has been successfully applied on some of our tiny and unique samples from objects, but for the lack of results for this additive, it is difficult to conclude: is the triton-like surfactant too low in concentration? Is there another additive that prevents the extraction of triton? Is there another type of surfactant? So many questions that point out the needs of next researches by collecting regularly, reference materials, information from manufacturers, and by sharing information through publications and networks. Maybe an approach with a multi-step procedure should be considered.

G PLASTICS

The identification of plastics is a growing issue because of their presence in many and varied museum collections (toys, fashion, design, fine-arts, and industrial heritage) and because unstable materials have been proven for most of these.

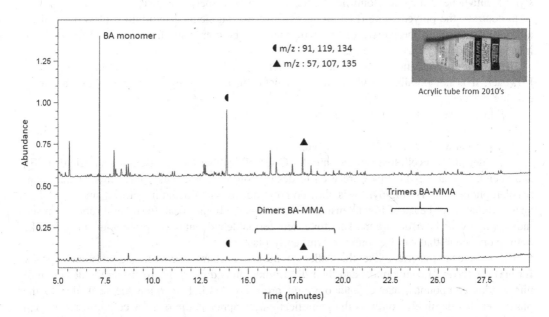

FIGURE 6.12 Triton additive (◀, ▲) is more visible in the water extracted fraction (top) than during analysis of paint sample (bottom).

The chemical families of plastics are numerous and what is most problematic for conservators and curators is that similar objects and similar properties can be obtained with different plastics, which over time will mean different conservation states.

A plastic is a complex mixture of a polymer and many additives that contribute to the properties of the material: plasticizer, reinforcing filler; and to its appearance: dyes, pigments, fillers for opacity. Some of them are required for the fabrication of objects, for example, the heat stabilizers added to PVC to avoid the degradation of the material during the extrusion molding process. Finally, additives are often incorporated in the formulations of plastics to extend the shelf life of the material.

The most frequent question is: what plastic is it? A clear sheet like the one used for animation film (called cells) can be made of cellulose nitrate (Celluloid), cellulose acetate (Rhodoïd), or polyester; foam can be made with polyethylene (Ethafoam), polystyrene (Styrofoam), PVC (Forex); polyurethane ester and ether are widely-used. Polypropylene, polycarbonate, and polyamide can be used for toys, packaging, fashion accessory, furniture, house appliance, etc. Among old objects, there are those made with plastics that are no longer part of the usual or most common materials, this involves Galalith (made with milk, casein formaldehyde), Bakelite (phenol-formaldehyde resin), urea, and melamine formaldehyde.

During the European project POPART (Preservation of Plastic Artefacts in museums) [79,80], a round-robin was made on a collection of plastics composed of industrial standards and pieces of everyday-life objects of unknown composition. Several analytical techniques—including the Py-GC/MS—were applied on a set of one hundred items. The goal of this work was to establish a list of markers compounds for each plastic, to monitor how the pyrolysis parameters can affect the result, and to verify if the addition of a chemical derivatization reagent can improve the detection of some molecules.

With a pyrolysis temperature set at 650 °C for 10 s at a rate of 10 °C/mS, and without derivatization reagent, most of the polymers give a specific fingerprint. This one can be made of very few peaks, like for poly(methyl methacrylate) (PMMA) that produces only one peak—the monomer methyl methacrylate. The polystyrene (PS) gives only three peaks, the monomer, the dimer, and the trimer of styrene, but the presence of butadiene or isocyanate peak indicates a styrene-butadiene and an acrylonitrile-styrene copolymer, respectively [81].

In contrast, the polyethylene (PE) produces some 60 fragments group into triplets of alkanes-alkenes-alkynes, both the pattern of the pyrogram and the composition of the triplet allow us to immediately identify this polyolefin.

In most cases, synthetic polymers yield about ten peaks that can be listed as marker compounds. However, it is not possible to list them all in detail in this writing, given the wide variety of existing plastics.

Building on several years of studies, plastics in museum collections are always somewhat the same and it is obvious that according to the type of collection of artifacts, some will be systematically present and others will be excluded.

The older plastics cellulose acetate, nitrate (CA, CN), rubber (NR), casein formaldehyde (CF), and phenol-formaldehyde (PF) were used to design combs, glass, fans, toys, films, switch, panhandler, phones, etc. During pyrolysis, they do not produce a significant fingerprint since it is often an accumulation of peaks of low intensity. Limonene is the bigger peak for natural rubber. Cresol, indole, phenol, and pyrrole are markers for casein formaldehyde, and for 'proteinous' material like horn, ivory, bone that can be confused with early plastic.

The cellulosic materials like the cellulose nitrate, acetate, butyrate, and propionate show furfuryl (or furfural) compounds, and their discrimination is done respectively by the detection of nitrogenous compounds, acetic, butanoic, and propanoic acid. It is often the peak due to the plasticizer that definitely confirms the identification: camphor is the one for cellulose nitrate and triphenyl phosphate is the one for cellulose acetate (Figure 6.13). Phthalates can also be added or substituted [81,82].

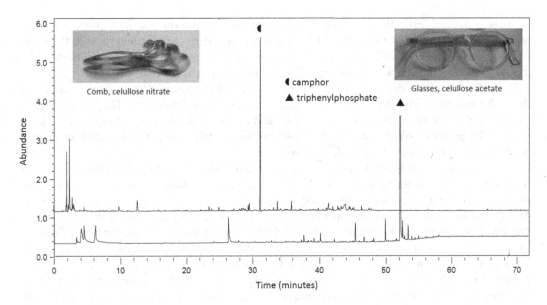

FIGURE 6.13 The main peak of cellulose nitrate (above) and acetate (below) is plasticizer camphor and triphenylphosphate respectively.

Other relevant plastics well represented in the collections are polyvinyl chloride (PVC), polyurethane (PUR), and polyamide (PA),

Flexible PVC, obtained by adding a high percentage of plasticizer to the polymer, is the most present form of PVC in a museum collection and it exhibits numerous conservation issues whose starting point is the loss of this additive. The objects become sticky—dust sticks and they yellow.

An in-depth study of plasticized PVC was carried out as part of a research on inflatable structures [83]. The selected objects had several origins: everyday life objects such as the Quasar's inflatable furniture line Aerospace from the 1960s, or inflatable Disney characters from the Cinémathèque Française collections (Paris). Larger pieces like the *Giant Ice Bag* (1969–1970) by Claes Oldenburg or *Dots* Obssession (2008) by Yayoi Kusama were also chosen because they were produced with more technical PVC and/or specialized companies that could involve different PVC formulations.

Their analysis reveals that this material can be plasticized with phthalates (diethylhexyl (DEHP), dioctyl (DOP), butyl (DBP), diisononyl (DINP)), adipate (Bis2-Ethylhexyl (DOA)), and triphenyl phosphate (TPP) sometimes in mixture.

During the pyrolysis at 650 °C where the plasticizer gives an intense peak (up to 50% of the area of the TIC), the PVC polymer gives a series of small peaks of aromatic compounds that would form from HCl elimination: toluene, ethylbenzene, styrene, indene, naphthalene, biphenyl, fluorine, and anthracene with the more intense being toluene (10% of the total area of the TIC). By changing the pyrolysis temperature and increasing it to 900 °C, the yield of the formation of these aromatic compounds increases (toluene has an area of 16% of the TIC); they can be considered as more relevant marker compounds of PVC [84].

Our experience on PUR is mainly focused on foams as, unfortunately, they are in poor condition and often exhibit specific conservation issues. Their degradation is due to hydrolysis for polyester-based polyurethane PUR(ES), whereas oxidation is the principal cause of degradation for polyether-based polyurethane PUR(ET). Around 45 samples of foams were gathered during a survey on design collections from several museums. Pyrolysis at 650 °C of unaged PUR(ES) and PUR(ET) foam leads to different marker compounds that differentiate them without ambiguity. The markers for the PUES are adipic ketone, diethylene glycol, toluene diisocyanate isomers

(TDI), and adipate derivatives—for the PUET, there are 1-propoxypropan-2-ol, propoxyacetone isomer, toluene diisocyanate isomers (TDI), and (G) glycol derivatives. Analysis with the same conditions of degraded PUET produce the same markers, whereas for PUES, in addition to the previously listed markers, a peak corresponding to the adipic acid is observed [85].

In 2016, a collection survey was carried out on several French design collections focusing on chairs and armchairs with foam inserts or made in foam with a paint coating. The studied objects come from the world and date between the 1950s and 2010s. Of the 30 samples analyzed, 25 were found to be polyether foam and only five were polyester—this material was widely suspected; maybe be the older objects have been restored.

In the polyamide family, the known element is certainly nylon invented in 1935 in the USA by a chemist from Du Pont de Nemours society. There are several nylons: the 6, 6-6, 6–9, 6-10, 11, and 12, where the figures refer to the number of carbon atoms in the amine and acid, respectively. As they are obtained from different monomers, their fragmentation leads to different fingerprints. The PA-6 is made by polymerization of caprolactam during the pyrolysis; the phenomena of unzipping occurs so the major peak, which is also the marker compound used to identify this material, is the one of the caprolactam. The PA 6-6 is made by polyaddition of adipic acid and 1-6 hexane diamine and its pyrolysis produces mainly cyclopentanone (cyclization of adipic acid under heat) and a smaller peak of hexanedinitrile that are considered as marker compounds for this material (Figure 6.14). Unlike the two materials mentioned below, we have rarely been faced with the study of degraded polyamide, so today we cannot determine if pyrolysis produces the same or different markers.

Our experience with polyamides mainly concerns the fiber form of this material. Nylon 6-6 was identified for the transparent filament used by Naum Gabo for *linear construction #2* (1949–1953) and for the carpet used by Daniel Tremblay to create a two-meters silhouette of a profile face (*sans titre, 1982*).

It is nylon 6 that goes into the composition of the carpet used by Patrick Saytour in the assembly named *Port Lympia* (1983), in the hair of several dolls from 1965, and conserve in the toys collection from the Decorative Art Museum of Paris.

The main problem with plastic is that when the degradation is visible, it is too late to expect a solution that will slow down and stop it. Analyses on plastics at different states of preservation

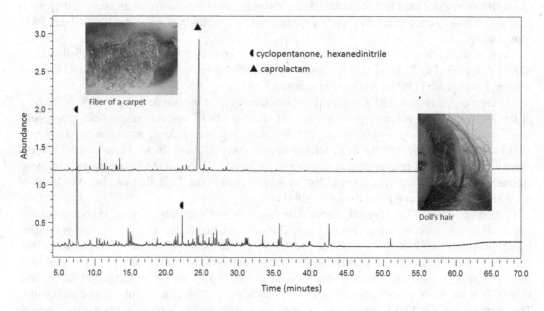

FIGURE 6.14 Comparison of pyrograms of nylon 6 (above) and nylon 6,6 (below).

(naturally or artificially aged) are carried out to check for the presence of chemical compounds that would indicate a degraded state of the material. As adipic acid indicates degradation for PUR ester (see above), a study of several pairs of glasses made of cellulose nitrate and in different states of preservation allowed to correlate the detection of levoglucosan with a degraded state of the material. Levoglucosan is formed by the breakdown of the nitrate ester groups (denitration) and it was detected on objects in perfect condition.

IV CONCLUSION

Pyrolysis-GC/MS is, without any hesitation, useful in the field of cultural heritage to verify hypotheses or explain the materials of pieces of art, to elucidate complex mixtures, and analyze macromolecules or cross-linking material due to aging—all this with only a tiny sample and without any previous treatment.

Ten years ago, the interpretation of a pyrogram might seem to be a challenging and time-consuming process but today, the abundant scientific literature, the on-going research, and networks under development make this task easier to grasp.

The main limitation is that Py-GC-MS in cultural heritage is not a fully quantitative technique. Indeed, it is impossible to establish all the calibration curves, as in most cases we do not know which compounds are in the sample, and the pyrolysis process can produce or destroy some molecules. It is a semi-quantitative result that can be obtained by comparing the areas of each pyrolysis product and express it as percentages of the total area. But it is inconvenient, as what is the relevance of the quantification of a unique and tiny sample coming from a handmade mixture like the ones made by old masters or from a mixture of around 20 ingredients like the ones of a modern paint.

REFERENCES

1. A. Lluveras-Tenorio, J. Mazurek, A. Restivo, M. P. Colombini, and I. Bonaduce, 2012. Analysis of plant gums and saccharide materials in paint samples: comparison of GC-MS analytical procedures and databases, *Chem Cent J.*, 6: 115. doi:10.1186/1752-153X-6-115.
2. E. Tammekivi, S. Vahur, O. Kekišev, et al., 2019. Comparison of derivatization methods for the quantitative gas chromatographic analysis of oils, *Anal Methods*, 11, 28: 3514–3522. doi:10.1039/C9AY00954J.
3. S. Wei, M. Schreiner, E. Rosenberg, H. Guo, and Q. Ma, 2011. The identification of the binding media in the Tang dynasty Chinese wall paintings by using Py-GC/MS and GC/MS techniques, *Int. J. Conserv. Sci.*, 2, 2: 77–88.
4. M. P. Colombini, A. Andreotti, I. Bonaduce, F. Modugno, and E. Ribechini, 2010. Analytical strategies for characterizing organic paint media using gas chromatography/mass spectrometry, *AccChemRes.*, 43, 6: 715–727. doi:10.1021/ar900185f.
5. G. Gautier and M. P. Colombini, 2007. GC–MS identification of proteins in wall painting samples: A fast clean-up procedure to remove copper-based pigment interferences, *Talanta*, 73, 1: 95–102. doi:10.1016/j.talanta.2007.03.008.
6. K. D. Jansson, C. P. Zawodny, and T. P. Wampler, 2007. Determination of polymer additives using analytical pyrolysis, *J. Anal. Appl. Pyrol.*, 79, 1: 353–361. doi:10.1016/j.jaap.2006.12.009.
7. M. F. Silva, M. T. Doménech-Carbó, L. Fuster-López, S. Martín-Rey, and M. F. Mecklenburg, 2009. Determination of the plasticizer content in poly(vinyl acetate) paint medium by pyrolysis–silylation–gas chromatography–mass spectrometry, *J. Anal. Appl. Pyrol.*, 85, 1–2: 487–491. doi:10.1016/j.jaap.2008.11.020.
8. S. Wei, V. Pintus, and M. Schreiner, 2012. Photochemical degradation study of polyvinyl acetate paints used in artworks by Py–GC/MS, *J.Anal. Appl. Pyrol.*, 97: 158–163. doi:10.1016/j.jaap.2012.05.004.
9. S. C. Moldoveanu, 1998. *Analytical Pyrolysis of Natural Organic Polymers*, Elsevier.
10. S. C. Moldoveanu, 2005. *Analytical Pyrolysis of Synthetic Organic Polymers*, Elsevier.
11. S. Tsuge, H. Ohtani, and C. Watanabe, 2001. *Pyrolysis – GC/MS Data Book of Synthetic Polymers: Pyrograms, Thermograms and MS of Pyrolyzates*, Elsevier.

12. P. Kusch, 2018. Pyrolysis–gas chromatography: mass spectrometry of polymeric materials, in *World Scientific (Europe)*. doi:10.1142/q0171. https://www.worldscientific.com/worldscibooks/10.1142/q0171 (Accessed March24, 2018).

13. I. Bonaduce and A. Andreotti, 2009. Py-GC/MS of organic paint binders, in *Organic Mass Spectrometry in Art and Archaeology*, John Wiley & Sons. https://onlinelibrary.wiley.com/doi/abs/10.1002/9780470741917.ch11 (Accessed March24, 2020).

14. J. Peris-Vicente, U. Baumer, H. Stege, K. Lutzenberger, and J. V. G. Adelantado, 2009. Characterization of commercial synthetic resins by pyrolysis-gas chromatography/mass spectrometry: application to modern art and conservation, *Anal. Chem., 81*, 8: 3180–3187. doi:10.1021/ac900149p.

15. J. La Nasa, G. Biale, F. Sabatini, I. Degano, M. P. Colombini, and F. Modugno, 2019. Synthetic materials in art: a new comprehensive approach for the characterization of multi-material artworks by analytical pyrolysis, *Herit. Sci., 7*, 1: 8. doi:10.1186/s40494-019-0251-4.

16. K. L. Sobeih, M. Baron, and J. Gonzalez-Rodriguez, 2008. Recent trends and developments in pyrolysis–gas chromatography, *J. Chromatogr. A, 1186*, 1: 51–66. doi:10.1016/j.chroma.2007.10.017.

17. F. Shadkami and R. Helleur, 2010. Recent applications in analytical thermochemolysis, *J. Anal. Appl. Pyrol., 89*, 1: 2–16. doi:10.1016/j.jaap.2010.05.007.

18. J. Poulin, 2017. Slow pyrolysis: Py-GC-MS analysis using a lower pyrolysis temperature and a slower heating rate, Lecture at the 8th MaSC meeting, University of Evora, Portugal, September 28–29, 2017.

19. N. Gallois, J. Templier, and S. Derenne, 2007. Pyrolysis-gas chromatography–mass spectrometry of the 20 protein amino acids in the presence of TMAH, *J. Anal. Appl. Pyrol., 80*, 1: 216–230. doi:10.1016/j.jaap.2007.02.010.

20. J. M. Challinor, 2001. Thermally assisted hydrolysis and methylation (THM) reactions, *J. Anal. Appl. Pyrol., 61*, 1: 1–2. doi:10.1016/S0165-2370(01)00152-8.

21. J. M. Challinor, 2001. Review: the development and applications of thermally assisted hydrolysis and methylation reactions, *J. Anal. Appl. Pyrol., 61*, 1: 3–34. doi:10.1016/S0165-2370(01)00146-2.

22. J. D. J. van den Berg and J. J. Boon, 2001. Unwanted alkylation during direct methylation of fatty (di)acids using tetramethyl ammonium hydroxide reagent in a Curie-point pyrolysis unit, *J. Anal. Appl. Pyrol., 61*, 1: 45–63. doi:10.1016/S0165-2370(01)00149-8.

23. D. Drechsel, K. Dettmer, and W. Engewald, 2003. Studies of thermally assisted hydrolysis and methylation-GC-MS of fatty acids and triglycerides using different reagents and injection systems, *Chromatographia, 57*, 1: S283–S289. doi:10.1007/BF02492117.

24. M. Mizumoto, E. Shimokita, T. Ona, T. Seino, Y. Ishida, and H. Ohtani, 2010. Rapid and direct characterization of total fatty acids in wood by thermochemolysis–gas chromatography–flame ionization detector/mass spectrometry with tetrabutylammonium hydroxide, *J. Anal. Appl. Pyrol., 87*, 1: 163–167. doi:10.1016/j.jaap.2009.11.004.

25. H. van Keulen, 2009. Gas chromatography/mass spectrometry methods applied for the analysis of a Round Robin sample containing materials present in samples of works of art, *Int. J. Mass Spectrom., 284*, 1: 162–169. doi:10.1016/j.ijms.2009.03.007.

26. O. Chiantore, C. Riedo, and D. Scalarone, 2009. Gas chromatography–mass spectrometric analysis of products from on-line pyrolysis/silylation of plant gums used as binding media, *Int. J. Mass Spectrom., 284*, 1: 35–41. doi:10.1016/j.ijms.2008.07.031.

27. K. B. Anderson and R. E. Winans, 1991. Nature and fate of natural resins in the geosphere. 1. Evaluation of pyrolysis-gas chromatography/mass spectrometry for the analysis of natural resins and resinites, *Anal. Chem., 63*: 2901–2908. doi:10.1021/ac00024a019.

28. V. Pintus, S. Wei and M. Schreiner, 2016. Accelerated UV ageing studies of acrylic, alkyd, and polyvinyl acetate paints: influence of inorganic pigments, *Microchem. J., 124*: 949–961. doi:10.1016/j.microc.2015.07.009.

29. F. Cappitelli, T. Learner and O. Chiantore, 2002. An initial assessment of thermally assisted hydrolysis and methylation-gas chromatography/mass spectrometry for the identification of oils from dried paint films, *J. Anal. Appl. Pyrol., 63*, 2: 339–348. doi:10.1016/S0165-2370(01)00164-4.

30. A. Piccirillo, D. Scalarone and O. Chiantore, 2005. Comparison between off-line and on-line derivatisation methods in the characterization of siccative oils in paint media, *J. Anal. Appl. Pyrol., 74*, 1: 33–38. doi:10.1016/j.jaap.2004.11.014.

31. A. Asperger, W. Engewald, and G. Fabian, 1999. Analytical characterization of natural waxes employing pyrolysis–gas chromatography–mass spectrometry, *J. Anal. Appl. Pyrol., 50*, 2: 103–115. doi:10.1016/S0165-2370(99)00031-5.

32. A. Asperger, W. Engewald, and G. Fabian, 1999. Advances in the analysis of natural waxes provided by thermally assisted hydrolysis and methylation (THM) in combination with GC/MS, *J. Anal. Appl. Pyrol., 52*, 1: 51–63. doi:10.1016/S0165-2370(99)00039-X.

33. A. Asperger, W. Engewald, and G. Fabian, 2001. Thermally assisted hydrolysis and methylation – a simple and rapid online derivatization method for the gas chromatographic analysis of natural waxes, *J. Anal. Appl. Pyrol., 61*, 1: 91–109. doi:10.1016/S0165-2370(01)00116-4.

34. I. Pastorova, K. J. van der Berg, J. J. Boon, and J. W. Verhoeven, 1997. Analysis of oxidised diterpenoid acids using thermally assisted methylation with TMAH, *J. Anal. Appl. Pyrol., 43*, 1: 41–57. doi:10.1016/S0165-2370(97)00058-2.

35. G. Chiavari, S. Montalbani, and V. Otero, 2008. Characterisation of varnishes used in violins by pyrolysis-gas chromatography/mass spectrometry, *Rapid Commun. Mass Spectrom., 22*, 23: 3711–3718. doi:10.1002/rcm.3785.

36. D. Scalarone and O. Chiantore, 2009. Py-GC/MS of natural and synthetic resins, in *Organic Mass Spectrometry in Art and Archaeology*, John Wiley & Sons. doi:10.1002/9780470741917.ch12.

37. G. Chiavari, N. Gandini, P. Russo, and D. Fabbri, 1998. Characterisation of standard tempera painting layers containing proteinaceous binders by pyrolysis (/methylation)-gas chromatography-mass spectrometry, *Chromatographia, 47*, 7: 420–426. doi:10.1007/BF02466473.

38. G. Chiavari, D. Fabbri and S. Prati, 2001. Gas chromatographic–mass spectrometric analysis of products arising from pyrolysis of amino acids in the presence of hexamethyldisilazane, *J. Chromatogr. A, 922*, 1: 235–241. doi:10.1016/S0021-9673(01)00936-0.

39. C. Riedo, D. Scalarone, and O. Chiantore, 2010. Advances in identification of plant gums in cultural heritage by thermally assisted hydrolysis and methylation, *Anal. Bioanal. Chem., 396*, 4: 1559–1569. doi:10.1007/s00216-009-3325-4.

40. C. Riedo, D. Scalarone, and O. Chiantore, 2013. Multivariate analysis of pyrolysis-GC/MS data for identification of polysaccharide binding media, *Anal. Methods, 5*, 16: 4060–4067. doi:10.1039/C3AY40474A.

41. L. Decq, V. Cattersel, D. Steyaert, et al., 2016. Optimisation of pyrolysis temperature for chromatographic analysis of natural resins, Conference poster, ChemCH2016, Brussels, July 5, 2016.

42. R. Falabella, 2013. Imitation amber beads of phenolic resin from the African trade, *BEADS: J. Soc. Bead Res. 28*, 1: 3–15.

43. J. Poulin and K. Helwig, 2014. Inside amber: the structural role of succinic acid in class Ia and class Id resinite,*Anal. Chem., 86*, 15: 7428–7435. doi:10.1021/ac501073k.

44. A. M. Shedrinsky, D. A. Grimaldi, J. J. Boon, and N. S. Baer, 1993. Application of pyrolysis-gas chromatography and pyrolysis-gas chromatography/mass spectrometry to the unmasking of amber forgeries, *J. Anal. Appl. Pyrol., 25*: 77–95. doi:10.1016/0165-2370(93)80034-W.

45. M. P. Colombini, E. Ribechini, M. Rocchi, and P. Selleri, 2013. Analytical pyrolysis with in-situ silylation, Py(HMDS)-GC/MS, for the chemical characterization of archaeological and historical amber objects, *Herit. Sci., 1*, 1: 6. doi:10.1186/2050-7445-1-6.

46. H. Khanjian, 2013. FTIR and Py-GC/MS Investigation of Archaeological Amber Objects from the J. Paul Getty Museum, *e-PRESERVATION Sci., 10*: 66–70. http://www.morana-rtd.com/e-preservationscience/Khanjian-28-08-2012.pdf (Accessed July28, 2020).

47. J. Park, E. Yun, H. Kang, J. Ahn, and G. Kim, 2016. IR and py/GC/MS examination of amber relics excavated from 6th century royal tomb in Korean Peninsula, *Spectrochim. Acta A Mol. Biomol. Spectrosc., 165*: 114–119. doi:10.1016/j.saa.2016.04.015.

48. I. D. van der Werf, A. Monno, D. Fico, G. Germinario, G. E. De Benedetto, and L. Sabbatini 2017. A multi-analytical approach for the assessment of the provenience of geological amber:the collection of the Earth Sciences Museum of Bari (Italy), *Environ. Sci. Pollut. Res. 24*, 3: 2182–2196. doi:10.1007/s11356-016-6963-z.

49. M. Havelcová, I. Sýkorová, K. Mach, and Z. Dvořák, 2014. Organic geochemistry of fossil resins from the Czech Republic, *Procedia Earth Planetary Sci., 10*: 303–312. doi:10.1016/j.proeps.2014.08.021.

50. M. Verbeeck-Boutin, 2007. Conservation d'une œuvre monumentale contemporaine de Jean-Pierre Pincemin, *CeROArt Conservation, Exposition, Restauration d'Objets d'Art*, 1. doi: 10.4000/ceroart.376. https://journals.openedition.org/ceroart/376 (Accessed July 28, 2020).

51. J.-F.-L. Mérimée, 1839. *The Art of Painting in Oil, and in Fresco: Being a History of the Various Processes and Materials Employed, from Its Discovery*, Whittaker & Company.

52. G. M. Languri, 2004. *Molecular Studies of Asphalt, Mummy and Kassel Earth Pigments: Their Characterisation, Identification and Effect on the Drying of Traditional Oil Paint*, Ph.D. thesis, University

of Amsterdam. https://dare.uva.nl/search?identifier=9576cd3b-89a2-456f-9cb2-42f4aa2ee8d1 (Accessed July 13, 2020).

53. M. Webb, 2000. *Lacquer: Technology and Conservation: A Comprehensive Guide to the Technology and Conservation of Asian and European Lacquer*, Oxford; Boston, Butterworth-Heinemann.

54. N. Niimura, T. Miyakoshi, J. Onodera, and T. Higuchi, 1996. Characterization of Rhusvernicifera and Rhus succedanea lacquer films and their pyrolysis mechanisms studied using two-stage pyrolysis-gas chromatography/mass spectrometry, *J. Anal. Appl. Pyrol., 37*, 2: 199–209. doi:10.1016/0165-2370(96)00945-X.

55. N. Niimura, T. Miyakoshi, J. Onodera, and T. Higuchi, 1996. Structural studies of melanorrhoeausitate lacquer film using two-stage pyrolysis/gas chromatography/mass spectrometry, *Rapid Commun. Mass Spectrom., 10*, 14: 1719–1724. doi:10.1002/(SICI)1097-0231(199611)10. 14<1719::AID-RCM706>3.0.CO;2-6 (Accessed July 28, 2020).

56. A. Heginbotham, H. Khanjian, R. Rivenc, and M. Schilling, 2008. A procedure for the efficient and simultaneous analysis of Asiand and European lacquers in furniture of mixed origin, ICOM Committee for Conservation, ICOM-CC, 15th Triennial Conference New Delhi, 22–26 September 2008, Preprints 1109–1114.

57. R. Lu, T. Honda, and T. Miyakoshi, 2012. Application of pyrolysis-gas chromatography/mass spectrometry to the analysis of lacquer film, *Advanced Gas Chromatography – Progress in Agricultural, Biomedical and Industrial Applications*. Published online March 21, 2012. doi:10.5772/32235.

58. M. R. Schilling, A. Heginbotham, H. van Keulen, and M. Szelewski, 2016. Beyond the basics: A systematic approach for comprehensive analysis of organic materials in Asian lacquers, *Stud. Conserv., 61*, sup3: 3–27. doi:10.1080/00393630.2016.1230978.

59. A.-S. Le Hô, M. Regert, O. Marescot, et al., 2012 Molecular criteria for discriminating museum Asian lacquerware from different vegetal origins by pyrolysis gas chromatography/mass spectrometry, *Anal. Chim. Acta, 710*: 9–16. doi:10.1016/j.aca.2011.10.024.

60. Recent Advances in Characterizing Asian Lacquer. https://www.getty.edu/conservation/our_projects/education/radical/radical_overview.html (Accessed July 14, 2020).

61. http://www.amdis.net/index.html (Accessed July 14, 2020).

62. H. van Keulen and M. Schilling, 2019. AMDIS &EXCEL: a powerful combination for evaluating THM-Py-GC/MS results from European lacquers, *Stud. Conserv., 64*, sup1: S74–S80. doi:10.1080/00393630.2019.1594580.

63. M. Baumeister, 2020. Jean Dunand: a french art déco artist working with Asian lacquer: 15. Wooden Artifacts Group Postprints. http://www.wag-aic.org/2002/02index.htm (Accessed July 14, 2020).

64. T. Learner, 2001. The analysis of synthetic paints by pyrolysis-gas chromatography-mass spectrometry (PyGCMS), *Stud. Conserv., 46*, 4: 225–241. doi:10.2307/1506773.

65. N. Khandekar, C. Mancusi-Ungaro, H. Cooper, et al., 2010. A technical analysis of three paintings attributed to Jackson Pollock, *Stud. Conserv., 55*: 204–215.

66. T. Learner, 2004. *Analysis of Modern Paints*, Getty Conservation Institute, Los Angeles.

67. P. Schossler, I. Fortes, JCD de F Júnior, F. Carazza, and L. A. C. Souza, 2013. Acrylic and vinyl resins identification by pyrolysis-gas chromatography/mass spectrometry: a study of cases in modern art conservation, *Anal. Lett., 46*, 12: 1869–1884. doi:10.1080/00032719.2013.777925.

68. J.-L. Andral, et al., 2011. *Picasso Express*, Musée Picasso, Antibes.

69. M. Kokkori, M.-O. Hubert, N. Balcar, G. Barabant, K. Sutherland, and F. Casadio, 2015. Gloss paints in late paintings by Francis Picabia: a multi-analytical study, *Appl. Phys. A, 122*, 1: 16. doi:10.1007/s00339-015-9532-2.

70. F. Cappitelli, 2004. THM-GCMS and FTIR for the study of binding media in Yellow Islands by Jackson Pollock and Break Point by Fiona Banner, *J. Anal. Appl. Pyrol., 71*, 1: 405–415. doi:10.1016/S0165-2370(03)00128-1.

71. C. Defeyt, M. Schilling, J. Langenbacher, J. Escarsega, and R. Rivenc, 2017. Polyurethane coatings in twentieth century outdoor painted sculptures. Part II: comparative study of four systems by means of Py-GC/MS, *Herit. Sci., 5*, 1: 15. doi:10.1186/s40494-017-0129-2.

72. S. Q. Lomax and T. Learner, 2006. A review of the classes, structures, and methods of analysis of synthetic organic pigments, *J. Am. Inst. Conserv., 45*, 2: 107–125. doi:10.1179/019713606806112540.

73. J. Russell, 2010. *A Study of the Materials and Techniques of Francis Bacon (1909–1992)*, Ph.D Thesis, Northumbria University, Newcastle, England. http://nrl.northumbria.ac.uk/3156/ (Accessed March 28, 2020).

74. N. Sonoda, 1999. Characterization of organic azo-pigments by pyrolysis–gas chromatography, *Stud. Conserv., 44*, 3: 195–208. doi:10.1179/sic.1999.44.3.195.

75. E. Ghelardi, I. Degano, M. P. Colombini, J. Mazurek, M. Schilling, and T. Learner, 2015. Py-GC/MS applied to the analysis of synthetic organic pigments: characterization and identification in paint samples, *Anal. Bioanal. Chem.*, *407*, 5: 1415–1431. doi:10.1007/s00216-014-8370-y.

76. J. Russell, B. W. Singer, J. J. Perry, and A. Bacon, 2011. The identification of synthetic organic pigments in modern paints and modern paintings using pyrolysis-gas chromatography–mass spectrometry, *Anal. Bioanal. Chem.*, *400*, 5: 1473. doi:10.1007/s00216-011-4822-9.

77. A. Rehorek and A. Plum, 2007. Characterization of sulfonatedazo dyes and aromatic amines by pyrolysis gas chromatography/mass spectrometry, *Anal. Bioanal. Chem.*, *388*, 8: 1653–1662. doi:10.1007/s00216-007-1390-0.

78. M. Recipon, 2017. Mise au point d'un protocole d'analyse pour la caracterisation d'additifs présents dans les peintures pour artiste et peintures industrielles par pyrolyse-chromatographie en phase gazeuse couplée a la spectrometrie de masse (an analysis protocol to characterize additives in latex paints usingpyrolysis-gas chromatography coupledwith mass spectrometry), MA thesis, Forensic Science, University of Lausanne.

79. Preservation of Plastic ARTefacts in museum collections | Popart. http://popart-highlights.mnhn.fr/ (Accessed July 14, 2020).

80. B. Lavédrine, A. Fournier, and G. Martin, 2012. *Preservation of Plastic Artefacts in Museum Collections*, Comité Des Travaux Historiques Et Scientifiques.

81. E. Richardson, M. Truffa Giachet, M. Schilling, and T. Learner, 2014. Assessing the physical stability of archival cellulose acetate films by monitoring plasticizer loss, *Polym. Degrad. Stab.*, *107*: 231–236. doi:10.1016/j.polymdegradstab.2013.12.001.

82. J. Salvant, K. Sutherland, J. Barten, C. Stringari, F. Casadio, and M. Walton, 2016. Two László Moholy-Nagy paintings on Trolit: insights into the condition of an early cellulose nitrate plastic. *e-PRESERVATION Sci.*, *13*: 15–22. http://www.morana-rtd.com/e-preservationscience/2016/ePS_2016_a3_Salvatn.pdf Khanjian-28-08-2012.pdf. (Accessed July 28, 2020).

83. H. Bluzat, 2013. Les œuvres gonflables en polychlorure de vinyle plastifié dans les collections patrimoniales (Inflatable works in plasticized polyvinylchloride in national collections), *Technè*, 38: 30–33.

84. N. Balcar and A. Colombini, 2009. Approche multi-analytique pour l'étude du PVC, Art d'aujourd'hui, patrimoine de demain Conservation-restauration des œuvres contemporaines (Art Today, Cultural properties of tomorrow Conservation of contemporary artwork)13è Journées de la SFIIC, Paris 24–26 june 2009, 46–155.

85. A. Lattuati-Derieux, S. Thao-Heu, and B. Lavédrine, 2011. Assessment of the degradation of polyurethane foams after artificial and natural ageing by using pyrolysis-gas chromatography/mass spectrometry and headspace-solid phase microextraction-gas chromatography/mass spectrometry, *J. Chromatogr. A*, *1218*, 28: 4498–4508. doi:10.1016/j.chroma.2011.05.013.

7 Environmental Applications of Pyrolysis

T. O. Munson[1], Karen D. Sam[2], and Alexandra ter Halle[3]
[1]Department of Math/Science, Concordia University, Portland, Oregon
[2]CDS Analytical, Oxford, Pennsylvania
[3]Paul Sabatier University, Toulouse, France

I INTRODUCTION

The stated purpose for this Analytical Pyrolysis Handbook includes "to be a practical guide to the application of pyrolysis techniques to various samples and sample types." The subset of this overall scope to be dealt with in this chapter comprises environmental applications of analytical pyrolysis.

A narrow view of this mission would include only the use of analytical pyrolysis techniques (e.g., pyrolysis-mass spectrometry (Py-MS) and pyrolysis-gas chromatography-mass spectrometry (Py-GC/MS)) to identify and measure "contaminants" in samples of outdoor air, soils, sediments, water, and biota. To include interesting and useful applications of analytical pyrolysis techniques that otherwise might not be mentioned in the other chapters of this handbook, a broader view of environmental applications will be used to include topics for the use of analytical pyrolysis in understanding natural environmental processes such as the conversion of plant materials into the soil, coal, and petroleum hydrocarbons. This subject was included in a review paper describing the use of analytical pyrolysis for environmental research [1]. Several other review papers contain references pertinent to environmental analysis [2–7].

No extensive effort has been made to uncover all published reports of pyrolysis use that might be classified as environmental. Instead, time has been devoted to identifying and including good examples of as many different types of environmental applications as possible. Because the intent of this Handbook is to be useful, another sensible criterion was to choose published reports that would be readily accessible to an interested reader.

The most useful format to display the gathered information did not readily present itself. An obvious choice was to arrange the applications according to the pyrolytic technique described in the report (e.g., Py-GC/MS, Py-MS, and so forth). This approach was not selected for two reasons: specific pyrolysis techniques are discussed in other chapters of this Handbook, and it seems appropriate for this chapter to emphasize the applications rather than the techniques. Following this sort of logic, applications in this chapter are sorted by the type of environmental media to which they apply (i.e. air, water, soil/sediment, and biota).

II APPLICATIONS RELATED TO AIR

Since the term pyrolysis as used in the context of this Handbook refers to the conversion of large, non-volatile organic molecules–such as organic polymers–to smaller volatile organic molecules, it should not be surprising that the reports discussed in this section fall into two general categories: the release of organic materials into the air by the pyrolytic decomposition of non-volatile material and the use of pyrolysis techniques for the examination of particulate/aerosol material recovered from air samples.

A Examination of Air Particulates/Aerosols

1 Smoke Aerosols

Tsao and Voorhees used Py-MS together with pattern recognition for the analysis of smoke aerosols from non-flaming [2] and flaming [3] combustion of materials such as Douglas fir, plywood, red oak, cotton wool, polyurethanes, polystyrene, polyvinyl chloride, nylon carpet, and polyester carpet. The driving force for these studies was to determine whether this technique could provide a means to assess the fuels involved in building fires from the analysis of the smoke aerosols formed. Apparently, this type of information can be useful in legal actions associated with catastrophic fires such as the one at the MGM Hotel in Las Vegas, NV.

During flaming combustion, smoke aerosols that contain — in addition to various other materials — an amorphous polymeric substance form, which the authors speculated should contain a great deal of the original polymer structure. Because Py-MS had been shown useful in differentiating, classifying, and sometimes characterizing non-volatile macromolecules, they speculated that Py-MS combined with pattern recognition procedures might be useful in classifying aerosol materials from combustion.

The sample preparation procedures were similar for both studies. Smoke aerosols of the materials of interest were produced in the laboratory under non-flaming [2] or flaming [3] combustion conditions, and the material produced collected on glass-fiber filters. The volatile organic fraction was removed from the collected material by gentle heating (50–55 °C) under a vacuum for 24–48 hours. A portion of the non-volatile material was then subjected to analysis by Py-MS using a Curie-point pyrolyzer (510°C) interfaced to a quadrupole mass spectrometer via an expansion chamber.

Upon pyrolysis, under both combustion conditions, the material collected from the smoke aerosols produced a unique Py-MS spectrum for each substance. In many cases, visual inspection of the spectra from simple mixtures could lead to the identification of the fuel materials. For more complex mixtures, however, extensive "crunching" of the data was necessary using pattern recognition techniques before judgments could be made as to what fuel mixture might have produced the smoke aerosol analyzed. The authors demonstrated considerable success in identifying some (or all the) fuel components of complex mixtures and speculated that with further experiments and refinement in the applications of chemometric techniques, this approach could become a useful tool in fire investigations.

Those interested in these studies might also be interested in a study of the products of non-flaming combustion of poly(vinyl chloride), in which 2 g samples of the granular polymer were decomposed in a tubular flow reactor under various temperature conditions and gas mixtures [4]. The evolved products were trapped in cold hexane or styrene and identified using GC and GC/MS. An earlier study along these lines, using phenol-formaldehyde resin foam, was reported by the same investigator [8].

There is a rich literature associated with studies of the breakdown and/or formation of specific chemical compounds and classes of chemical compounds during burning processes. A study of the thermal decomposition of pentachlorobenzene, hexachlorobenzene, and octachlorostyrene in the air contains many citations [9]. In this study, nearly pure 10–20 mg samples of the cited chemicals were decomposed in a vertical combustion furnace, and the decomposition product trapped on cooled XAD-4 resin followed by charcoal tubes. The adsorbed components were desorbed with toluene and analyzed using capillary GC and GC/MS. The decomposition products formed depended upon the applied temperature, the oxygen concentration, and the residence time in the hot zone of the combustion chamber.

2 Airborne Particulates

Voorhees et al. [10] reported a study of the insoluble carbonaceous material in airborne particulates from vehicular traffic using Py-MS and Py-GC/MS (in addition to TGA, elemental analysis, and radiocarbon analysis). The solvent-soluble organic compounds separated from atmospheric particulate matter (which encompasses a broad spectrum of solid and liquid particles generally ranging in size from several hundred angstroms to several hundred micrometers) had been extensively studied by numerous investigators. Due to its complexity, the insoluble carbonaceous material (ICM) in urban and rural particulate material had not been studied in depth.

Apparently, this complexity had caused earlier attempts to associate ICM with possible formation sources to be highly speculative. In this study, the authors proposed to characterize ICM collected under conditions which would ensure a known source — vehicular traffic.

Air particulate samples were collected on glass microfiber filters in the Eisenhower tunnel on Interstate 70 under conditions designed to minimize particulates other than those formed by the vehicular traffic passing through the tunnel. Water and volatile organics were removed from the samples by drying the filters at 45 °C under vacuum for 18 hours. The soluble organics were removed from the particulates by successive 18-hour Soxhlet extractions with methanol, acetone, methylene chloride, and cyclohexane.

Py-GC/MS of the ICM was accomplished using a Pyroprobe (725°C set temperature, 550 °C actual temperature, for 20 s with the pyrolyzate trapped with liquid nitrogen at the head of a 15-m long by 0.32-mm ID bonded-phase capillary GC column). The Py-MS was accomplished as described above for the study of smoke aerosols.

The capillary column pyrogram shown in Figure 7.1 — and the peak identification list shown in Table 7.1 from the analysis of ICM that had not been sorted by particle size — demonstrate the immense power of this analytical technique. A 0.5 mg sample of ICM containing only a fraction of that amount of organic material (perhaps about 25 μg) was separated into 134 GC peaks, about 90 of which were presumptively identified (that is, by comparison of the unknown spectra to computer library spectra of known compounds). Py-GC/MS analysis of ICM that was size-sorted into four size fractions (<0.6, 0.6–2.7, 2.7–10.4, and >10.4 μm, respectively) showed pyrograms of similar composition to the unsorted material.

Py-GC/MS of samples (prepared as were the ICM samples) of the four materials thought to be the major sources of the tunnel air particulates — tire rubber, road salt, diesel exhaust, and gasoline exhaust — showed that road salt gave no pyrolysis products, and the other three materials gave pyrograms of complexity similar to the ICM samples. The overlap of compound types and the absence of unique "marker" compounds demonstrated the relative contribution of each of the three sources to the total impractical using the Py-GC/MS profiles.

Figure 7.2 shows the comparison of the Py-MS spectra obtained from the ICM (referred to as tunnel particles in the figure) and the three most likely contributor materials. While there are many ions in common, the patterns of ions in the spectra of the three contributor materials are distinctly

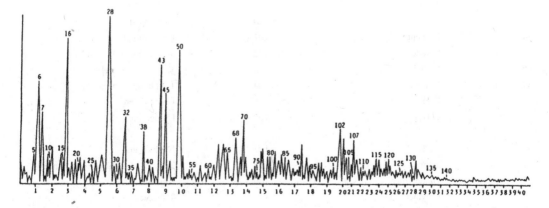

FIGURE 7.1   550 °C pyrogram of insoluble carbonaceous material from vehicular traffic (from Ref. 10).

TABLE 7.1
Peak Identification: Tunnel 2

Peak No.	Compound	Peak No.	Compound
1	1,3-Butadiene	68	l-Ethyl-3-methylbenzene
2	1,3-Butadiene	69	1-Decene
3	Pentane	70	1,2,3-Trimethylbenzene
4	2-Pentene	71	Decane
5	Pentane	72	4- or 5-Decene
6	Acetone	73	Unidentified
7	4-Methyl-2-pentene	74	Unidentified
8	2,3-Dimethyl-2-butene	75	Unidentified
9	Methylpentene	76	C4-Alkylbenzene
10	Hexane	77	1-Methyl-4-(1-methyl ethyl)benzene
11	2-Methylfuran	78	Unidentified
12	Hexene	79	Unidentified
13	2,3-Dimethylbutadiene	80	2-Propenylbenzene
14	1,3,5-Hexatriene	81	Unidentified
15	2-Hexene	82	Propynylbenzene
16	Benzene	83	Methylpropylbenzene
17	Hexadiene	84	C_4-Alkylbenzene
18	Methylcyclopentene	85	Trimethyloctane
19	1-Heptene	86	1-Methyl-2-propylbenzene (or 1,3)
20	Methylhexadiene	87	1-Methyl-2-(propenyl) benzene
21	2-Methylhexadiene	88	Unidentified
22	2,5-Dimethylfuran	89	Unidentified
23	Unidentified	90	C_4-Alkenylbenzene
24	Unidentified	91	Undecene
25	Trimethylcyclohexane	92	Undecene
26	1,3,5-Heptatriene	93	Undecene
27	2,4-Heptadiene	94	Ethyldimethylbenzene
28	Toluene	95	C_4-Alkenylbenzene
29	2-Methylthiophene	96	Tetramethylbenzene
30	Heptadiene	97	C_4-Alkenylbenzene
31	1-Octene	98	C_4-Alkenylbenzene
32	Trimethylbutene	99	1-Methyl-2-(1-methylethyl)benzene
33	Unidentified	100	Dimethylstyrene
34	Unidentified	101	Unidentified
35	Octene	102	1- or 3-Methyl-lH-indene
36	Octadiene	103	1,4-Divinylbenzene
37	Unidentified	104	1,3-Divinylbenzene
38	C_8H_{12}	105	Unidentified
39	C_8H_{12}	106	Unidentified
40	C_8H_{12}	107	Naphthalene
41	1-Ethylcyclohexene	108	3-Methyl-1,2-dihydro naphthalene
42	Unidentified	109	Unidentified
43	Ethylbenzene	110	Tridecene

(Continued)

TABLE 7.1 (Continued)

Peak No.	Compound	Peak No.	Compound
44	Unidentified	111	Unidentified
45	1,2-or 1,4-Xylene	112	C_5-Alkenylbenzene
46	2,4-Dimethylthiophene	113	Unidentified
47	Unidentified	114	Unidentified
48	Unidentified	115	C_6-Alkylbenzene
49	Unidentified	116	C_6-Alkylbenzene
50	Styrene	117	C_5-Alkenylbenzene
51	Unidentified	118	C_6-Alkenylbenzene
52	Unidentified	119	C_{13}-Alkene
53	Unidentified	120	Alkylbenzene
54	1,2,3-Trimetylcyclohexane	121	Unidentified
55	Unidentified	122	C_{14}-Alkene
56	Unidentified	123	Unidentified
57	Unidentified	124	Unidentified
58	1-Methylethylbenzene	125	Unidentified
59	Unidentified	126	Unidentified
60	Dimethyloctane	127	Unidentified
61	2,4- or 3,4-dimethylpyridine	128	Alkylbenzene
62	2-Methylstyrene	129	Alkylbenzene
63	Propylbenzene	130	Alkylbenzene
64	Methylethylbenzene	131	Branched C_{14}-Alkene
65	1,3,5-Trimethylbenzene	132	Branched C_{14}-Alkene
66	Unidentified	133	Branched C_{15}-Alkene
67	Unidentified	134	Alkylbenzene

(Adapted from Table 3, ref 10)

different from each other. Using a statistical method for the comparison of the Py-MS spectra of the three contributor materials to the spectrum for the ICM, the authors estimated that diesel exhaust particulates, gasoline exhaust particulates, and tire wear particulates accounted for 64%, 26%, and 10%, respectively of the tunnel insoluble carbonaceous material.

3 Organics in Urban Airborne Particles

Curie-point-Py-GC/MS was used to examine the organic compounds in urban aerosols [11]. Summer and winter samples (1999) were taken from a series of sampling sites in Leipzig, Germany. Size segregated aerosol samples were collected using a five-stage low-pressure cascade impactor set up in such a way that particulate samples were deposited directly on small pieces of Curie-point foil (590°C). The authors stated that the organic compounds that were measured by the GC/MS were not pyrolysis products but evaporated directly from the particulate matter on the foils during the rapid heating at the Curie-point. Various alkanes, polynuclear aromatic hydrocarbons (PAHs), and oxygenated PAHs were observed.

4 Bioaerosols

An analytical problem of considerable interest to the military concerns the detection of bioaerosols — specifically the need for early warning of a foreign biological presence at perimeters of military posts,

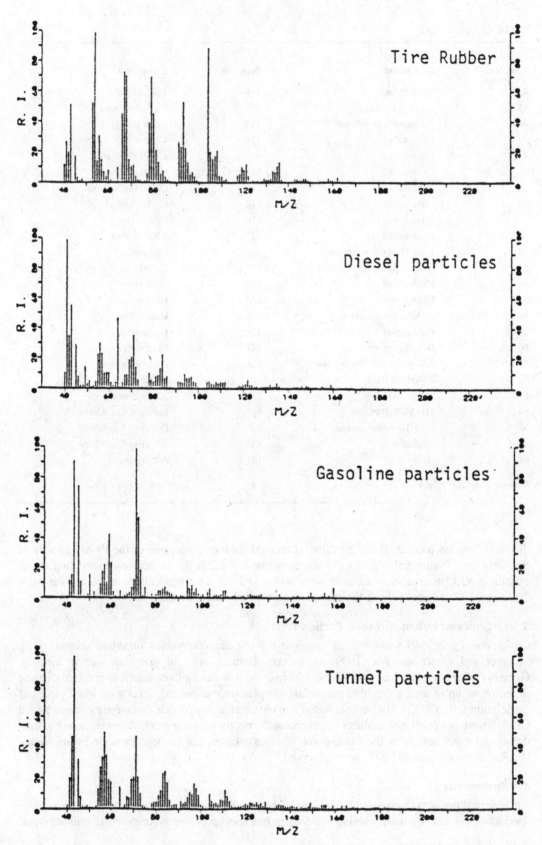

FIGURE 7.2 Py-MS spectra of the tunnel particles and various contributors to that sample (from Ref. 10).

stations, airfields, and battlefields. Especially desirable would be an analytical device compact enough to be resident upon military vehicles such as tanks and personnel carriers yet sophisticated enough to provide useful information about the nature of the biological material detected. A study testing the ability of a pyrolysis-gas chromatography-ion mobility spectrometry (Py-GC-IMS) biodetector system to respond to aerosolized biological substances in a Canadian prairie environment showed promising results for this technique [12]. In a series of 42 trials, this Py-GC-IMS biodetector was shown not only able to distinguish between biological and nonbiological aerosols, but also able to discriminate between aerosols of a gram-positive spore, a gram-negative bacterium, and a protein. The reproducible limits of detection were less than one bacterial analyte-containing particle per two liters of air.

For the biodetector to provide discriminatory data at this level of detection, large (2000-L) air samples were required. A 916 L/min XM-2 aerosol collector/concentrator (25 × 17 × 13 inch, 66 lb) interfaced to the Py-GC-IMS biodetector (12 × 16 × 5 inch, 15 lb, including the computer) almost continuously collected air particulates, providing a cumulative sample to the biodectector at about 3.5 min intervals. No particulate collection took place during the 0.7 min required for the transfer of the collected particulates from the fourth stage of the XM-2 to the quartz microfiber filter in the pyrolysis tube. At the end of the sample transfer time, the XM-2 resumed sampling and the biodectector processed the transferred sample (drying followed by pyrolysis at 350°C, GC separation of the components of the pyrolyzate, and characterization of the components by IMS). The report provides a wealth of information about the data collected during these field trials.

5   Chloroorganic Compounds in Precipitation

Studies of organic compounds in rain and snow that showed that only a small portion (about 0.1%) of the chlorinated organic material was volatile enough for examination directly by GC techniques prompted a study of the nonvolatile organic compounds in such samples by Py-GC/MS and Py-GC/AED [13]. The use of AED (atomic emission detection) was particularly interesting because this detector system allowed specific monitoring for halogenated analytes.

The samples that were presented to the pyrolyzer were some (less than one mg) of the solid residue remaining after evaporation to dryness (using a rotary evaporator at 40°C) of 3-5 L of rainwater, or a similar amount of the solid residue obtained when 3-5 L of rainwater or melted snow was passed through a polyacrylate resin (XAD-8) column, the column eluted with acetonitrile, and the eluate evaporated to dryness. Although the pyrograms showed an abundance of chlorinated organic compounds, the interpretation of the results was complicated by the observation (using ^{35}Cl-enriched sodium chloride) that, in the presence of inorganic chloride, many of the observed chlorinated compounds were formed to some extent during pyrolysis of the samples. However, while varying the pyrolysis conditions, it was discovered that, when heated to 150°C prior to pyrolysis, three of the five samples examined (snow from Gdansk, Poland, and rainwater from Sobieszewo, Poland and Linköping, Sweden) evolved the chlorinated flame retardant, tris(2-chloroethyl) phosphate, clearly not a pyrolysis artifact. In addition, the Py-GC/MS data suggested that carbohydrates and nitrogen-containing biopolymers were abundant in the samples.

B   EXAMINATION OF AIR POLLUTANTS FROM THERMAL DECOMPOSITION

1   Pyrolysis of Plastic Wrapping Film

Pyrolysis has been used to examine the types of air pollutants that might be generated from various types of waste plastic materials during incineration or thermal decomposition in landfills. One such study [14] used GC and GC/MS to examine the products that were formed when two kinds of widely used plastic wrapping film (made of poly(vinylidene chloride)) were pyrolyzed.

Samples of the wrapping material were pyrolyzed at 500°C in an airflow of 300 mL/min and the products evolved trapped in ice-cold hexane. Capillary column GC and GC/MS then analyzed the concentrated extracts. Figure 7.3 shows a typical capillary gas chromatogram of the pyrolysis products from one of the wrapping materials. The numbers on the peaks correspond to the numbers of the compounds identified and shown in Table 7.2. The identifications were carried out by comparison of the mass spectra and retention index of the sample peaks with mass spectra and retention index of authentic compounds. The original form of this table includes the retention indices (calculated using alkanes as references) and average amounts formed calculated from eight runs.

Pyrolyzing the material of interest separately from the analysis allows a greater latitude in the selection of the pyrolysis conditions. For instance, in this case, the gas flow that contained about 25% oxygen would have been incompatible with the capillary GC column. On the other hand, direct coupling of the pyrolysis step to the analytical step (Py-GC and Py-GC/MS) would have allowed examination of the most volatile products formed which could not be done here due to losses during the concentration of the hexane, and the huge hexane solvent peak in the early portion of the chromatogram. A modern pyrolysis-concentrator unit allows the best of both arrangements by providing a gas stream other than the GC carrier gas during pyrolysis (while trapping the pyrolyzate) and then passing the pyrolyzate to the GC column under optimum conditions.

2   Waste Plastic Processing

An ideal way to recycle waste plastics is to re-melt them and form new products directly, but this process is usually only possible in the case of rather homogeneous waste streams. With heterogeneous plastic containing waste streams, however, because other processing steps would be required, landfilling or incineration are the most favored disposal methods from an economic point

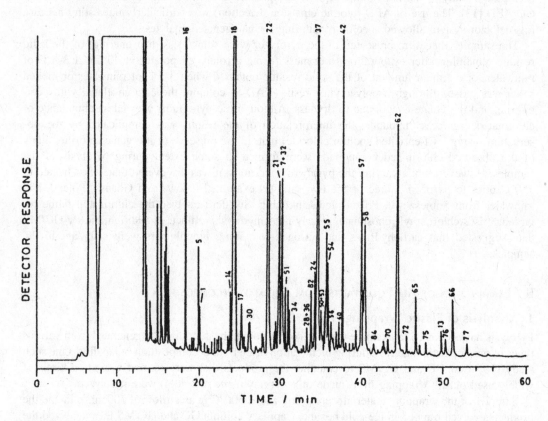

FIGURE 7.3   Typical gas chromatogram of pyrolysis products from a plastic wrapping film (from Ref. 14).

TABLE 7.2

Compounds Identified from Pyrolysis of Plastic Wrap

Peak No.	Compound	Peak No.	Compound
1	Trichlorobutadiene	43	Tetrachlorostyrene
2	Trichlorobutadiene	44	Tetrachlorostyrene
3	Pentachlorobutadiene	45	Tetrachlorostyrene
4	2,2,3-Trimethyloxetan	46	Tetrachlorostyrene
5	Styrene	47	2-Chlorophenol
6	Phenylacetylene	48	2,5-Dichlorophenol
7	Naphthalene	49	2,6-Dichlorophenol
8	Biphenyl	50	3,5-Dichlorophenol
9	Methylphenylacetylene	51	Dichlorophenylacetylene
10	Phenanthrene	52	Dichlorophenylacetylene
11	Phenol	53	Trichlorophenylacetylene
12	Benzaldehyde	54	1-Chloronaphthalene
13	Dibutyl phthalate	55	2-Chloronaphthalene
14	Benzofuran	56	Chlorotetrahydronaphthalene
15	Dibenzofuran	57	Dichloronaphthalene
16	Chlorobenzene	58	Dichloronaphthalene
17	1,2-Dichlorobenzene	59	Dichloronaphthalene
18	1,3-Dichlorobenzene	60	Dichloronaphthalene
19	1,4-Dichlorobenzene	61	Trichloronaphthalene
20	1,2,3-Trichlorobenzene	62	Trichloronaphthalene
21	1,2,4-Trichlorobenzene	63	Trichloronaphthalene
22	1,3,5-Trichlorobenzene	64	Trichloronaphthalene
23	1,2,3,4-Tetrachlorobenzene	65	Trichloronaphthalene
24	1,2,3,5- or 1,2,4,5-Tetrachlorobenzene	66	Tetrachloronaphthalene
25	4-Chlorotoluene	67	Tetrachloronaphthalene
26	Dichlorotoluene	68	2-Chlorobiphenyl
27	Dichlorotoluene	69	2,4- or 2,5-Dichlorobiphenyl
28	Trichlorotoluene	70	2,3'-Dichlorobiphenyl
29	Tetrachlorotoluene	71	3,5-Dichlorobiphenyl
30	Chlorostyrene	72	3,4-or 3,4'-Dichlorobiphenyl
31	Chlorostyrene	73	2,3,6- or 2,3',6-Tri chlorobiphenyl
32	Dichlorostyrene	74	2,3,5- or 2',3,5-Tri chlorobiphenyl
33	Dichlorostyrene	75	2,3',4- or 2,3',5-Tri chlorobiphenyl
34	Dichlorostyrene	76	3,4',5-Trichlorobiphenyl
35	Dichlorostyrene	77	2,3',5,5'-Tetrachlorobiphenyl
36	Dichlorostyrene	78	2,3',4,5'-Tetrachlorobiphenyl
37	Trichlorostyrene	79	2,2',3,5',6-Pentachlorobiphenyl
38	Trichlorostyrene	80	Chlorobenzofuran
39	Trichlorostyrene	81	Chlorobenzofuran
40	Trichlorostyrene	82	Dichlorobenzofuran
41	Trichlorostyrene	83	Dichlorobenzofuran
42	Tetrachlorostyrene	84	Tetrachlorobenzfuran

(Adapted from Table II, Ref. 14).

of view. These disposal processes can lead to environmental problems, especially in the case of halogen-containing plastics such as PVC — poly(vinyl chloride).

Because PVC is widely used in combination with aluminum — and because ABS (a blend of acrylonitrile, butadiene, and styrene) and PET (polyethylene terephthalate) may well be also in the future — a study was undertaken to examine the use of pyrolysis to process combinations of these plastics and aluminum [15].

The thermal stability of the plastics (separately and in combination with aluminum, and under various gas atmospheres) was investigated using thermogravimetry. With this pyrolysis technique, the decomposition of a material is measured as a function of applied temperature by monitoring the weight of the sample as the temperature is increased at a programmed rate. For these experiments, the temperature was programmed from 300 to 1000 K at 0.083 K/s. If the data from such a test is plotted as the normalized mass change per unit time (based on the initial sample mass) as a function of temperature, the resulting differential thermogravimetric (DTG) curve provides a profile of the decomposition of the material as a function of temperature. The DTG curves for virgin PVC in different atmospheres (Figure 7.4) clearly show that the thermal decomposition of virgin PVC is a two-step process.

In addition to thermogravimetry experiments, batch pyrolysis experiments continuously monitoring HCl formation were performed with PVC to determine the optimum temperature for HCl formation. Under conditions of maximum HCl formation, PVC was pyrolyzed with and without oxygen in a fluidized bed reactor and the formation of polychlorinated dibenzodioxins (PCDDs) and polychlorinated dibenzofurans (PCDFs) measured. Avoiding the formation of these highly toxic compounds would be a critical element in any waste stream processing scheme.

In addition to these techniques, analytical pyrolysis experiments were performed where 0.5 mg samples were batch pyrolyzed under flowing helium gas in a tube furnace connected to a liquid nitrogen cold trap. By warming the cold trap, the pyrolysis products were transferred directly to a GC/MS system for identification.

Several important design guidelines for processing aluminum scrap, plastic waste, and combinations of the two were elucidated by these studies. For instance, thermal decomposition of PVC in the presence of oxygen (combustion) generated 10-1,000 times as much PCDD and PCDF as did thermal decomposition in the absence of oxygen. Because anaerobic thermal decomposition of PVC does not cause PCDD and PCDF emission problems, simpler emissions scrubbing steps can be employed. Most of the organic fraction of aluminum scrap can be removed by pyrolysis below the melting point of the aluminum. If the plastic fraction of the aluminum scrap contains a considerable amount of PVC, it is more economical to incinerate the pyrolysis products on-site in

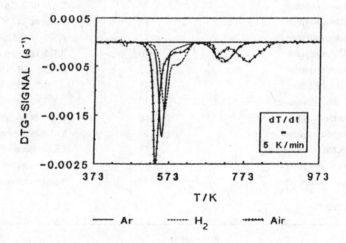

FIGURE 7.4 DTG curves of virgin PVC in different atmospheres (from Ref. 15).

combination with HCl removal and heat recovery. In the case of other plastics, it is possible to get a valuable pyrolysis oil.

PVC is acceptable for two-stage recycling: first dehydrochlorination, and then degradation of the residue. Hydrochloric acid is highly reactive, and components that react with it will reduce dehydrochlorination, limiting the recycling of the residue. Separating PET from PVC in mixed plastics cannot be done inexpensively; separation based on density is not possible because of the density similarities, and other separations are expensive. Offline pyrolysis was carried out in a stainless-steel reactor by German et al., and volatile products were condensed and analyzed using GC-MS. Samples were first heated to below 350°C to dehydrochlorination PVC and then heated to 450°C [16].

By interpreting mass spectra, terephthalic acid, 2-chloroethylene diester, 2-chloroethyl monoester, methyl-2-chloroethyl ester, and ethyl-2-chloroethyl ester were detected. Reactions between these two polymers are not helpful for PVC wastes recycling as it increases the chlorine content in the final residue. If the chlorine fraction is too high, it cannot be used by petrochemical plants [16].

The findings in this publication can also be demonstrated by pyrolyzing PVC with PET with Py-GC/MS. Here, PVC was pyrolyzed with PET, revealing a chloroester of terephthalic acid (Figure 7.5).

3   Pyrolysis of Sewage Sludge

Disposal of sewage sludge has become a major environmental problem, particularly in areas of dense population. The increase in metal content in soil and the potential presence of toxic organic compounds has thrown the traditional utilization of sewage sludge on farmlands into question. Thermal decomposition of sewage sludge, followed by sanitary landfill disposal of the resultant slag and ashes, represents an alternative to agricultural use, but one must then be concerned with the fate of the inorganic and organic constituents during the thermal treatment and final disposal. A laboratory-scale pyrolysis reactor was used to determine the process conditions for a minimum flux of metals to the environment from the anaerobic thermal decomposition (pyrolysis) of sludge, and subsequent disposal of the pyrolysis products [17].

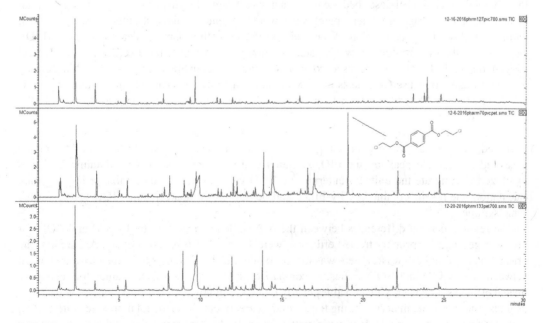

FIGURE 7.5 Pyrolysis-GC/MS of PET (top), PET with PVC (center), and PVC (bottom), peak for chloroester of terephthalic acid, a reaction product, is marked with its structure.

Sewage sludge containing less than 20% water was dried in 600 g batches to less than 1% water. A fraction of each batch (30–40 g) was analyzed for organic matter and metals (Cr, Ni, Cu, Zn, Cd, Hg, and Pb), and the rest was pyrolyzed at selected temperatures (350°C, 505°C, 625°C, 750°C) for one hour with the off-gases being passed through traps and filters.

Analysis of the solid residue (char) and the various traps and filters showed that, in all temperatures, more than 97% of the Hg was evaporated completely from the sludge. At 505 °C, all the Cd remained in the char, but at 750°C, all the Cd was volatilized. At 625°C, the partitioning of Cd between the gas phase and the solid phase depended on the residence time at the pyrolysis temperature. Essentially, all the other metals remained in the char at all pyrolysis temperatures.

This study demonstrated that the best temperature for the pyrolysis of sewage sludge with respect to the content of the metals is in the range of 500°–600°C. Higher temperatures, which leave less char, are less desirable because, at $T > 600°C$, Cd and other metals with relatively high vapor pressures are transferred to the gas phase. Since Hg is very difficult to scrub from the off-gas stream and is completely volatilized at even the lowest temperatures, the most economical approach with respect to Hg seems to be to limit the input of Hg to the sewer system.

In another study [18], an apparatus consisting of a primary pyrolyzer (Pyroprobe 1000) combined with a second reactor was used to study the thermal decomposition of three different chemical sewage sludges. The pyrolysis gases were swept directly into a gas chromatograph for analysis. Yields of twelve pyrolysis products were determined (methane, ethylene, ethane, propylene, propane, methanol, acetic acid, acetaldehyde, C_4-hydrocarbons, CO, CO_2, and water). The temperatures could be adjusted in the two-stage process such that nearly all the organic material was converted to CO, CO_2, and water at temperatures that retained the heavy metals (except for Cd and Hg) in the final residue.

III APPLICATIONS RELATED TO WATER

A WASTEWATER

1 Measuring the Organic Carbon in Wastewater

In 1970, Nelson et al. [19] described the use of a newly designed instrument for measuring the total organic load in sewage treatment plant wastewater samples. This instrument employed flash pyrolysis followed by quantitation of the total pyrolyzate with a flame ionization detector (FID). Two tests of the organic load in wastewater — chemical oxygen demand (COD) and biological oxygen demand (BOD) — are used to provide data to adjustment processes within sewage treatment plants. Because COD tests take two hours and BOD tests can take up to five days, this pyrolysis-FID (Py-FID) technique was proposed as a rapid alternative procedure to provide the needed information.

The design of the pyrolysis reactor was to convert all organic material, either suspended or dissolved in a 50-μL wastewater sample, to volatile organic compounds that would then be measured as a single peak by the FID. A standard mixture of glucose and glutamic acid was pyrolyzed to calibrate the unit. By entering the BOD value of the standard mixture into the final calculation, the results obtained by Py-FID could be directly compared to the BOD values for the same samples.

The relative percent differences between the organic loads measured by Py-FID and BOD for raw sewages, and for primary treated effluents, were 4.1% and 5.6%, respectively. Apparently, the organic material in both wastewaters was mostly biodegradable, judging from the close agreement between the Py-FID and BOD values. A comparison of Py-FID and BOD values for secondary treated effluents, however, yielded values nearly twice as high by Py-FID. Apparently, the bacterial degradation of organic material during the secondary treatment shifts the total organic material to a higher ratio of nonbiodegradable organic matter. Since Py-FID measures the total organic material and BOD only the biodegradable, this large difference is understandable.

Among other techniques, Py-GC/MS was used to study the nature of the residual organic matter in wastewater from a sewage treatment plant [20]. As in the earlier study, it was found that as the total load of organic material decreased during biological treatment of the waste stream, the percent of the total remaining organic material that was resistant to chemical or biological hydrolysis increased. The analysis of this "refractory" organic matter by Py-GC/MS did not lead to a clear understanding of the chemical nature or origin of this material but provided the basis for some interesting speculations.

2 Analysis of Pulp Mill Effluents Entering the Rhine River

Van Loon et al. [21] used Py-MS and Py-GC/MS as analytical techniques for examining the high molecular weight, dissolved organic fraction from pulp mill effluents entering the Rhine River. The study had the following as its objectives: (1) to optimize the analytical method of ultrafiltration combined with Py-MS (which had been reported earlier); (2) to analyze qualitatively the chlorolignins and lignosulphonates in pulp mill effluents entering the Rhine River to find structurally specific pyrolysis products for quantitative analysis in river water; and (3) to obtain an overview of the amounts of adsorbable organic halogens discharged by these pulp mills.

Portions of the effluent samples were treated to remove the non-dissolved materials and then subjected to ultrafiltration to remove the materials with molecular weights less than 1000 Da. With repetitive treatments, water and lower molecular weight organic and inorganic materials were removed, effectively concentrating and de-salting the high molecular weight, dissolved organic carbon (DOC) fractions. Volatile organic fractions were collected by vapor stripping followed by trapping on charcoal tubes, and nonpolar and moderately polar organic compounds were collected by adsorption/elution using XAD-4 resin columns.

The high molecular weight DOC samples were analyzed by Py-MS using the in-source platinum filament pyrolysis technique. The filament, bearing a 1–20 µg sample, was heated at 15°C/s to a final temperature of 800°C. The mass spectrometer collected one scan/s over the m/z range 20–800 amu for 1.5 min.

The Py-GC/MS analysis was performed with 20-µg samples, a Curie-point temperature of 610°C, and a pyrolysis time of 4 s. The resulting pyrolyzate was then swept onto a capillary GC column (50-m long by 0.32-mm I.D. with a 1 µM film of methyl silicones) heated from 30° to 300°C at 4°C/min for a total analysis time of about one hour.

A discussion of the many interesting and important topics presented in this paper is well beyond the scope of this Handbook. However, the inclusion, in detail, of some of the Py-MS, Py-GC/MS, and GC/MS data might provide some thought-provoking material for those interested in the usefulness of these techniques for the examination of a complex organic matrix such as pulp mill effluent.

Figure 7.6, for instance, shows the Py-MS spectra of the high molecular weight DOC isolated from samples from three pulp mills. A distinctive feature of the spectra is the presence of intense signals from small pyrolysis products containing sulfur or chlorine (m/z 34, hydrogen sulfide; 48, methyl sulfide; 62, dimethyl sulfide; 64, sulfur dioxide; 76, carbon disulfide; 50, methyl chloride; and 36, hydrogen chloride. Also noteworthy are the relatively low intensities of lignin pyrolysis products. These features are explained as due to the highly sulfonated, chlorinated, and oxidized nature of the chlorolignins and lignosulfonates, which leads to macromolecules with relatively low aromatic content and high aliphatic functional group content. Comparing these spectra to the Py-MS spectra of standard materials (sodium lignosulfonate, lignosulfonic acid, sodium polystyrene sulfonate, and polystyrene sulfonic acid) provided some insights into mechanisms for the formation of the small pyrolysis products containing sulfur or chlorine.

Figure 7.7 presents the Py-GC/MS profile of the high molecular weight DOC from pulp mill effluent B. The peak numbers refer to the compound identifications shown in Table 7.3, a listing of all the compounds found in the pulp mill effluents. Most likely, the peak identifications in Table 7.3 were presumptive identifications obtained by matching spectra against known spectra in a computer

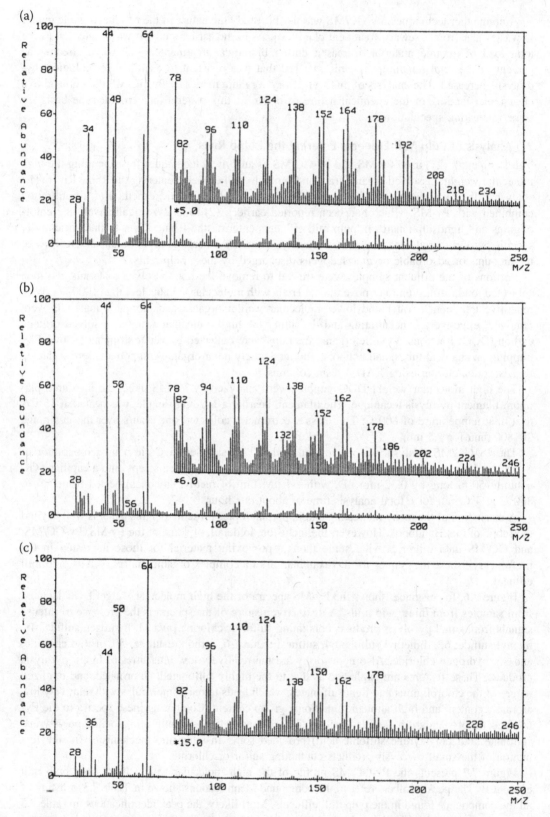

FIGURE 7.6 Pyrolysis-mass spectra of high molecular weight material (MW > 1000) isolated from pulp mill effluents A, B, and D (from Ref. 21).

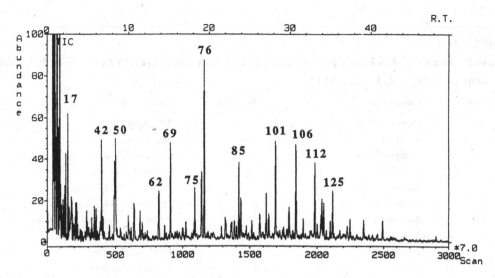

FIGURE 7.7 Curie-point pyrolysis-GC-MS total ion current profile of high molecular weight material (MW>1000) from pulp mill effluent B. (from Ref. 21).

database, and likely some of the identifications were incorrect; for instance, peaks 78 and 86 were both identified as 4-methyl guaiacol. Nevertheless, the types of compounds identified — and the relative amounts of the various types of compounds — allowed the authors to reach some useful conclusions about the chemical composition of the high molecular weight DOC. The original table presented considerably more information than shown here in Table 7.3 (such as relative retention times, the m/z of the base peak and molecular ions in each spectrum, and the pulp mill effluents found to contain each compound).

It is interesting to compare the compounds found in the pyrolyzate of the high molecular weight DOC (Table 7.3) with the types of compounds identified by GC/MS analysis of the low molecular weight DOC (Table 7.4). Many of the same compounds are present, leading to the observation that pyrolytic and chemical degradation can lead to monomeric products with a high structural similarity.

This paper presents a good example of how the two techniques, Py-MS and Py-GC/MS, can be used in a complementary fashion to elucidate various aspects of a very complex analytical problem. Each method has its strengths and weaknesses. Py-MS, for instance, does not provide a wealth of detail concerning the various individual compounds that constitute the pyrolyzate of the material being studied but has the great advantage of speed and simplicity. If a feature of the Py-MS spectrum represents a marker for a component of interest, many samples can be examined for this marker in a relatively short time. The authors suggest that sulfur dioxide may be a marker used to determine lignosulfonates in river water. (For further reading in this area, consult references [21–25].

B SURFACEWATER: ANALYSIS OF NATURAL ORGANIC MATTER

The 1980s and early 1990s have been referred to as the golden age of environmental analytical chemistry in the United States [26]. This era saw the development of relatively inexpensive, very powerful GC/MS systems, and the widespread use of GC/MS in laboratories that formerly had been using only GC. During this same period, Py-GC/MS became one of the routine analytical tools applied to the study of natural organic material (NOM) in water. An extensive discussion of the use of Py-GC/MS for the examination of NOM is well beyond the scope of this chapter but the studies cited below will lead the interested reader into studies of NOM in surface waters around the

TABLE 7.3
Curie-Point PY-GC-MS Pyrolysis Products of High Molecular Weight Material in Pulp Mill Effluent of Mills A, B, C, D, and E[a]

Peak No.	Compound	Peak No.	Compound
1	Carbon dioxide	75	2-Methylphenol
2	Sulfur dioxide	76	4-Methylphenol
3	Methyl chloride	77	Guaiacol(2-methoxyphenol)
4	Acetaldehyde	78	4-Methylguaiacol
5	Butene	79	Dimethylphenol-isomer
6	Methanethiol	80	Dimethylphenol-isomer
7	Acetonitrile	81	1-Methyl-l-formyl-lH-irnidazole
8	2-Propenal	82	1-Methyl-lH-indene
9	2-Propanone	83	Ethylphenol-isomer
10	Furan	84	Ethylphenol-isomer
11	Dimethylsulfide	85	Dimethylphenol-isomer
12	Acetic acid, methyl ester	86	4-Methylguaiacol
13	Carbon disulfide	87	Monochloroguaiacol-isomer
14	Cyclopentadiene	88	Naphthalene
15	1-Butanal	89	Methylguaiacol
16	2-Methyl-2-propenal	90	1,2-Dihydroxybenzene
17	2,3-Butanedione	91	Monochloroguaiacol-isomer
18	2-Butanone	92	4-Vinylphenol
19	2-Methylfuran	93	Dimethylguaiacol
20	3-Methylfuran	94	Methoxydihydroxybenzene
21	Propionic acid, methyl ester	95	Methyldihydroxybenzene-isomer
22	2-Butenal	96	Methyldihydroxybenzene-isomer
23	3-Methylbutanal	97	Monochloroguaiacol-isomer
24	Acetic acid, methyl ester	98	4-Ethylguaiacol
25	2-Methylbutanal	99	Monochloroguaiacol-isomer
26	Benzene	100	Methyldihydroxybenzene
27	Thiophene	101	Monochloroguaiacol-isomer
28	Hydrochloric acid	102	4-Vinylguaiacol
29	Methylisothiocyanate	103	Chloro-4-methylguaiacol
30	2,3-Pentadione	104	2-Methylnaphthalene
31	4-Thiol-3-butyn-2-one	105	Syringol(2,6-dimethoxyphenol)
32	1-Thiol-2-propanone	106	4-(1-propenyl)guaiacol
33	2,5-Dimethylfuran	107	4-Formylguaiacol
34	Acetic acid	108	Chloro-l,2-dimethoxybenzene
35	(Methylthio)methanol	109	Chloro-4-methylguaiacol
36	2-Vinylfuran	110	4-(2-propenyl)guaiacol
37	1-Methyl-lH-pyrrole	111	Chloro-l,2-dihydroxybenzene
38	2,3-Dihydro-3-metylfuran	112	4-Methylsyringol
39	2-Methylpyrole	113	4-(2-Propenyl)guaiacol
40	Pyridine	114	1,2-Dimethoxy-4-benzaldehyde
41	Dimethyldisulfide	115	Chloro-4-ethylguaiacol
42	2-Butenoic acid, methyl	116	4-Propylguaiacol ester
43	Toluene	117	Dichlorguaiacol-isomer
44	3,4-Dihydro-2H-pyran	118	4-Ethanalguaiacol

(*Continued*)

TABLE 7.3 (Continued)

Peak No.	Compound	Peak No.	Compound
45	Methylthiophene	119	4-C_3H_3-guaiacol
46	2,3-Dihydrofurfural	120	4-C_3H_3-guaiacol
47	3-Furaldehyde	121	4-(1-Propenone)guaiacol
48	3-Acetylfuran	122	Chloro-4-vinylguaiacol-isomer
49	2,5-Dimethyl-l,4-dioxane	123	4-(3-Hydroxypropenyl)guaiacol
50	Dimethylsulfoxide	124	4(2-Propanone)guaiacol
51	2-Furaldehyde	125	4-Ethylsyringol
52	Methyl-lH-pyrrole	126	4(1-Propanone)guaiacol
53	Methyl-lH-pyrrole	127	Chloro-4-vinylguaiacol-isomer
54	4-Methylpyridine	128	Tetrachloro-compound
55	Ethylbenzene-isomer	129	Trichloroguaiacol-isomer
56	Ethylbenzene-isomer	130	4-(Propane-1,2-dione)guaiacol
57	Dimethylbenzene-isomer	131	Trichloroguaiacol-isomer
58	2,4-Dimethylthiophene	132	4-(2-Chloro-2ethanone)guaiacol
59	CH_3-S-CO-SH	133	4-Vinylsyringol
60	3-Methyl-2-cyclopenten-l-	134	4-(1-propenyl)syringolone
61	Dimethylbenzene-isomer	135	Chloro-4-propenylguaiacol
62	2-Acetylfuran	136	4-Ethyoxyguaiacol
63	5-Methyl-2-furaldehyde	137	4-Formylsyringol
64	2,3-Dimethylpyrazine	138	3-Methoxy-2-naphthalenol
65	Methoxybenzene	139	Chloro-4-(2-propanone)guaiacol
66	CH_3-SO-CH_2-CO-CH_3	140	4-C_3H_3-syringol
67	3-Methyl-2-cyclopenten-l-	141	Biphenol-isomerone
68	5-Methyl-2-furaldehyde	142	C_3H_3-syringol
69	2-Furancarboxylic acid,	143	Trichloroguaiacol-isomer methyl ester
70	Phenol	144	4-(1-Propenyl)syringol
71	Benzofuran	145	4-(Prop-l-en-3-al)guaiacol
72	Trimethylbenzene	146	4-ethanalsyringol
73	2-Hydroxy-3-Methyl-2-	147	4-(2-Propanone)syringolcyclopenten-l-one
74	1H-indene	148	C_{16}-fatty acid methyl ester
		149	1,2-Diguaiacylethene

[a]Pulp mill identities are provided in the text.
(Adapted from Ref. 21).

world: rivers in France [27]; nine different unspecified water sources in Norway [28]; comparison of the Mediterranean with the Black Sea [29]; estuarine water on the border between The Netherlands and Germany [30]; lake water in Germany [31]; Keddara dam reservoir water in Algeria [32]; reservoir water in Australia [33]; lake water in Finland [34]; and in the United States, freshwater wetlands water in Florida [35], and 17 drinking water sources in Alaska [36].

In addition, the NOM in seawater from off the coast of Florida was examined using Py-MS [37], and the NOM in the lake water in Germany was investigated using pyrolysis-field ionization mass spectrometry (Py-FIMS) and pyrolysis-gas chromatography-combustion-isotope ratio mass spectrometry (Py-GC-C-IRMS) [38].

NOM in water is generally organic material that derives from nonanthropomorphic sources. One of the complexities in examining this material is that the organic materials range in size from relatively simple, organic molecules to quite complex biopolymers, and these materials may be

TABLE 7.4

Low Molecular Weight Compounds Identified with GC/MS in Effluent from Pulp Mills A, B, C, and D[a]

Peak No.	Compound	Peak No.	Compound
1	Benzene	26	4-Formylguaiacol
2	Pentanol	27	Syringol(2,6-dimethoxyphenol)
3	Toluene	28	1,7,7-Trimethyl-bicyclo[2.2.1]heptan-2-ol
4	Dimethylfuran	29	C_{11}-Alkane
5	2-Furancarboxaldehyde	30	1,1,1-Trichloro-2-propanone
6	Dimethylsulfide	31	4-Propenylguaiacol
7	2-Methanolfuran	32	4-Ethanalguaiacol
8	Hexanal	33	4-Acetylguaiacol
9	2-Acetylpyrole	34	C_{12}-alkane
10	5-Methyl-2-furancarboxaldehyde	35	4-(2Propanone)guaiacol
11	Trichloromethane	36	4-(Ethanoic acid)guaiacol
12	C_3-Benzene	37	C_{13}-Alkane
13	Guaiacol(2-methoxyphenol)	38	Dichloroguaiacol(3 isomers)
14	2-Acetylfuran	39	1,1,3,3-Tetrachloro-2propanone
15	C_4-Benzene	40	Trichlorophenol
16	4-Propylphenol	41	C_{14}-Alkane
17	C_{10}-Alkane	42	4-(2-Propanone)syringol
18	Dichlorobenzene	43	C_{15}-Alkane
19	4-Vinylguaiacol	44	4-(Propanedione)syringol
20	4,6,6-Trimethyl-bicyclo-[3.3.1]hept-3-en-2-one	45	C_{16}-Alkane
21	1,7,7-Trimethyl-bicyclo-	46	C_{14}-Fatty acid [2.2.1]heptan-2-one
22	1,7,7-Trimethyl-bicyclo-	47	C_{17}-Alkane [2.2.1]heptan-2-ol
23	1,3,3-Trimethyl-bicyclo-	48	C_{18}-Alkane [2.2.1]heptan-2-one
24	2,6,6-Trimethyl-bicyclo-	49	Tetrachloroguaiacol [3.1.1]heptan-3-one
25	4-Ethylguaiacol		

[a] Pulp mill identities are provided in the text.
(Adapted from Ref. 21).

dissolved and/or suspended in the water and/or adsorbed upon inorganic suspended particulates in the water. The raw water samples can be processed in a variety of ways to yield a variety of organic fractions that can be examined by a variety of techniques, one of which is frequently Py-GC/MS. In the studies cited above, specific compounds found in the pyrolyzates are used to infer the presence of various classes of biopolymers in the original sample (e.g., polysaccharides, proteins, N-acetylamino sugars, polyhydroxy-aromatics, and lignins).

In one example, Py-GC-MS was performed on ultrafiltered DOM (UDOM) obtained from headwater streams across a climatic gradient to better understand Dissolved Black Carbon (DBC). DBC consists of organic residues of incomplete combustion including charcoal and soot, which has a significant role in the global C cycle [39].

C   ANALYSIS OF GROUNDWATER CONTAMINATION

Voorhees et al. [40] described an environmental application that is not a pyrolysis application, *per se*, but rather an environmental application of pyrolysis equipment which seemed too clever not to mention here.

Groundwater contamination by organic chemicals is frequently measured by drilling wells into aquifers to obtain samples of the groundwater for subsequent purge-and-trap GC/MS analysis. When used in a reconnaissance mode, this process can be very expensive. These organic chemicals can migrate vertically to the surface as vapors, but their concentrations are often too low for direct measurement. The trapping procedure described by these authors, however, provides a viable approach.

Static traps were prepared by coating about 1 cm of the end of 358°C Curie-point pyrolysis wires with finely powdered activated charcoal. After being precleaned by being heated to the Curie point under vacuum, the traps were transported to the field test site in sealed culture tubes. At the field test site (a site known for groundwater contamination by tetrachloroethylene, PCE), the static traps were placed in 25- to 35-cm deep holes, covered with inverted aluminum cans, and the soil replaced in the holes. After three days, the traps were removed for analysis.

Analysis was performed by desorbing the organics from the traps with a Curie-point pyrolyzer unit in series with a quadrupole mass spectrometer. The data produced had similarities when compared to Py-MS data, although quite likely, thermal desorption was taking place rather than pyrolysis. The typical mass spectrum obtained from the contaminated areas was dominated by the major ions of PCE. Table 7.5 shows the various compounds that were identified in spectra obtained from the 25 samples spaced around the contaminated area. Table 7.6 shows the compounds identified by static trapping from a specific location and by purge-and-trap GC/MS analysis of water from an adjacent well.

This static trapping technique could be an excellent reconnaissance technique for many volatile organic compounds. In this study, the ion counts for the PCE were proportional to the surface fluxes from the plume contamination, falling off sharply at the plume edges. An attractive feature of the technique is the short analysis time — that is, a few minutes per sample. However, as pointed out by the authors, the technique is limited in some instances by the lack of separation of the adsorbed components prior to the measurement by mass spectrometry. Apparently, the overlapping ions from the mixture of components can make some components unidentifiable in the mixed spectrum. This can be a serious limitation in some cases; for instance, in the comparison shown in Table 7.6, a very high chloroform concentration in the purge-and-trap GC/MS analysis was undetected in the static trapping/MS analysis due to a high background of ions from other substances masking the chloroform ions. The authors suggest that combining this Py-MS technique with Py-GC or Py-GC/MS could be an improvement.

IV ANALYSIS OF SOIL AND SEDIMENT

A THERMAL DISTILLATION-PYROLYSIS-GC

The use of thermal distillation-pyrolysis-GC (TD-Py-GC) and GC/MS to examine marine sediments and suspended particulates for anthropogenic input was reported by Whelan et al. [41,42].

TABLE 7.5

Compounds Identified by Static Trapping/MS Analysis for a Denver Industrial Site

Benzene	Dichlorobenzene
Toluene	Chloroform
Xylene	Trimethylbenzene
Phenol	Naphthalene
Trichloroethylene	Carbon tetrachloride
Tetrachloroethylene	

(Adapted from Table I, Ref. 40).

TABLE 7.6

Compounds Identified by Static Trapping/MS and by Purge-and-Trap GC/MS

Compound	Static Trapping/MS above Well 23179 (Ion Counts)	Purge-and Trap GC/MS Well 23179 (µg/L)
Dichloroethylene	ND[a]	12
Benzene	ND	22.1
Toluene	1052	<2
Xylene	853	4.3
Trichloroethylene	352	9.5
Tetrachloroethylene	616	75.2
Chloroform	_[b]	3440
Carbon tetrachloride	110	<3

[a]ND = not detected.
[b] other compounds interfered with m/z 83 and 85; identification could not be made.
(Adapted from Table II, Ref. 40).

The TD-Py-GC technique differed from typical Py-GC in that the temperature of the sample was raised slowly (e.g., 60°C/min) to 800°C, evolving two distinct peaks of organic material, well separated in time. The first, a low-temperature peak (P_1), in the range of 100–200°C, contained unaltered, absorbed volatile organic compounds; and the second pyrolyzate peak (P_2), which emerged in the 350–600°C range, consisted of compounds thermally "cracked" from high molecular weight organic materials. The TD-Py-GC apparatus is shown diagrammatically in Figure 7.8.

As described in the first report [41], a 0.5- to 50-mg sample of wet (or frozen) sediment was placed in a quartz tube, which was then placed in the platinum coil of a pyrolysis probe. The probe was inserted into the cooled interface chamber of the reaction system, which was then purged with helium carrier gas. The interface chamber was heated to 250°C and the pyrolysis probe temperature ramp started. About 10% of the helium carrier stream was split off to a thermal conductivity-flame ionization detector series that provided profiles of products being evolved from the sample. The main portion of the helium carrier flow was directed via switchable multiport valves through a pair of Tenax traps for the appropriate time periods to trap the P_1 and P_2 groups of evolved organic materials.

The GC-FID analysis of the Tenax-trapped portions of P_1 and P_2 was accomplished sequentially. By flash desorbing the adsorbed organics from each of the Tenax traps in turn, P_1 and P_2 were sent sequentially to the GC for analysis. One column (a micropacked column — 3-m long with a 3% OV-17 stationary phase coated on a 160-180 mesh support material) of a dual-column, dual-FID GC connected to the system was used for this analysis. To improve chromatographic peak shape, the desorbed organic materials were cryo-focused with liquid nitrogen prior to being transferred to the GC column. Qualitative analysis of the low-molecular-weight portions of P_1 and P_2 that were not trapped by the Tenax traps (e.g., C_1-C_5 hydrocarbons) was accomplished by sweeping the helium effluent from the interface chamber through the Tenax traps (via switchable multi-port valves) to a liquid nitrogen trap on the first coil of the other column of the GC (n-octane/ Porasil C, 6-ft long by 0.085-in ID).

Rather than analysis by GC-FID, the P_1 and P_2 fractions could be analyzed by GC/MS by being trapped on small Tenax traps. These traps would then be desorbed by being placed in the heated inlet of the GC/MS system, with the desorbed organics being cryo-focused with liquid nitrogen at the head of the GC column.

In the later work [41], the micropacked GC column was replaced with a 50-m long, narrow-bore capillary column with probably more than ten times as many theoretical plates. Comparing the P_2

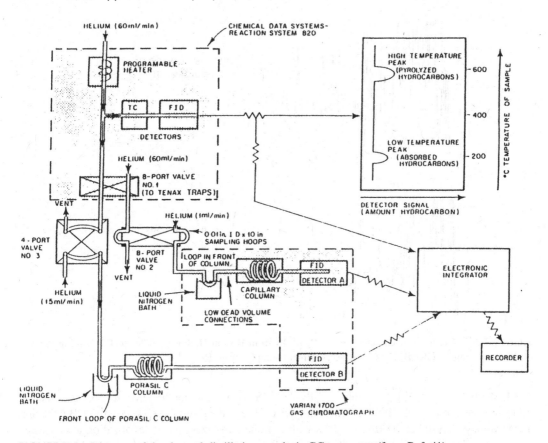

FIGURE 7.8   Diagram of the thermal distillation-pyrolysis-GC apparatus (from Ref. 41).

profile (micropacked GC column) in Figure 7.9 with the P_2 profile (narrow-bore capillary GC column) in Figure 7.10 illustrates the tremendous additional amount of detail that can be obtained by using high-resolution capillary columns for the separation of pyrolyzates.

Whether they were generated using a packed column or a narrow-bore capillary column, the GC profiles from the TD-Py-GC technique proved quite useful for examining the distribution of organic materials in the various "compartments" of the marine ecosystem. Due to the relatively small size sample required, the method was especially useful for examining organic matter in several types of samples, such as marine particles and parts of small marine animals. Analysis of sediments and various particles, including sewage sludge from the Boston Harbor area, provided interesting insights into sediment deposition processes in this area and the fate of the anthropogenic organic matter.

B   FLASH PYROLYSIS-GC AND GC/MS

Several groups of researchers noticed that nonpolymeric organic compounds that are reasonably volatile at elevated temperatures do not fragment upon pyrolysis but simply volatilize [43,44]. They realized that pyrolysis could be a rapid way of extracting these compounds from a complex sample matrix and transferring them directly into the analytical system, thus providing a rapid screening method and avoiding the frequently lengthy extraction and cleanup procedures. One group [44] referred to this procedure as flash evaporation pyrolysis (EV-Py). This technique differs from TD-Py, described above, in that the sample is elevated to the pyrolysis temperature in milliseconds rather than the slow temperature ramp used in TD-Py.

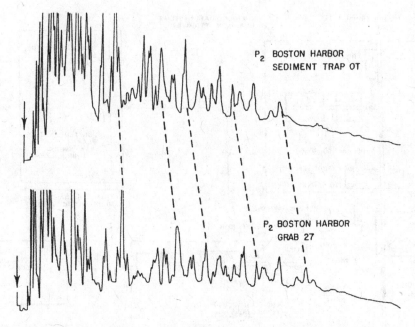

FIGURE 7.9   Micropacked GC analyses of P_2 peaks from Boston Harbor particles. OT is a midwater sediment trap sample; GRAB 27 is a surface sediment sample (from Ref. 41)

McMurtrey et al. [43] evaluated EV-Py-GC/MS as a rapid method for screening soils for polychlorinated biphenyls (PCBs). 5-mg samples of dried lake sediment spiked with PCB (Aroclor 1254) were pyrolyzed at 1000°C for 10 s with a probe pyrolyzer and the evolved organics swept onto an 80°C packed GC column (3% OV-1, 6-ft long by 2-mm I.D.) that was temperature programmed to 275°C. This method proved easily capable of demonstrating the presence of the spiked PCB at the 10ppm level using the MS in the full scan mode. Although they did not do the experiments, the authors theorized that the detection limit could be reduced considerably by operating the MS in the selected ion monitoring mode. They realized that Py-GC using the much more sensitive electron-capture detector would be a more attractive screening method for PCBs than Py-GC/MS, but their preliminary attempts in that direction did not seem very promising because pyrolysis of sediments released so much electron-capturing material that the peak patterns of the PCBs were obscured and, hence, unrecognizable.

De Leeuw et al. [44] demonstrated the usefulness of EV-Py-GC/MS as a rapid screening technique for polycyclic aromatic hydrocarbons (PAHs), halogenated organics, aliphatic hydrocarbons, heteroaromatics, elemental sulfur, cyanides, and pyrolysis products of synthetic polymers. A 200-µg sample (either polluted soil or sediment) was heated to 510°C with a Curie-point pyrolyzer and the evolved organics swept onto a high-resolution capillary column (CP-SIL 5, 25-m long by 0.22-mm I.D.) at 0°C. The GC temperature was programmed to 275°C at 3°C/min. The MS scanned the range 50–550 amu at 1.5 s/scan.

Figure 7.11 displays a total ion current profile (TICP) of EV-Py-GC/MS analysis of a polluted soil sample. A central portion of the TICP has been expanded 8× to emphasize the wealth of compounds present (upper trace). The numbers in the mass chromatograms (lower third of the figure) represent the *m/z* values indicative of the following classes of compounds: 104, 118, 132 (C_0-C_2 styrenes); 116, 130 (C_0-C_1 indenes); 128, 142, 156, 170 (C_0-C_3 naphthalenes); 134, 148, 162 (C_0-C_2 benzo[b]thiophenes; 154 (biphenyl and acenaphthene); 168, 182, 196 (C_1-C_2 biphenyls and C_0-C_2 dibenzofurans); 166,180 (C_0-C_1 fluorenes and 9-fluorenone); 184 (dibenzothiophene); 202 (fluoranthene and pyrene); 228 (benzo[c]phenanthrene, benz[a]anthracene, chrysene, and

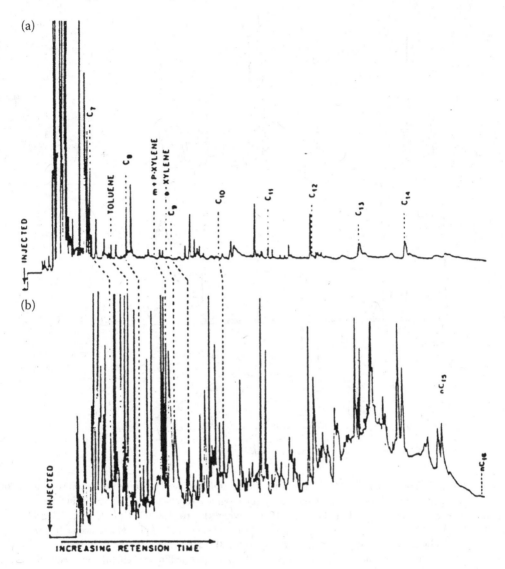

FIGURE 7.10 P_2 capillary GC patterns: seston (A) and sediment trap particles (B) (from Ref. 42).

triphenylene); and 252 (benzo[e]pyrene and benzo[a]pyrene). The x-axes of the mass chromatograms correspond exactly with the appropriate parts of the TIC x-axis directly above them (e.g., the major peak in the mass chromatogram of m/z 168 corresponds with peak 41 in the total ion current profile). The peak numbers in Figure 7.11 correspond to the peak numbers of the presumptively identified compounds listed in Table 7.7. Relating the response for phenanthrene (peak 63) and benzo[a]pyrene (peak 73) to the 200 μg applied to the pyrolysis wire, it was estimated that the detection limit for PAHs using this method was about 5 ppm.

The authors pointed out that, because no pre-treatment of the samples was carried out, the peaks present in the TICP reflected components generated by pyrolysis of primary sample compounds ("real pyrolysis products"), and components that were present as such in the sample and simply evaporated ("free products"). In this soil sample, for instance, they saw four different groups of anthropogenic compounds: HCN and dicyanogen (pyrolysis products), elemental sulfur (present as such), PAH (mainly unsubstituted, present as such), and styrenes and phenyl ethers (pyrolysis products). Then, they went on to speculate about the previous industrial activities that took place in

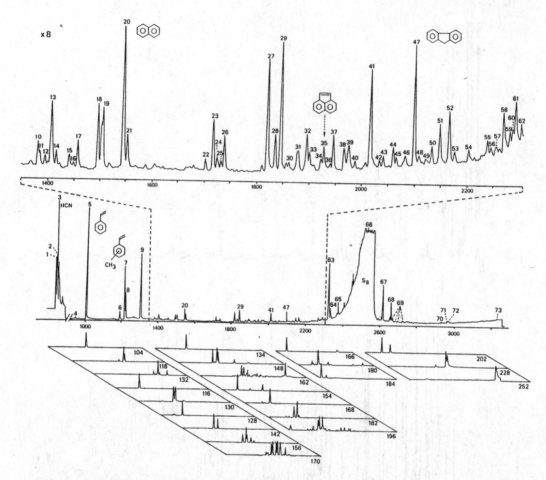

FIGURE 7.11 TIC of EV-Py-GC-MS analysis of Polluted Soil Sample. The upper trace represents a part of the TIC magnified 8 times. The number in the mass chromatograms represents the m/z values indicative of the following classes of compounds: 104, 118, 132 (C_0-C_2 styrenes); 116, 130 (C_0-C_1 indenes); 128, 142, 156, 170 (C_0-C_3 naphthalenes); 134, 148, 162 (C_0-C_1 benzo[b]thiophenes; 154 (biphenyl and acenaphthene); 168, 182, 196 (C_1-C_2 biphenyls and C_0-C_2 dibenzofurans); 166,180 (C_0-C_1 fluorenes and 9-fluorenone); 184 (dibenzothiophene); 202 (fluoranthene and pyrene); 228 (benzo[c]phenanthrene, benz[a]anthracene, chrysene, and triphenylene); and 252 (benzo[e]pyrene and benzo[a]pyrene). The x-axes of the mass chromatograms correspond exactly with the appropriate parts of the TIC x-axis directly above them (e.g., the major peak in the mass chromatogram of m/z 168 corresponds with peak 41 in the total ion current profile). (from Ref. 44).

the area where the soil sample was collected, which might have led to the suite of compounds identified by the EV-Py-GC/MS analysis.

The authors presented an analytical scheme (Figure 7.12) to use flash evaporation pyrolysis with a variety of detectors — in addition to MS — to provide greater selectivity and/or sensitivity for common classes of pollutants, but only data for MS were presented in this report.

Over the years, this technique (usually referred to simply as Py-GC or Py-GC/MS) has proven extremely useful in the examination of soils and/or sediments to determine the nature of the natural organic materials or the degree of contamination by anthropomorphic organic materials [45–52]. One study [53] analyzed river sediments and sewage sludges by pyrolysis gas chromatography using an atomic emission detector (Py-GC/AED) and compared the results with Py-GC/MS analysis. Multi-element detection focused on C, N, S, O, and Cl was shown to be a useful tool for specifying the distribution of organic heteroatom compounds.

TABLE 7.7

Identified Evaporation and Pyrolysis Products of the Soil Sample

Peak No.	Compound	Peak No.	Compound
1	H_2S, CO_2, CO	37	(C_1-phenyl)ethyl tertbutyl ether (tentative)
2	Dicyanogen	38	(C_1-phenyl)ethyl tertbutyl ether (tentative)
3	Hydrogen cyanide	39	Acenaphthene + 4methylbiphenyl
4	Ethylbenzene	40	3-Methylbiphenyl
5	Styrene	41	Dibenzofuran
6	α-Methylstyrene	42	C_3-Naphthalene
7	3-Methylstyrene	43	C_3-Naphthalene
8	4-Methylstyrene	44	C_3-Naphthalene
9	Indene	45	C_3-Naphthalene
10	α,3-Dimethylstyrene	46	C_3-Naphthalene
11	3-Ethylstyrene	47	Fluorene
12	α,4-Dimethylstyrene	48	C3-Naphthalene + dimethylbiphenyl
13	3,5-Dimethystyrene	49	Dimethylbiphenyl
14	α,2-, or 2,5- or 2,4-Dimethylstyrene	50	Unknown organic sulfurcompound
15	Phenyl ethyl ether	51	Methylbenzofuran
16	2,3-Dimethylstyrene	52	Methylbenzofuran
17	3,4-Dimethylstyrene	53	Methylbenzofuran
18	Methylindene	54	α-1-Phenyl ethyl ether(tentative)
19	Isomeric methylidenes	55	2-Methylfluorene
20	Naphthalene	56	1-Methylfluorene
21	Benzo[b]thiophene	57	C_2-Benzofuran
22	Methylbenzo[b]thiophene	58	9-Fluorenone
23	2-Methylnaphthalene	59	C_2-Benzofuran
24	Methylbenzo[b]thiophene	60	C_2-Benzofuran
25	Methylbenzo[b]thiophene	61	Dibenzothiophene
26	1-Methylnaphthalene	62	C_2-Benzofuran
27	1-Phenylethyl tert-butylether (tentative)	63	Phenanthrene
28	Biphenyl	64	Anthracene
29	Unknown organic sulfurcompound	65	Bis(1-phenylethyl)thioether (tentative)
30	1-Ethylnaphthalene + dimethylbenzo[b]thiophene	66	Elemental sulfur
31	2,6- and/or 2,7- dimethylnaphthalene	67	Fluoranthene
32	1,3-Dimethylnaphthalene	68	Pyrene
33	1,7- and/or 1,6- Dimethylnaphthalene	69	Isomeric naphthobenzofurans
34	2,3- and/or 1,4- Dimethylnaphthalene	70	Benzo[c]phenanthrene
35	Acenaphthalene	71	Benzo[a]anthracene
36	1,2-Dimethylnaphthalene	72	Chrysene + triphenylene
		73	Benzo[a]pyrene + benzo[e]pyrene

(Adapted from Table I, Ref. 44).

The usefulness of py-GC/MS as a screening tool for sediment contamination was studied by Kruge et al. [54]. Pyrolyzing sludge at 600°C using a CDS 120 Pyroprobe interfaced to a GC/MS, they were able to detect marker compounds for sewage sludge in a sludge deposit offshore of Barcelona, Spain. Py-GC/MS was able to provide detection of different sewage marker compounds, such as linear alkylbenzenes (LABs) and trialkyl amines for surfactant marker compounds; other organonitrogen compounds from amino acids and proteins in the biomass of sludge. The fecal sterol coprostanol was

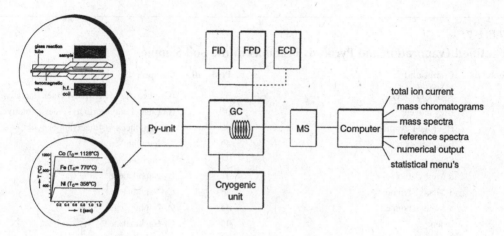

FIGURE 7.12   Instrumental setup for screening analysis by evaporation/pyrolysis gas chromatography (from Ref. 44).

also a marker. Additionally, while sterenes are also markers for petroleum contamination, a sterene carbon number distribution pattern was proposed as a signature for sewage pollution.

C   Pyrolysis of Sewage Sludge to Produce Biochar

Diamadopoulos et al. studied the possibility of producing biochar from sewage that can improve soil productivity and remediate contaminated soils. They conducted pyrolysis of 20 g of sludge between 300° and 500°C in a muffle furnace under a flow of nitrogen, also studying the potential increase of surface area by adding K_2CO_3 and H_3PO_4. These agents were chosen to take advantage of the potential leaching of nutrients K and P to the soil. Leaching tests were also performed to study the release of heavy metals, and kinetic experiments were performed to study the ability to absorb AS(V) (known for causing adverse human and environmental effects) and CR(III) (to study the possible role in cation immobilization). They determined that pyrolysis at 300°C increase biochar yield, and that the addition of K_2CO_3 increased the biochar surface area, but H_3PO_4 decreased the surface area. Non-impregnated biochar was best at retaining heavy metals in the leaching study, and it also was the most effective at adsorbing cations and anions [55].

D   Soil Organic Matter (SOM)

SOM is an essential part of ecosystem dynamics. Soil is the largest terrestrial carbon sink, so pyrolysis GC/MS is used to understand the organic matter content of soils. Examples of Py-GC-MS used to study SOM include differences related to climatic regions [56], the effects of wildfires [57], and changes related to land usage [58].

E   TD-Py-FID Applied to Marine Sediments

Kennicutt et al. [59] described the use of Py-FID in assessing the areal distribution of drilling fluids in surficial marine sediments around drilling platforms. Unlike the Py-FID that utilized flash pyrolysis (described earlier in Section IIIC), their technique subjected the samples to a 30°C/min temperature ramp from ambient to 700°C, with the evolved organic compounds being swept into an FID for measurement. This technique would better be described as thermal distillation pyrolysis-FID (TD-Py-FID) rather than Py-FID.

Freeze-dried, ground, and sieved surficial sediment samples (upper few centimeters) were heated from ambient temperature to 700°C at 30°C/min in a helium atmosphere, and the evolved organics swept into an FID with the area under the resulting peak (or peaks) integrated and digitized. The resulting FID area units were "normalized" by being divided by the organic carbon content (measured by combustion in an oxygen atmosphere — after carbonate removal — and measurement of the evolved CO_2) of the sample, expressed as nanograms, to produce values referred to as pyrolysis ratios (PRs).

The data presented clearly demonstrated that different materials — such as planktonic debris, ancient shales, leaves, and wood — show widely different PRs. Just what the ratio between the amount of organic carbon as measured by TD-Py-FID and the amount of organic carbon as measured by an oxidative technique represented was not at all clear. Nevertheless, the isopleths showing PRs as a function of the distance from drilling platforms indicated that PRs could be used to delineate the areal extent of materials added to surficial sediments by drilling operations.

F ROCK-EVAL PYROLYSIS

Rock-Eval Pyrolysis from its inception in 1977 (referred to in Ref. [60]) has become a widely used technique for geochemical characterization of organic matter in sediments. Two references have been included here that will lead interested persons into the rich literature in this' area [61]. One of these studies, in particular [60], describes the method in some detail.

Rock-Eval Pyrolysis is similar to Py-FID (described in Section III C) in that the organic material from the sample passes directly to an FID for measurement without separation by GC, and similar to TD-Py-GC (described in Section IVA) in that two organic profiles are obtained: one from a low-temperature treatment of the sample and one from the subsequent high-temperature pyrolysis of the sample. As described in reference [60], "In this technique, bulk dried samples are heated in an inert helium atmosphere, whereupon thermo-vaporization and pyrolysis, hydrocarbons are quantified by flame-ionization detection. Compounds occurring as free gases and liquids in sediments are separated from those that occur in bound form, or as particulate organic matter by temperature control. The S_1-detector signal records free hydrocarbons, which are thermo-vaporizable at 300°C, and the S_2-detector signal measures those compounds liberated during programmed pyrolysis from 300 to 550°C."

Manipulation of the S_1 and S_2 data provides a measure of the hydrogen richness and oxygen content of the organic matter, parameters that can be used to assess the biological origin of the organic matter and the degree of microbial reworking or abiotic oxidation.

G TMAH-PYROLYSIS-GC/MS

A report from a forensic science laboratory in 1989 described a technique to pyrolyze synthetic polymer samples and simultaneously chemically derivatize (methylate) the pyrolysis products prior to analysis by capillary GC and GC/MS — a technique that was referred to as simultaneous pyrolysis methylation capillary gas chromatography (SPM-GC) and SPM-GC/MS [62]. The methylation is caused to take place *in situ* by the simple expedient of adding a few microliters of methanol containing tetramethylammonium hydroxide (TMAH) to the sample in the sample holder of the pyrolysis device. When applied to sediment samples for characterization of the organic matter, this technique is referred to as TMAH-Py-GC/MS [63,64]. The methylation procedure quite likely allows the measurement of many compounds that otherwise would pass undetected. Table 7.8 shows a list of compounds that were identified in programs of river and lake sediments subjected to analysis by TMAH-Py-GC/MS [64].

TABLE 7.8
Identification of peaks from TMAH-Py-GC/MS

Peak No.	Compound Name	Peak No.	Compound Name
1	Phosphonic acid, dimethyl ester	46	Pentadecanoic acid, methyl ester
2	Stryene	47	1-Octadecene
3	Phosphoric acid, trimethyl ester	48	1-(Dimethoxyphenyl)-1,2,3-trimethoxypropane
4	Diethylene glycol dimethyl ether	49	Tetradecanoic acid, 12-methyl-, methyl ester
5	Benzaldehyde	50	1,2,4-Benzenetricarboxylic acid, trimethyl ester
6	Methylstyrene	51	2-Propenoic acid, 3-(3,4-dimethoxyphenyl)-, methyl ester
7	(Methoxymethyl)benzene	52	Hexadecanoic acid, methyl ester
8	1-Methoxy-4-methylbenzene	53	N,N-Dimethyloctadecylamine
9	2-Butendioic acid, dimethyl ester	54	Heptadecanoic acid, methyl ester
10	Butanoic acid, methyl ester	55	Octadecenoic acid, methyl ester
11	Benzoic acid, methyl ester	56	N,N-Dimethyloctadecylamine
12	Octanoic acid, methyl ester	57	Octadecanoic acid, methyl ester
13	1,4-Dimethoxybenzene	58	Tetracosane
14	Decanal	59	Eicosanoic acid, methyl ester
15	Triethylene glycol dimethyl ether	60	Dehydroabietic acid?
16	3,4-Dimethoxytoluene	61	Pentacosane
17	4-Methoxybenzaldehyde	62	Heneicosanoic acid, methyl ester
18	1-Methylindole	63	Hexacosane
19	Lignine derivative (m/z-98)	64	Dicosanic acid, methyl ester
20	Decanoic acid, methyl ester	65	Heptacosane
21	Proteinaceous derivative (m/z = 98)	66	Tricosanoic acid, methyl ester
22	1,2,4-Trimethoxybenzene	67	Docosanoic acid, 2-methoxy-, methylester
23	4-Methoxybenzoic acid, methyl ester	68	Octacosane
24	2-Propenoic acid, 3-phenyl-, methyl ester	69	Tetracosanoic acid, methyl ester
25	Dodecanal	70	Tricosanoic acid, 2-methoxy-, methyl ester
26	3-Methoxy-4,7-dimethyl-1H-isoindole	71	Sterol 1
27	1,3,5-Trimethoxybenzene	72	Sterol 2
28	1,2-Benzenedicarboxylic acid, dimethyl ester	73	Nonacosane
29	3,4-Dimethoxybenzaldehyde	74	Tetracosanoic acid, 2-methoxy, methyl ester
30	1,4-Benzenedicarboxylic acid, dimethyl ester	75	Triacontane
31	Dodecanoic acid, methyl ester	76	Hexacosanoic acid, methyl ester
32	Nonanedioic acid, dimethyl ester	77	5β-Cholestanol
33	Ethanone, 1-(3,4-dimethoxyphenyl)-	78	C_{27}-stanol
34	3,4-Dimethoxybenzoic acid, methyl ester	79	3 β-Methoxycholest-5-ene
35	1-Hexadecene	80	Sterol 3
36	3,4,5-Trimethoxybenzaldehyde	81	5α-Cholestanol
37	Benzeneacetic acid, 3,4-dmethoxy-, methyl ester	82	C_{28}-stanol
38	4,5-Dimethoxy-2-(2-propenyl)phenol	83	Octaconsanoic acid, methyl ester
39	Ethanone, 1-(3,4,5-trimethoxyphenyl)-	84	Sterol 4
40	2-Propenoic acid, 3-(4-methoxyphenyl)-, methyl ester	85	C_{29} 5β,3β,3α-stanol
41	3,4,5-trimethoxybenzoic acid, methyl ester	86	C_{29} stanol
42	Tetradecanoic acid, methyl ester	87	24S-stigmast-5-en-3.beta.-ol

(*Continued*)

TABLE 7.8 (Continued)

Peak No.	Compound Name	Peak No.	Compound Name
43	2-Propenoic acid, 3-phenyl-, ethyl ester	88	Sterol 5
44	4-Acetyl-2,3,6-trimethoxytoluene	89	Sterol 6
45	1-(Dimethoxyphenyl)-1,2,3-trimethoxypropane		

(Adapted from Ref. 64).

V OTHER ENVIRONMENTAL APPLICATIONS

A Ancient Limestone – Examination by Py-GC and Py-GC/MS

Py-GC and Py-GC/MS analysis played an interesting role in the characterization of a green layer of material 1 mm below the surface of the limestone on the south-facing exterior wall of a fourteenth-century building in Tongren, Belgium [65]. This green layer consisted of a discrete band (of about 0.2- to 0.5-mm thickness) beneath the surface of the limestone, following the surface contours of the stone. Various bits of evidence suggested that this green layer comprised a cryptoendolithic ecological niche being filled by moss, and two species of cyanobacteria. Presumably, these photosynthetic organisms gained some protection from the covering limestone surface but were able to use the moisture and light which passed through it.

A small portion of the green layer was ground to a powder, applied to the wire of a Curie-point pyrolysis unit, and heated within 0.1 s to 610°C for 10 s. The resulting pyrolysis products were separated on a 25-m long by 0.32-mm I.D. capillary GC column temperature programmed from 0°C to 310°C at 3°C/min. The mixture of compounds volatilized from the sample and/or generated during the pyrolysis was resolved into more than 150 separate peaks by the capillary GC, yielding a pyrogram (not shown here) not greatly dissimilar in appearance from some of those shown earlier in this chapter. Table 7.9, however, lists the compounds presumptively identified from this mixture by Py-GC/MS. The list of compounds, deriving from the pyrolysis of living material, is quite unlike the lists shown in the earlier tables in this chapter.

The substances identified to support the assertion that the green layer contained biogenic material because a vast majority of the evaporation/pyrolysis products obtained were related to polysaccharides, proteins, and lipids — substances that are the main cellular components of organisms such as cyanobacteria and mosses.

B Outdoor Bronze Monuments – Examination of Corrosion Patinas by Py-GC/MS

Samples of the corrosion layers from six outdoor Italian bronze monuments were examined by Py-GC/MS, leading to the identification of organic compounds deriving from both environmental pollution and from protective organic coatings [66]. In addition to direct pyrolysis, some of the samples were subjected to SPM-Py-GC/MS (described in Section IVE). For SPM, 5 µL of 25% tetramethylammonium hydroxide in methanol was added to the dry sample inside the quartz pyrolysis tube. This methylation procedure was especially useful in improving the chromatographic behavior of fatty acids present in the sample. Analytical results varied considerably among monuments, variously showing the presence of nitrogen compounds (pyridine, benzonitrile, cyanopyridine, dicyanobenzene), oxygenated compounds (phenol, benzoic acid, phthalic anhydride), fatty acids, and alkanes.

The ease with which the sample analysis could be accomplished allowed the collection of multiple samples from each monument. The comparison of results from protected portions of monuments to other areas more directly in contact with wind and rain allowed some interesting speculations about the weathering process.

TABLE 7.9

Compounds Identified from Pyrolysis of a Green Layer in Limestone from a Building in Tongeren, Belgium

Peak No.	Compound	Peak No.	Compound
1	Sulfur dioxide	72	n-Dodec-l-ene
2	1,3-Butadiene + but-l-ene	73	n-Dodecane
3	Methanethiol	74	Methylbenzylcyanide
4	trans-But-2-ene	75	VAL-VAL sequence + C_3-alkylphenol
5	Acetone	76	C_3-alkylphenol + C_2-alkylindene
6	Furan	77	VAL-VAL sequence
7	Cyclopentadiene	78	C_2-alkylindene
8	2-Methylpropanal	79	Quinoline
9	2,3-Butanedione	80	Methylnaphthalene
10	2-Butanone	81	$C_{5:1}$-alkylbenzene
11	2-Methylfuran	82	C_3-alkylphenol
12	3-Methylfuran	83	$C_{5:1}$-alkylbenzene + Methylnaphthalene
13	2-Butenal	84	$C_{5:1}$-alkylbenzene
14	3-Methylbutanal	85	n-Tridec-l-ene
15	Benzene	86	n-Tridecane
16	2-Methylbutanal	87	VAL-ILEU sequence
17	Pentane-3,4-dione	88	ILEU-VAL sequence
18	Cyclopentanone	89	VAL-LEU sequence + $C_{5:1}$-alkylbenzene
19	Pentane-2,3-dione	90	VAL-ILEU sequence
20	2,5-Dimethylfuran	91	VAL-LEU sequence
21	2,4-Dimethylfuran	92	ILEU-VAL sequence
22	Vinylfuran	93	LEU-VAL sequence
23	N-Methylpyrrol + pyridine	94	C6:1-alkylbenzene
24	Toluene	95	LEU-VAL sequence
25	3-Furaldehyde	96	$C_{6:1}$-alkylbenzene
26	2-Furaldehyde	97	Methylindole
27	2-Methyl-2,3-dihydrofuran-3-one	98	$C_{6:1}$-alkylbenzene
28	Ethylbenzene	99	C_{14}-aldehyde
29	m/p-Xylene	100	n-Tetradec-l-ene
30	1-Acetoxypropan-2-one	101	LLEU-ILEU sequence
31	Styrene	102	n-Tetracecane
32	o-Xylene	103	LLEU-LEU sequence
33	2-Methyl-2-cyclopenten-l-one	104	LLEU-ILEU sequence
34	C_3-alkylfuran	105	LLEU-LEU sequence + LEU-ILEU sequence
35	C_3-alkylfuran	106	LEU-LEU sequence
36	C_1-alkylfuran	107	LEU-ILEU sequence
37	C_3-alkylfuran	108	LEU-LEU sequence
38	Benzaldehyde	109	C_8-alkylbenzene
39	C_3-alkylbenzene	110	Levoglucosane
40	5-Methyl-2-furaldehyde	111	C_x-methylketone
41	3-Methyl-2-cyclopenten-l-one	112	1,2-Diphenylethane
42	C_3-alkylbenzene	113	n-Pentadec-l-ene
43	C_3-alkylbenzene	114	n-Pentadecane
44	a-Methylstryene	115	Isoprenoid hydrocarbon

(Continued)

TABLE 7.9 (Continued)

Peak No.	Compound	Peak No.	Compound
45	Benzofuran	116	Methyl-pentadecane
46	C_3-alkylbenzene	117	Dodecanoic acid
47	n-Dec-l-ene	118	n-Hexadec-l-ene
48	Phenol	119	n-Hexadecane
49	n-Decane	120	1,3-Diphenylpropane
50	C_3-alkylbenzene	121	Methy-lhexadecane
51	$C_{3:1}$-alkylbenzene	122	Biphenyl
52	C_3-alkylbenzene	123	9,10-Dihydroanthracene
53	Indene + 2-hydroxy-3-methyl-2-cyclopenten-1-one	124	1,3-Diphenyl-3-methyl-lpropene
54	Cyanopyridine	125	n-Heptadec-l-ene
55	C_4-alkylbenzene	126	n-Heptadecane
56	C_4-alkylbenzene + $C_{4:1}$-alkylbenzene	127	Prist-l-ene
57	Tolualdehyde	128	7-Methylheptadecane
58	m/p-Cresol	129	Tetradecanoic acid
59	C_4-alkylbenzene	130	Pristane
60	p-Cresol	131	n-Octadec-l-ene
61	n-Undec-l-ene	132	n-Octadecane
62	n-Undecane	133	Dialkyl phthalate
63	Benzylcyanide + C_4-alkylbenzene	134	Phytadiene
64	C_4-alkylfuran	135	phyt-l-ene
65	Methylindene	136	Pentadecanoic acid
66	Methylindene	137	Phytadiene
67	Ethylphenol	138	Phytadiene
68	C_2-alkylphenol	139	n-Nonadecane
69	C_4-alkylbenzene	140	Dialkyl phthalate
70	Naphthalene	141	Hexadecanoic acid
71	C_2-alkylphenol	142	n-Eicosane

(Adapted from Table 3 of Ref. 65).

C DIGESTED AND UNDIGESTED POLLENS – DISCRIMINATION BY PY-MS

The so-called "yellow rain" occurrences in southeast Asia in the early 1980s led to this application of Py-MS. A high pollen count was reported for some of the yellow samples and it was suggested that pollens might be used as a support or carrier for the distribution of chemical agents. As part of the investigations, it became important to be able to distinguish between pollen and bee feces (undigested and digested pollen, respectively). Because Py-MS had been used extensively for the characterization of biological and other polymeric materials — and, combined with pattern recognition techniques, had been successful for providing compositional information for a number of different types of materials — a study was undertaken to evaluate whether Py-MS analysis coupled with pattern recognition data analysis procedures could distinguish between bee feces and pollen samples [67].

The US Army Chemical Research and Development Center provided 11 southeast Asian samples for this study as unknowns — that is, with no sample history or identification known to the analysts at the time of analysis. The analytical task was to determine whether these samples were pollen, bee feces, or some other unrelated material. For purposes of comparison, a set of known materials was constructed consisting of bee feces, pollens, beeswax, and honey ——20 samples of different origins obtained from various sources.

The samples were applied to Curie-point wires as methanol suspensions of about 10 mg/mL and pyrolyzed at 610°C with a rise time of 100 ms. The pyrolyzer was interfaced to a quadrupole mass spectrometer set to accumulate low-energy (15 eV) electron-impact ionization spectra across the scan range of 45–245 amu with a scan speed of 1200 amu/s. For the subsequent data analysis, 30 spectra were summed together to produce a Py-MS spectrum for each sample. The 31 samples were run three times, as three sets of 31, with each set being arranged in random order before the beginning of the analysis.

The four Py-MS spectra — for pollen, bee feces, and two of the unknowns — presented as a figure in this report (but not shown in this chapter) clearly showed distinct differences, but the authors found that the complexity present in the 93 spectra made visual classification of the samples impossible. A discussion of the data analysis techniques used for the classification of these samples is beyond the scope of this chapter, but a summary of the results can be made. (Those readers with interest in statistical methods for the analysis of data generated by analytical pyrolysis mass spectrometry should find this paper interesting and may also want to read more recent work in this area [68,69]).

Unsupervised learning (i.e. factor analysis and non-linear mapping) showed that the Py-MS data contained enough distinguishing chemical information that digested and undigested pollens could be differentiated. Also, samples that were quite different from pollen, such as waxes and honeys, were identified as non-pollen samples through unsupervised learning techniques. Discriminant analysis was the most successful supervised statistical procedure used. Employing blind digested and undigested pollen standards, a classification success rate of 95% was achieved. Of the 11 Southeast Asian samples submitted blind, four of the known compositions were correctly classed as either digested or undigested pollen. Of the five "yellow rain" samples in the set, three were classified as digested pollens (bee feces) and two as undigested pollen.

D SPRUCE NEEDLES – EXAMINATION BY PY-FIMS

Field ionization mass spectrometry (FIMS) was initially developed for molecular weight determinations and mixture analyses. This soft ionization technique is particularly well suited for the mass spectrometric examination of the extremely complex mixture of compounds resulting from the pyrolysis of macromolecules because predominantly characteristic, high-mass molecular ions are produced [70].

As part of a research project on forest damage, spruce needles were examined by pyrolysis-field ionization mass spectrometry (Py-FIMS) using high-resolution mass spectrometry, time-resolved high-resolution mass spectrometry, and Curie-point Py-GC/MS [71]. Subsequent chemometric analysis of the data using pattern recognition techniques led to the conclusion that in the geographical region studied, the impact of acids and water stress were the major causes of the observed tree damage [1]. Presentation and discussion of the data obtained from the elegant analytical mass spectrometric techniques utilized for these studies are beyond the scope of this chapter. However, for those with access to the necessary instrumentation, Py-FIMS appears to offer unique insights into difficult environmental problems.

E FOREST SOILS – EXAMINATION BY PY-MBMS

Impressed by the information provided by Py-FIMS, another research group developed Pyrolysis Molecular Beam Mass Spectrometry (Py-MBMS) by coupling a quartz pyrolysis chamber to the inlet of a triple-quadrupole mass spectrometer [72]. This group wished to study the organic matrix of forest soils and saw two limitations for the application of Py-FIMS to their study: first, that the very small sample used in Py-FIMS might not be representative of the macro soil sample; and second, that the time required for each sample analysis by Py-FIMS was relatively long if one wished to examine and compare a large number of soil samples.

In their Py-MBMS instrument, about 0.1 g of each soil sample was heated to 500°C in the pyrolysis chamber containing flowing helium gas, and the pyrolysis vapors passed into the ion source of the mass spectrometer (the sampling orifice of the MS was located inside the end of the pyrolysis chamber). Although details of the sample introduction process were sketchy, a "typical" profile of soil pyrolysis product evolution for eight samples run in a period of 14 min was shown.

As with the Py-FIMS technique, presentation, and discussion of the data obtained by Py-MBMS is beyond the scope of this chapter, but the resulting pyrolysis mass spectra were incredibly complex and information-rich. With the use of chemometric techniques, the data obtained from whole soil samples taken from four sites at three depths successfully characterized changes in the chemical composition of the soil organic matter.

F   Analysis of Intractable Environmental Contamination

In the United States, most of the field investigation, remediation, and monitoring of environmental chemical contamination utilizes specific analytical methodologies, primarily those set forth by the US Environmental Protection Agency. These methods are directed almost entirely to measuring the presence and amounts of the several hundred chemical elements and compounds on the various USEPA lists of chemicals of concern. Most of the methods for organic compounds involve a primary extraction followed by a concentration step, or dissolution followed by dilution, with subsequent separation of the organic mixtures by gas chromatography prior to measurement with any one of a variety of detectors.

These methods not only fail to identify or measure most synthetic polymer materials but are also greatly hampered by the presence of these materials within the sample matrix. Probably every environmental laboratory has suffered from this problem, although it may have gone unrecognized. The experience of having a single sample suddenly "trash" an analytical system, rendering it incapable of meeting analytical quality assurance criteria without a major system overhaul, unfortunately, is not uncommon. Many of these occurrences are probably due to synthetic polymers that do not pass through the analytical system as the volatile and semi-volatile analytes do but, rather, "coat" the system and change its analytical performance characteristics.

Py-GC and Py-GC/MS could find useful places in the routine environmental analytical laboratory to help with the non-routine samples. Abandoned drums of "goo," or the collected containers of household waste chemicals which no longer bear labels possibly could be handled with some sort of thermal vaporization/pyrolysis analytical scheme.

Using analytical pyrolysis for such samples would be just one component of a systems approach to multimedia analysis, a multifaceted approach to the examination of environmental samples centered upon the use of a purge and trap/headspace/pyrolysis/gas chromatography system utilizing multiple detection systems. Although the development of such an analytical scheme was under discussion more than ten years ago [73], no analytical method using Py-MS, Py-GC, or Py-GC/MS has yet been promulgated.

G   Plastic Pollution

Plastic pollution is a growing environmental concern. Research efforts intensified around the mid-2000s to evaluate the potential harmful consequences of plastic on the ecosystems [74]. Yet, there are still fundamental questions unanswered about plastic fate once released into the environment. Upon solar light and mechanical stresses, plastic debris undergoes size reduction upon reaching the micrometric and even the nanometric scale. The presence of microscopic plastic particles in the environment is important from an ecological context because their size allows them to pass across biological barriers and to enter cells, while the high surface area to volume ratios enhance their reactivity [75]. The detection and characterization of these microscopic plastic particles is challenging. While infrared and Raman spectroscopy are key technologies to characterize the macroscopic plastic debris, the transition toward sub-millimetric detection becomes challenging. An automated image analysis using micro-Fourier transform infrared microscopy was developed to quantify and characterized plastic debris down to 30 μm after deposition on an appropriate filter [76]. After several steps of sample purification, the method was efficient with complex environmental matrices [77].

Py-GC/MS was only seldom used to detect plastic in environmental samples. In most studies, particles suspected to be plastic were manually isolated to be analyzed by Py-GC/MS [78–80]. The challenge lies in the detection of microscopic particles without handling of the particles. Presented here are two examples that use Py-GC/MS, illustrating the challenge of detecting microscopic plastic particles. The first study describes a semi-quantitative method developed for fish stomach samples [81]. The method was developed for eight polymers: polyethylene (PE), polypropylene (PP), polystyrene (PS), polyethylene terephthalate (PET), polyvinyl chloride PVC, poly(methyl methacrylate (PMMA), polycarbonate (PC), and polyamide 6 (PA6). The calibration was obtained with the introduction of a fine-grained polymer. After the animal tissues were enzymatically and chemically digested, the content was filtered on an Anodisc™ filter membrane that was dried, milled, and transferred to the pyrolysis target. The pyrolysis was performed using a Curie-point pyrolyzer (590°C) after the addition of TMAH. Spiked fish samples demonstrated the feasibility of the method. For non-spiked samples, the detected signals proved evidence of the presence of plastic particles, but the signal was too small for quantification.

The second method was the first demonstration of the presence of plastic particles at the nanometric scale in environmental samples [82]. The samples were collected in one of the five accumulation areas in oceans where floating plastic debris is gathered by oceanic currents. These samples were collected in the North Atlantic subtropical gyre. One liter of seawater was filtered on a 1.2 μm polyethersulfone membrane to remove every particle larger than the micrometer. The seawater was then concentrated by a factor of 200 by ultrafiltration (10 kDa polyethersulfone membrane). The presence of colloids was evidenced by light diffusion scattering experiments, but this technique does not render the chemical nature of the colloids. The concentrated seawater was evaporated dry and the resulting salts transferred to the pyrolysis target for pyrolysis at 700°C. The molecules usually derived from the pyrolysis of natural marine organic matter were not detected, probably because of the low volume of seawater analyzed. The colloids isolated presented the chemical fingerprint of a mixture of polymers: PE, PS, PVC, and PET.

This chapter was updated from the original work of T.O. Munson. G. Plastic Pollution was authored by Alexandra ter Halle.

REFERENCES

1. N. Simmleit and H.-R. Schulten, *J. Anal. and Appl. Pyrol.*, *15*: 3 (1989).
2. R. Tsao and K. J. Voorhees, *Anal. Chem.*, *56*: 368 (1984).
3. K. J. Voorhees and R. Tsao, *Anal. Chem.*, *57*: 1630 (1985).
4. A. Alajbeg, *J. Anal. Appl. Pyrol.*, 12: 275 (1987).
5. T. P. Wampler, *J. Anal. Appl. Pyrol.*, *16*: 291 (1989).
6. T. P. Wampler, *J. Chrom.*, *842*: 207(1999).
7. S. C. Moldovenau, *J. Microcolumn. Sep.*, *13*: 102 (2001).
8. A. Alajbeg, *J. Anal. Appl. Pyrol.*, *9*: 255 (1986).
9. W. Klusmeier, P. Vogler, K.-H. Ohrback, H. Weber, and A Kettrup, *J. Anal. Appl. Pyrol.*, *14*: 25 (1988).
10. K. J. Voorhees, W. D. Schulz, L. A. Currie, and G. A. Klouda, *J. Anal. Appl. Pyrol.*, *14*: 83 (1988).

11. A. Plewka, K. Möller, and H. Herrmann, *Proceedings from the EUROTRaC-2 Symposium 2000*, Springer-Verlag Berlin, Heidelberg, 2001.
12. A. P. Snyder, A. Tripathi, W. M. Maswadeh, J. Ho, and M. Spence, *Field Anal. Chem. Technol.*, *5*: 190 (2001).
13. K. Laniewski, H. Boren, A. Grimvall, and M. Ekelund, *J. Chrom. A*, *826*: 201 (1998).
14. A. Yasuhara and M. Morita, *Environ. Sci. Technol.*, *22*: 646 (1988).
15. A. B. J. Oudhuis, P. De Wit, P. J. J. Tromp, and J. A. Moulijn, *J. Anal. Appl. Pyrol.*, *20*: 321 (1991).
16. K. German and K. Kulesza, *J. Anal. Appl. Pyrol.*, *67*:123–124 (2003).
17. R. C. Kistler, F. Widmer, and P. H. Brunner, *Environ. Sci. Technol.*, *21*: 70 (1987).
18. J. A. Caballero, R. Front, A. Marcilla, and J. A. Canesa, *J. Anal. Appl. Pyrol.*, *40-41*: 433 (1997).
19. K. H. Nelson, I. Lysyj, and J. Nagano, *Water & Sewage Works*, *117*: 14 (1970).
20. M.-F. Dignac, P. Ginestet, D. Rybacki, A. Bruchet, V. Urbain, and P Scribe, *Water Res.*, *34*: 4185 (2000).
21. W. M. G. M. van Loon and J. J. Boon, *Anal. Chem.*, *65*: 1728 (1993).
22. D. van de Meent, J. W. de Leeuw, and P. A. Schenck, *J. Anal. Appl. Pyrol.*, *2*: 249 (1980).
23. E. R. E. van der Hage, M. M. Mulder, and J. J. Boon, *J. Anal. Appl. Pyrol.*, *25*: 149 (1993).
24. M. Kleen and G. Lindblad, *J. Anal. Appl. Pyrol.*, *25*: 209 (1993).
25. W. M. G. M. van Loon and J. J. Boon, *Trends Anal. Chem.*, *13*: 169 (1994).
26. W. L. Budde, *Analytical Mass Spectrometry-Strategies for Environmental and Related Applications*, Oxford University Press, UK (2001).
27. L. Cotrim da Cunha, L. Serve, F. Gadel, and J.-L. Bazi, *Sci. Total Environ.*, *256*: 191 (2000).
28. A. A. Christy, A. Bruchet, and D. Rybacki, *Environ. Int.*, *25*: 181 (1999).
29. Y. Coban-Yildiz, D. Fabbri, D. Tartari, S. Tugrul, and A. F. Gaines, *Org. Geochem.*, *31*: 1627(2000).
30. J. D. H. van Heemst, L. Megens, P. G. Hatcher, and J. W. de Leeuw, *Org. Geochem.*, *31*: 847 (2000).
31. H.-R. Schulten and G. Gleixner, *Water Res.*, *33*: 2489 (1999).
32. A. Aouabed, R. Ben Aim, and D. E. Hadj-Boussaad, *Environ. Technol.*, *22*: 597 (2001).
33. D. W. Page, J. A. van Leeuwen, K. M. Spark, and D. E. Mulcahy, *Mar. Freshwat. Res.*, *52*: 223 (2001).
34. N Paaso, J. Peuravuori, T. Lehtonen, and K. Pihlaja, *Environ. Int.*, *28*: 173 (2002).
35. X. Q. Lu, N. Maie, J. V. Hanna, D. L. Childers, and R. Jaffe, *Water Res.*, *37*: 2599 (2003).
36. D. M. White, D. S. Garland, J. Narr, and C. R. Woolard, *Water Res.*, *37*: 939 (2003).
37. B. Little, and J. Jacobus, *Org. Geochem.*,*8*: 27 (1985).
38. H.-R. Schulten and G. Gleixner, *Water Res.*, *33*: 2489 (1999).
39. J. Kaal, S. Wagner, R. Jaffe, *J. Anal. Appl. Pyrol*, *118*, 181–191 (2016).
40. K. J. Voorhees, J. C. Hickey, and R. W. Klusman, *Anal. Chem.*, *56*: 2604 (1984).
41. J. K. Whelan, J. M. Hunt, and A. Y. Huc, *J. Anal. Appl. Pyrol.*, *2*: 79 (1980).
42. J. K. Whelan, M. G. Fitzgerald, and M. Tarafa, *Environ. Sci. Technol.*, *17*: 292 (1983).
43. K. D. McMurtrey, N. J. Wildman, and H. Tai, *Bull. Environ. Contam. Toxicol.*, *31*: 734 (1983).
44. J. W. de Leeuw, E. W. B. de Leer, J. S. Sinninghe Damsté, and P. J. W. Schuyl, *Anal. Chem.*, *58*: 1852 (1986).
45. M. A. Kruge, P. K. Mukhopadhyay, and C. F. M. Lewis, *Org. Geochem.*, *29*: 1797 (1998).
46. E. A. Guthrie, J. M. Bortiatynski, J. O. H. van Heemst, J. H. Richman, K. S. Hardy, E. M. Kovach, and P. G. Hatcher, *En. Sci. Technol.*, *33*: 119 (1999).
47. D. M. White, H. Luong, and R. L. Irvine, *J. Cold Reg. Eng.*, *12*: 1 (1998).
48. D. Fabbri, C. Trombini, and I. Vassura, *J. Chrom. Sci.*, *36*: 600 (1998).
49. D. S. Garland, D. M. White, and C. R. Woolard, *J. Cold Reg. Eng.*, *14*: 24 (2000).
50. D. Fabbri, J. Anal. Appl. *Pyrol.*, *58-59*: 361 (2001).
51. B. Tienpont, F. David, F. Vanwalleghem, and P. Sandra, *J. Chrom. A*, *911*: 235 (2001).
52. Y. Zegouagh, S. Derenne, C. Largeau, P. Bertrand, M.-A. Sicre, A. Saliot, and B. Rousseau, *Org. Geochem.*, *30*: 101 (1999).
53. P. Faure, F. Vilmin, R. Michels, E. Jarde, L. Mansuy, M. Elie, and P. Landais, *J. Anal. Appl. Pyrol.*, *62*: 297 (2002).
54. M. Kruge, A. Permanyer, J. Serra, D. Yu, *J. Anal. Appl. Pyrol.*, *89*: 204–217 (2010).
55. E Diamadopoulos, E. Agrafioti, G. Bouras, and D. Kalderis, *Journal of Analytical and Applied Pyrolysis*, 2013, 101, 72–78.
56. K. Vancampenhout, K. Wouters, B. De Vos, P. Buurman, R. Swennen, and J. Deckers, *Soil Biol. Biochem.*, *41*: 568–579 (2009).
57. J. De la Rosa, S. Faria, M. Varela, H. Knicker, F. González-Villa, J. González-Pérez, and J. Keizer, *Geoderma*, *191*:24–30 (2012).

58. I. Yassir and P. Buurman, *Geoderma, 173-174*: 94–103 (2012).

59. M. C. Kennicuitt II, W. L. Keeney-Kennicutt, B. J. Bresley, and F. Fenner, *Environ. Geol., 4*: 239 (1983).

60. G. Luniger, and L. Schwark, *Sedimentation Geol., 148*: 275 (2002).

61. M. Yamamoto, H. Kayanne, and M. Yamamuro, *Geochem. J., 35*: 385 (2001).

62. J. M. Challinor, *J. Anal. and Appl. Pyrol., 16*: 323 (1989).

63. A. P. Deshmukh, B. Chefetz, and P. G. Hatcher, *Chemoshere, 45*: 1007 (2001).

64. L. Mansuy, Y. Bourezgui, E. Garnier-Zarli, E. Jarde, and V. Reveille, *Org. Geochem., 32*: 223 (2001).

65. C. Saiz-Jimenez, J. Garcia-Rowe, M. A. Garcia Del Cura, J. J. Ortega-Calvo, E. Roekens, and R. Van Grieken, *Sci. Total Environ., 94*: 209 (1990).

66. G. Chiavari, A. Colucci, R. Mazzeo, and M. Ravanelli, *Chromatographia, 49*: 35 (1999).

67. S. J. DeLUca, K. J. Voorhees, and E. W. Sarver, *Anal. Chem., 58*: 2439 (1986).

68. R. S. Sahota, and S. L. Morgan, *Anal. Chem., 65*: 70 (1993).

69. R. Goodacre, A. N. Edmonds, and D. B. Kell, *J. Anal. Appl. Pyrol., 26*: 93 (1993).

70. H.-R. Schulten, N. Simmleit, and R. Müller, *Anal. Chem., 59*: 2903 (1987).

71. H.-R. Schulten, N. Simmleit, and R. Müller, *Anal. Chem., 61*: 221 (1989).

72. K. A. Magrini, R. J. Evans, C. M. Hoover, C. C. Elam, and M. F. Davis, *Environ. Pollut., 116*: S255 (2002).

73. J. E. Bumgarner, Personal Communication (1993).

74. Sedlak, D. Three lessons for the microplastics voyage, *Environ. Sci. Technol., 51*, 14: 7747–7748 (2017).

75. T. S. Galloway, M. Cole, and C. Lewis, Interactions of microplastic debris throughout the marine ecosystem, *Nat. Ecol. Evol., 1*, 5 (2017).

76. S. Primpke, C. Lorenez, R. Rascher-Friesenhausen, and G. Gerdts, An automated approach for microplastics analysis using focal plane array (FPA) FTIR microscopy and image analysis, *Analytical Methods* 2017, 9, 1499.

77. S. M. Mintenig, I. Int-Veen, M. G. J. Loder, S. Primpke, and G. Gerdts, Identification of microplastic in effluents of waste water treatment plants using focal plane array-based micro-Fourier-transform infrared imaging, *Water Res., 108*, 365–372 (2017).

78. E. Fries, J. H. Dekiff, J. Willmeyer, M. T. Nuelle, M. Ebert, and D. Remy, Identification of polymer types and additives in marine microplastic particles using pyrolysis-GC/MS and scanning electron microscopy, *Environ. Sci. Proc. Imp., 15*, 10: 1949–1956 (2013).

79. M. T. Nuelle, J. H. Dekiff, D. Remy, E. Fries, A new analytical approach for monitoring microplastics in marine sediments, *Environ. Pollut., 184*: 161–169 (2014).

80. J. H. Dekiff, D. Remy, J. Klasmeier, and E. Fries, Occurrence and spatial distribution of microplastics in sediments from Norderney, *Environ. Pollut., 186*: 248–256 (2014).

81. M. Fischer and B. M. Scholz-Bottcher, Simultaneous trace identification and quantification of common types of microplastics in environmental samples by pyrolysis-gas chromatography-mass spectrometry. *Environ. Sci. Technol., 51*, 9: 5052–5060 (2017).

82. A. Ter Halle, L. Jeanneau, M. Martignac, E. Jarde, B. Pedrono, L. Brach, and J. Gigault, Nanoplastic in the North Atlantic Subtropical Gyre, Environ. *Sci. Technol.*, 51, 23: 13689–13697 (2017).

8 Examination of Forensic Evidence

John M. Challinor[1], David A. DeTata[1], Kari M. Pitts[1], and Céline Burnier[2]
[1]ChemCentre, Bentley, Western Australia
[2]Ecole des Sciences Criminelles, Université de Lausanne, Switzerland

I INTRODUCTION

A THE ANALYSIS CHALLENGE

The chemical characterization of forensic evidence requires highly sensitive and discriminatory analytical techniques due to the trace level of material available, and to ensure maximum discrimination from other material in the same class. Forensic laboratories are often equipped with an extensive instrument suite that can characterize various forms of material found at crime scenes as part of routine procedures. To ensure sample integrity, items should be analyzed "as received" if possible, and sample modification should be minimized. Pyrolysis gas chromatography (Py-GC) and pyrolysis gas chromatography-mass spectrometry (Py-GC-MS) are have been proved to be effective means of satisfying these requirements in many forensic science laboratories [1–3].

B TYPES OF EVIDENCE

Py-GC-MS is an appropriate characterization technique for crime scene materials such as paint from burglaries and hit/run traffic accidents or adhesives from insulation tapes and improvised explosive devices. Rubbers in auto tires, foams in carpets and clothing, fibers from garments, or plastics from household goods and auto parts are examples of commodities that may be identified and compared. Other applications are as diverse as the examination of bloodstains [4], propellants in ammunition [5], human hair [6], chewing gum [7], aging documents [8], and soils [9]. The numerous uses of analytical pyrolysis, including those with forensic applications, have been previously reviewed [10,11]. The technique is essential for the effective operation of any crime laboratory required to examine these and other organic materials found as evidence.

C DEVELOPMENT OF PY-GC

Py-GC was adopted by forensic scientists in the 1970s [12]. Packed Carbowax phase GC columns were accepted as the standard in many forensic science laboratories [13]. Reproducible pyrograms enabled a reliable database to be established. However, packed columns had a limited life span and were unsatisfactory for the chromatographic examination of very polar and higher molecular weight compounds, which are often diagnostic for many polymers of forensic interest. However, as a comparative technique, Py-GC was excellent. [2,14–16]. The limitations of packed column GC prompted the use of capillary GC columns, particularly high-resolution vitreous silica types. Pyrolysis capillary column GC has become the standard method adopted in most forensic science laboratories and discussion of the applications in this chapter will be restricted to this technique.

The crucial feature of pyrolysis is the need to undertake rapid and highly reproducible heating rates and final temperatures, which can be achieved using three main modes of heating: inductive (Curie-point), resistive (filament), and microfurnace [17]. While each technique has its own unique advantages [18,19], there exists a significant degree of comparability and reproducibility between the techniques [17].

D "Pyrolysis Derivatization" – modifications to the pyrolysis process

To obtain greater information about the composition of polymers and macromolecular material, modifications to the analytical pyrolysis process have been developed. These have included on-line pyrolysis hydrogenation of polyolefins using hydrogen carrier gas and a palladium catalyst [20]. Polyacetals have been pyrolyzed in the presence of a cobalt catalyst to produce cyclized compounds, which afford greater information about the structure of the polymer [21].

A further development in pyrolysis technique was developed by Challinor that involves high-temperature chemolysis and derivatization reactions, which provide more information about the structure of polymers [22]. In the "pyrolysis derivatization" process, a tetraalkylammonium hydroxide—in particular, tetramethylammonium hydroxide (TMAH)—when reacted at elevated temperatures with polymers having hydrolyzable groups, gives alkyl derivatives that reflect the composition of the polymer. Resins used in drying oil-modified alkyd enamels and saturated or unsaturated polyesters are particularly amenable to this procedure. The chromatographic profiles are generally simplified and are easier to interpret in terms of chemical composition. The precursors of the polymer or other resin components may be identified, providing data about the polymer that are often unattainable by conventional Py-GC. The reaction has been termed thermally-assisted hydrolysis and methylation (THM).

The pyrolysis products in conventional Py-GC may sometimes be obscure, particularly in the pyrolysis of some step-growth polymers, and are not easy to interpret in terms of the polymer composition. However, THM results in the formation of products that clearly identify the composition of these polymers and adds another dimension to the analytical capability of the pyrolysis technique.

II PAINT

Paint can provide strong evidence in a forensic investigation due to its variability and complexity [11,23]. There are, for example, numerous different monomers used in acrylic paint formulations that are difficult to identify by other analytical techniques such as infrared spectroscopy. Py-GC is an effective method in identifying and differentiating the organic resin/binder of paint that may vary widely within a class and, in some cases, additives may also be detected and identified. The Py-GC examination of paint, in the context of an overall forensic analysis of this type of evidence, has been described [24].

A Automotive Paint

Motor vehicle-related crime can include hit/run, wilful damage, and homicide incidents. Automotive paint binder types can be identified on microgram-sized samples of topcoat [25–27]. Some examples of typical pyrograms of acrylic lacquer, acrylic enamel, and alkyd enamel types found in original paint systems are shown in Figure 8.1.

The variability in the pyrolysis profiles of the different classes of coatings is self-evident. The interpretation of the composition revealed is as follows: the acrylic lacquer (General Motors) is a methyl methacrylate/methacrylic acid copolymer plasticized with dibutyl-, butyl cyclohexyl-, and butyl benzyl phthalates. The acrylic enamel (Ford) is a styrene/ethylhexyl acrylate/methyl

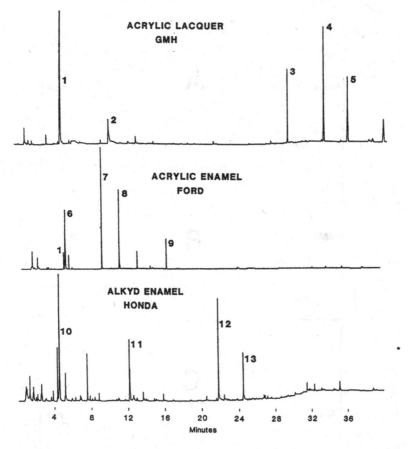

FIGURE 8.1 Pyrograms of an acrylic lacquer, acrylic enamel and an alkyd enamel used in some original automotive topcoat paint systems. 1 = methyl methacrylate, 2 = methacrylic acid. 3 = dibutyl phthalate, 4 = butyl cyclohexyl phthalate, 5 = butyl benzyl phthalate, 6 = butanol, 7 = styrene, 8 = butyl methacrylate, 9 = 2-ethylhexyl acrylate, 10 = isobutanol, 11 = vinyltoluene, 12 = phthalic anhydride, and 13 = phthalimide.

methacrylate terpolymer. The alkyd enamel (Honda) pyrolysis profile indicates that the paint resin is an orthophthalic alkyd containing a butylated-amino resin crosslinking component.

Repainted Vehicles

Acrylic lacquer paints are frequently used for refinishing vehicles after accident damage. The formulations are often based on methyl methacrylate monomer and are plasticized by the incorporation of monomers that produce "softer" polymers such as butyl methacrylate and/or external plasticizers such as the phthalates. The pyrograms of three different refinishing acrylic lacquers are shown in Figure 8.2.

Alkyd-Based Enamels

Alkyd enamels occurring as original baked enamels or spraying enamels may be identified by the THM pyrolysis derivatization modification of the Py-GC technique [28]. These alkyd polyesters are converted to methyl derivatives of their polyol, polybasic acid, and drying oil. More chemical structure information can be gained from the THM-GC analysis of alkyd resins than from conventional Py-GC. For example, Figure 8.3 shows the THM chromatogram of a blue paint smear found on a crowbar that had been allegedly used to damage a blue automobile. The paint is a baked

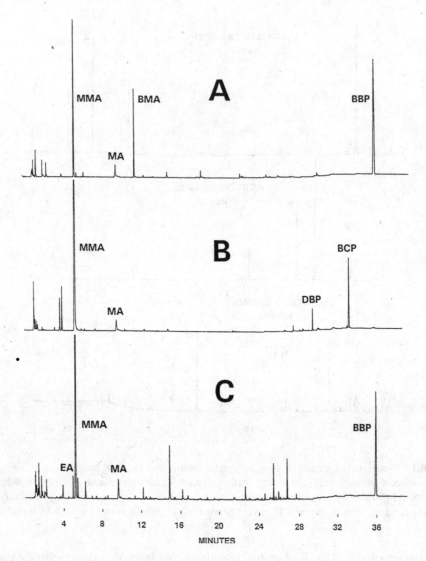

FIGURE 8.2 Pyrograms of three refinishing acrylic lacquers of different composition. A is a methyl methacrylate (MMA) - butyl methacrylate (BMA) - methacrylic acid (MA) terpolymer plasticized with butyl benzyl phthalate (BBP). B is an MMA – MA copolymer plasticized with a mixture of dibutyl phthalate (DBP) and butyl cyclohexyl phthalate (BCP) and C is an MMA – EA (ethyl acrylate) – long chain methacrylate type polymer plasticized with BBP.

alkyd enamel. From the THM chromatogram, it is evident that the paint is a pentaerythritol-orthophthalic acid baked alkyd enamel having a coconut non-drying oil crosslinked with a butylated melamine formaldehyde resin.

B ARCHITECTURAL PAINT

Break-in tools used to gain access to premises may carry traces of paint that have abraded from painted surfaces at the point of entry. Commonly encountered household paint types include polyvinyl acetate (PVA), acrylics, alkyd enamels, epoxies, and chlorinated rubbers. Py-GC can distinguish between these different classes (Figure 8.4).

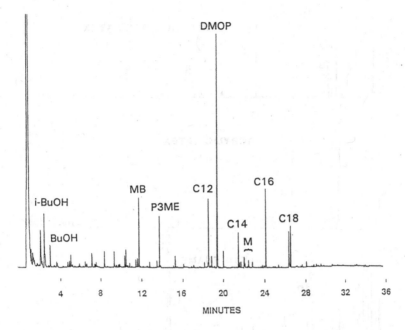

FIGURE 8.3 THM chromatogram of the baked alkyd enamel paint smear. i-BuOH = isobutanol, BuOH = butanol, MB = methyl Benzoate, P3ME = pentaerythritol trimethyl ether, DMOP = dimethyl orthophthalate, M = triazines from melamine, C12, C14, C16 and C18 = respective fatty acid methyl esters.

Furthermore, their polymer class and composition may be determined. The PVA type has a vinyl acetate/2-ethylhexyl acrylate copolymer binder. The acrylic is a methyl methacrylate/butyl acrylate copolymer. The alkyd enamel is an orthophthalic alkyd type and the epoxy is a bisphenol-A type.

In some cases, paint additives may be identified. The presence of a latex coalescing agent, trimethyl pentanediol mono-isobutyrate (Texanol) may also be detected (Peak 5 in Figure 8.4).

Studies on the differentiation between alkyd enamels by conventional Py-GC have been carried out [29,30]. As discussed previously, more structural information about alkyd enamels may be obtained by the THM technique [28]. Polybasic acid, polyhydric alcohol, drying oil composition, oil length, degree of cure, and any rosin modification can be determined. The discrimination between alkyd enamels is, therefore, improved. For example, the alkyd enamel whose pyrogram is shown in Figure 8.4 gives the following THM profile, shown in Figure 8.5.

In this case, the paint can be identified as a pentaerythritol o-phthalic alkyd consistent with having a high proportion of a soya bean drying oil (from the C16.0: C18.0 ratio). The total absence of unsaturated C18 acid methyl esters and significant azelaic acid indicates that the paint has a high degree of cure. No rosin or other modification is detected.

C INDUSTRIAL PAINTS

Industrial coatings are generally used on goods manufactured in a factory production line. They include domestic appliances, furniture, tools, and products derived from sheet metal. Their organic binders are either heat cured, air dry, catalyzed air-dry, or radiation cured. Heat cured coatings or baking and stoving enamels are the most commonly found industrial coatings. These binders can be alkyd enamels, thermosetting acrylics, polyesters, epoxies, silicone acrylics, vinyl and heat resistant polyimides and fluorocarbons. The air-dry coatings include acrylic, vinyl and nitrocellulose lacquers, alkyd resins, urethanes, and latex types. Polyurethanes and epoxy resins are examples of catalyzed air-dry binders.

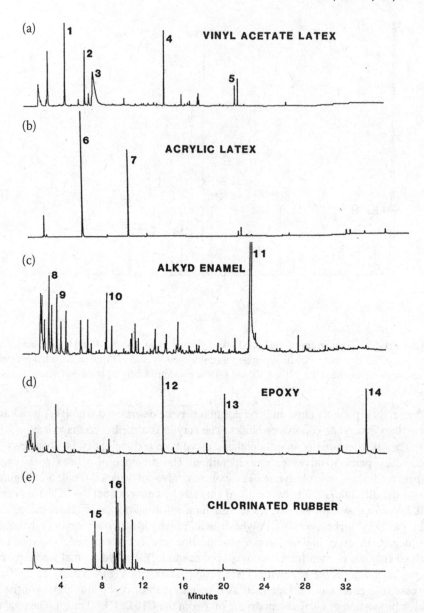

FIGURE 8.4 Pyrograms of vinyl acetate, acrylic, alkyd enamel, epoxy and chlorinated rubber type architectural paints. 1 = benzene, 2 = isooctene, 3 = acetic acid, 4 = 2-ethylhexyl acrylate, 5 = 2,2,4-trimethyl 1,3-pentanediol mono-isobutyrate, 6 = methylmethacrylate, 7 = butylmethacrylate, 8 = acrolein, 9= methacrolein, 10 = hexanal, 11 = phthalic anhydride, 12 = phenol, 13 = isopropenylphenol, 14 = bisphenol-a, 15 = xylenes, 16 = trimethylbenzenes.

Circumstances where these coatings occur as evidence involve items such as crowbars, tire levers, baseball bats, domestic appliances, and furniture.

Typical examples of Py-GC characterization of two of the wide variety of these coatings using THM are shown in Figure 8.6. It can be concluded that the beverage can coating is a methylmethacrylate/ethylhexyl acrylate copolymer modified polyester. The polyester is an adipic acid modified neopentyl glycol iso-/orthophthalic acid type. The electrical appliance coating is a butylated amino resin cross linked pentaerythritol-orthophthalic thermosetting alkyd enamel.

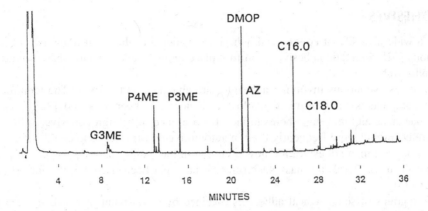

FIGURE 8.5   THM chromatogram of an architectural alkyd enamel. G3ME = glycerol trimethyl ether, P4me = pentaerythritol tetramethyl ether, P3me = pentaerythritol trimethyl ether, Dmop = dimethyl orthophthalate, Az = methyl azelate, C16.0 = methyl palmitate, C18.0 = methyl stearate.

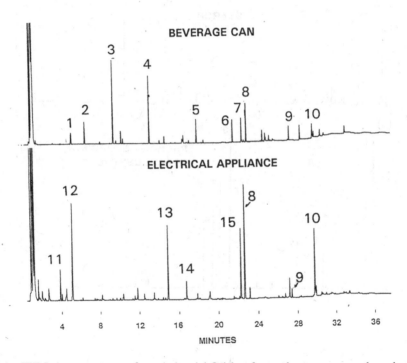

FIGURE 8.6   THM chromatograms of two industrial finishes from a beverage can and an electrical appliance. 1 = methylmethacrylate, 2 = neopentylglycol dimethyl ether, 3 = neopentylglycol monomethyl ether, 4 = ethylhexyl alcohol, 5 = dimethyl adipate, 6 = methyl laurate, 7 = dimethyl orthophthalate, 8 = dimethyl isophthalate, 9 = methyl palmitate, 10 = methyl stearate, 11 = isobutanol, 12 = *n*-butanol, 13 = methyl benzoate, 14 = pentaerythritol trimethyl ether, 15 = *N*-methyl phthalimide.

Conventional Py-GC of these coatings gives misleading diagnosis or data from which it is impossible to obtain useful structural information. THM provides more useful data about their composition [31]. Other applications of Py-GC to synthetic resins include the examination of photocopy toners and surface coatings on currency notes.

III ADHESIVES

There is a wide diversity of commercial adhesives available to the retail trade and for industrial applications [32]. Synthetic adhesives tend to replace animal- and vegetable-based glues such as casein and starch types.

Forensic case situations involving adhesives are abundant and various. The identification of adhesive components in improvised explosive devices has been reported [33]. Fraudulently resealed packages and mail may be examined for presence of foreign adhesive. Adhesives and sealers are becoming more frequently used in automotive construction and could occur in traffic-related crime. Hot melt adhesives are now common in the sealing of several types of packaging and domestic uses. Pyrolysis mass spectrometry has also been used for the identification of adhesives [34].

The pyrograms of some typical adhesives obtained by Curie point pyrolysis using a medium polarity phase vitreous silica column are shown in Figure 8.7.

The major diagnostic pyrolysis products found in common adhesives are tabulated as follows:

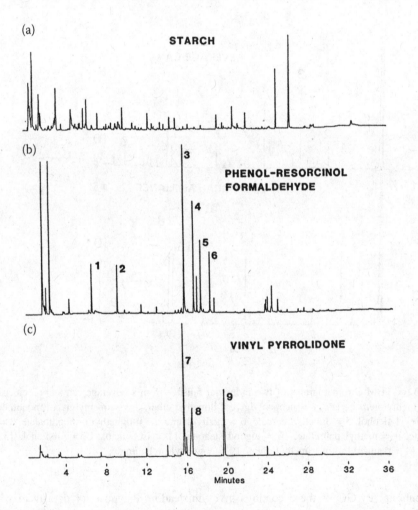

FIGURE 8.7   Pyrograms of (a) starch-based, (b) phenol-resorcinol formaldehyde and (c) vinylpyrrolidone type adhesives. 1 = toluene, 2 = xylene, 3 = phenol, 4 = ortho-cresol, 5 = para-cresol, 6 = xylenol, 7 = vinyl pyrrolidone, 8 = pyrrolidone, 9 = caprolactam.

Adhesive Resin	Diagnostic Pyrolysis Products
Vinyl acetate	Acetic acid
Acrylic	Ethyl acrylate, methylmethacrylate, butyl acrylate, 2-ethylhexyl acrylate, 2-ethylhexylalcohol, isooctene
Epoxy	Phenol, isopropenyl phenol, bisphenol A
Polyindene	Indene, vinyl toluene
Polyisoprene	Dipentene, isoprene, dimethylvinylcyclohexene
Neoprene-Novolac	Chloroprene oligomers, substituted phenols
Phenol formaldehyde	Phenol, cresols, dimethyl phenols
Silicone	Methyl substituted cyclic siloxanes
Starch	Furans, laevoglucosan
Vinylpyrollidone	Vinylpyrrolidone, caprolactam

III A INTRACLASS DIFFERENTIATION

Within-class discrimination is a crucial factor in the forensic examination of materials, and Py-GC is an appropriate technique to distinguish closely related polymers. Vinyl acetate polymers, for example, may be plasticized internally or externally. Copolymers include ethylhexyl acrylate, while phthalate plasticizers—such as dibutyl or di-isooctyl phthalate—may be used as external plasticizers in commercial products.

IV RUBBERS

Rubbers have physical characteristics and a chemical composition that precludes their successful identification by infrared spectroscopy due to their inherent elasticity and highly filled composition. In contrast, no such difficulties are encountered with Py-GC. Crime scene rubber evidence from automotive tires and rubber vehicle components is found in hit/run cases and in the soles of shoes worn by offenders in offences against property. Discrimination of vehicle bumper rubbers by Py-GC has been reported [35]. Volatile and polymeric components of rubbers and other polymers have been analyzed by Py-GC and the inorganic residue recovered for subsequent analysis [36]. The technique may also be used to quantitate rubber blends by measuring the ratios of characteristic pyrolysis products. Figure 8.8 shows examples of the pyrograms of three common types of rubber.

Peak ratios of 2 (from butadiene), 3 (from styrene), and 4 (from polyisoprene) reflect the relative proportions of butadiene, styrene, and isoprene in the rubber blend. Differences in the relative proportions of these compounds were used to obtain classification and discrimination of tires [37].

V PLASTICS

The modern world has a vast exposure to plastic materials. Motor vehicles, fire scene debris, mobile electronics, and household utensils are just a few examples of materials that contain plastics. Their identification and comparison are often necessary to further criminal investigation. A wide range of thermoplastic and thermosetting plastics may be encountered as crime scene evidence. These include polyolefines, polyacrylates, polyamides, polyesters, epoxies, and vinyl polymers. Pyrograms of some of these types are shown in Figure 8.9. The pyrogram of polyethylene was obtained on a non-polar methyl silicone phase capillary column.

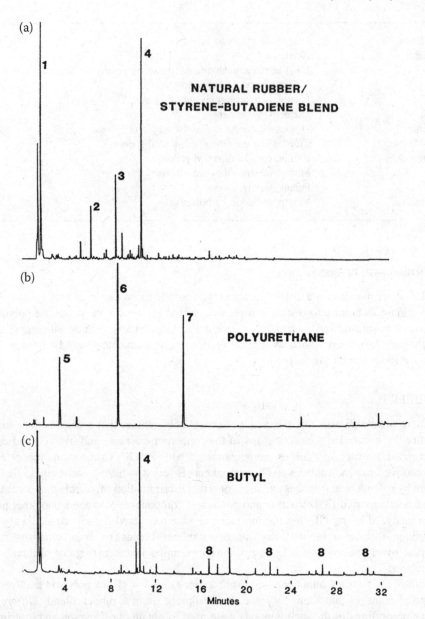

FIGURE 8.8 Pyrograms of (a) natural rubber/styrene butadiene blend, (b) polyurethane and (c) butyl rubbers. 1 = isoprene, 2 = vinylcyclohexene, 3 = styrene, 4 = dipentene, 5 = tetrahydrofuran, 6 = cyclopentanone, 7 = butanediol, 8 = isobutene oligomers.

Additional structural information can also be derived by Py-GC. Information on stereoisomerism, crystallinity, and sequence distribution data are sometimes able to be obtained. Isotactic and syndiotactic polypropylene may be identified by the ratio of oligomeric components [38]. High- and low-density polyethylene can also be determined by the proportion of branch chain alkane pyrolysis products [20] using a novel pyrolysis hydrogenation technique.

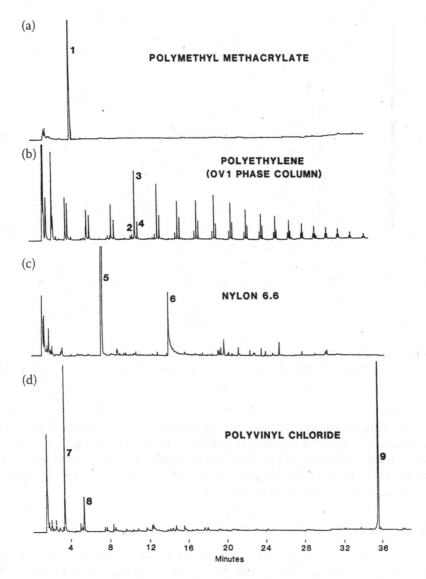

FIGURE 8.9 Pyrograms of (a) polymethyl methacrylate, (b) polyethylene, (c) nylon 6.6 and (d) polyvinyl chloride. 1 = methyl methacrylate, 2 = a, w-decadiene, 3 = 1-decene, 4 = decane, 5 = cyclopentanone, 6 = hexamethylene diamine, 7 = benzene, 8 = toluene, 9 = butyl benzyl phthalate.

VI MOTOR VEHICLE BODY FILLERS

Body fillers from repainted vehicles in traffic accidents may be left at the crime scene. Identification and subsequent comparison to material from the vehicle of the offender may establish that the vehicle was at the accident. Auto body fillers are usually a mixture of easily sanded inorganic fillers and styrenated unsaturated polyester resins. The composition of the resins may vary according to the application, resin supplier, and the relative cost of raw materials at the time of manufacture. The resins may vary in polyol, polybasic acid, or additive composition. The THM-GC procedure provides the most diagnostic information about the structure of the resin. A pyrogram of a typical body filler is shown in Figure 8.10. From this data, it is apparent that the resin in the body filler is a styrenated diethylene glycol-isophthalic acid polyester cross-linked with maleic anhydride/fumaric acid and modified with adipic acid to give the resin flexibility. It might

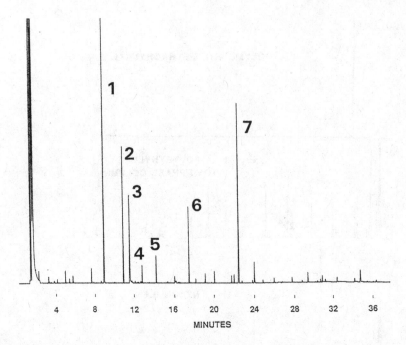

FIGURE 8.10   THM chromatogram of an automotive body filler. 1 = styrene, 2 = diethylene glycol dimethyl ether, 3 = diethylene monomethyl ether, 4 = dimethyl adipate, 5 = methyl benzoate, 6 = dimethyl adipate, 7 = dimethyl isophthalate.

be possible to determine the degree of cure by monitoring the concentration of residual maleic anhydride/fumaric acid, detected as dimethyl maleate in the THM procedure.

In a survey of body fillers [39], it was found that the resin composition varies between products in the proportion of styrene, the phthalic acid isomer, and the presence of adipic acid. Most resins contained diethylene glycol and, in some cases, propylene glycol.

VII   FIBERS

Textile fibers can provide compelling evidence in criminal activity because of the wide variety in color, dye, and organic structure found in their composition. Fibers may be transferred between garments in person-to-person contact during assaults, transferred to vehicles in hit-run traffic incidents, and encountered as ropes or twine in cases of kidnapping or deprivation of liberty.

The forensic examination of natural and synthetic fibers employs optical and electron microscopy for characterizing color, morphology, surface features, and elemental composition. The organic composition of fibers can be determined by Fourier-Transform infrared spectroscopy and Py-GC. A single fiber is often sufficient to obtain an identification of the polymer class by both techniques. The Py-GC technique has been criticized for lack of sensitivity. However, with contemporary pyrolyzers and efficient gas chromatographic systems, it is often possible to identify submicrogram quantities of fibers. Using THM and selected ion monitoring-mass spectrometry techniques, it is possible to lower the limit of detection, with modern systems capable of greater sensitivity compared to those in previous years.

The application of Py-GC to the forensic examination of textile fibers has been reviewed [40]. Examples of the selectivity of Py-GC to several fiber types were described. The limitations in sensitivity of the technique and the action to be taken to alleviate this restriction and possible solutions were outlined.

Py-GC can, therefore, be used as an effective means of determining the chemical composition of man-made homopolymer and copolymer fibers, natural fibers, fiber blends, or partly degraded fibers. Pyrograms of polyester, acrylic, and cotton fibers show how fibers can readily be identified (Figure 8.11).

Polyester fibers are composed of linear chains of polyethylene terephthalate (PET), which produces benzene, benzoic acid, biphenyl, and vinyl terephthalate on pyrolysis. Acrylic fibers comprise chains made up of acrylonitrile units, usually copolymerized with less than 15% by weight of other monomers like methyl acrylate, methyl methacrylate, or vinylpyrrolidone. Thermolysis results in the formation of acrylonitrile monomer, dimers, and trimers with a small amount of the copolymer or its pyrolysis product. In this case, the acrylic is Orlon 28, which contains methyl vinyl pyridine as comonomer. Residual dimethyl formamide solvent from the manufacturing process may also be found in the pyrolysis products. Cotton, which is almost pure cellulose, comprises chains of glucose units. The pyrolysis products of cellulose identified by GC-MS include carbonyl compounds, acids, methyl esters, furans, pyrans, anhydrosugars, and

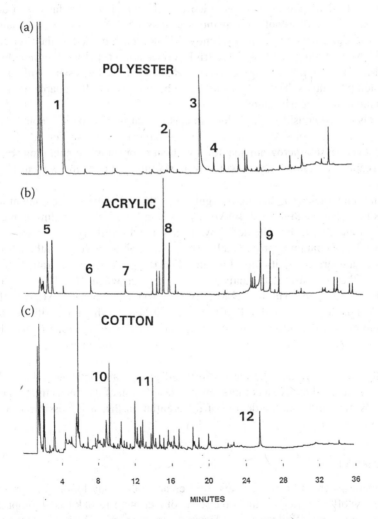

FIGURE 8.11   Pyrograms of (a) polyester, (b) acrylic and (c) cotton fibers. 1 = benzene, 2 = vinyl benzoate, 3 = benzoic acid, 4 = biphenyl, 5 = acrylonitrile, 6 = dimethyl formamide, 7 = methyl vinyl pyridine, 8 = acrylonitrile dimers, 9 = acrylonitrile trimers, 10 = 2-furaldehyde, 11 = dihydromethylfuranone, 12 = levoglucopyranose.

hydrocarbons. The major pyrolysis products are levoglucosan (1,6-anhydro-B-D-glucopyranose) and substituted furans.

Further, Py-GC examination of synthetic polymer fibers can often provide more data than other techniques, in cases where there are minor differences in composition within a class. In contrast, fibers that are chemically similar are difficult to differentiate by IR and Py-GC. Cotton vs. viscose rayon, polyesters based on PET, and wool vs. regenerated protein are examples of the use of these methods.

The pyrolysis derivatization approach may also be used. THM results in the formation of dimethyl terephthalate when PET fibers are subjected to the procedure. Tetrabutyl ammonium hydroxide may be used to replace TMAH in the reaction to confirm the presence of vinyl acetate in acrylonitrile-vinyl acetate copolymer acrylic fibers. The derivatized product is butyl acetate [22].

VIII   OILS AND FATS

Vegetable oils and animal fats are found alone or combined with other ingredients in many proprietary products. They are usually identified and compared as their fatty acid derivatives or their triglycerides. The well-established methods usually depend on GC as a means of identification. THM procedures developed more recently [31] provide a related method for characterizing these materials. Microgram quantities of the triglycerides are reacted with tetramethylammonium hydroxide (TMAH) at high temperature to yield fatty acid methyl esters without the need to employ multistep procedures. The methods are have been proved to be reliable, reproducible, and particularly suitable for forensic samples.

Vegetable oils can be considered to belong in classes that include oleic, linoleic, linolenic, and ricinoleic rich types. Examples of THM-GC of these types indicate the degree of discrimination that can be achieved on submicrogram quantities without prior sample preparation (Figure 8.12). A high polarity capillary column and 100°C oven starting temperature was used for the GC separations.

Olive oil and other oleic rich oils are recognized by their relatively high content of oleic acid, detected as its fatty acid methyl ester (FAME) (peak 3 in Figure 8.12). Palmitic and stearic acid FAMES (peaks 1 and 2) are also detected. Soya bean oil and, similarly, safflower, sunflower, and dehydrated castor oils contain approximately 60% linoleic acid, detected as the FAME (peak 4). Linseed oil is characterized by its high linolenic FAME (peak 5) concentration. Some thermal isomerization of the polyunsaturated fatty acids takes place when TMAH is employed for the THM procedure (peak 5a/b). These artefacts are not produced when TMAH is replaced by the commercially available reagents MethPrep 1 and MethPrep 2, which are trifluoromethylphenyl trimethyl ammonium hydroxide solution in water and methanol. Castor oil, which contains a high proportion of ricinoleic acid (9-hydroxy octadecenoic acid) in triglyceride form, may be identified by the presence of its free hydroxy FAME (peak 7), or as the methoxy derivative (peak 6), where both the hydroxy and the carboxylic acid groups are both methylated. While these examples are not intended to give a comprehensive description of the determination of fatty acids in vegetable oils, they indicate the degree of differentiation that it is possible to obtain by the technique.

IX   COSMETICS

Traces of cosmetics such as lipstick smears, face creams, and body lotions can provide valuable evidence to link an offender to a victim at the scene of a crime. Lipstick bases comprise a blend of waxes, oils, and emollients. While dyes and pigments present may be identified by techniques such as FTIR, Raman spectroscopy, high performance liquid chromatography and/or electron microscopy, Py-GC techniques may be used to characterize lipstick bases [41]. THM-GC is useful for the identification of triglycerides and polyols formulated into the product. Fatty acids in the

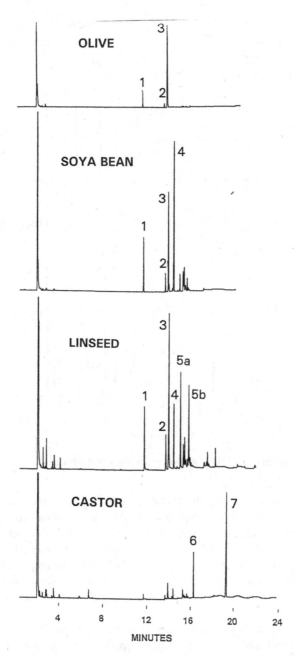

FIGURE 8.12 THM-GC profiles of olive oil (oleic rich), soya bean oil (linoleic rich), linseed oil (linolenic rich) and castor oil (ricinoleic rich) using a cyanopropyl (50%)-methyl silicone phase capillary column.

triglycerides are converted to methyl esters, while aliphatic alcohols are converted to their methyl ethers. The triglyceride of ricinoleic acid is the major component of castor oil—the most common medium found in lipstick bases. Saponification and derivatization of the triglyceride by the THM process is necessary to facilitate its confirmation. Using the pyrolysis apparatus to volatilize the organic components of the cosmetic product into the GC, Py-GC facilitates the identification of mineral oils, aliphatic alcohols, and simple fatty acid esters used as emollients. Figure 8.13 shows the Py-GC and THM-GC profiles of a typical lipstick base.

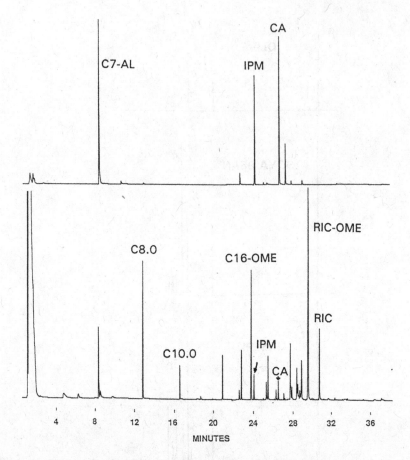

FIGURE 8.13   Py-GC (upper) and THM-GC (lower) profiles of a lipstick base.

Volatilization of the lipstick using the Py-GC process indicates the presence of cetyl acetate (CA) and isopropyl myristate (IPM). Heptanal (C-7AL), a pyrolysis product of castor oil, is a major product. The THM profile in this case is more complex and gives more information about the composition of the product. Cetyl acetate is converted to the methyl ether of cetyl alcohol (C-16-OME), while IPM is partly converted to the methyl ester. The major components, RIC-OME and RIC, are identified as the methyl ether of ricinoleic acid methyl ester and ricinoleic acid methyl ester having a free OH group, respectively. These compounds result from hydrolysis and complete—or partial—methylation of the major component of castor oil. The presence of coconut oil in the formulation is confirmed by the detection of the C8.0 and C10.0 fatty acid methyl esters. This product was differentiated from more than 60 lipsticks examined in the study, based on the composition of the organic components.

Body lotions are usually comprised of oil/water emulsions that may contain long chain fatty alcohols, glycerol, long chain fatty acids, vegetable oils, mineral oils, and emollients. Approximately 50 body lotions have been characterized by these techniques [35] and it has been possible to differentiate all these products. Figure 8.14 shows a THM-GC profile of a body lotion, identified from traces found at a crime scene.

The fatty acid methyl esters C16.0, C18.0, and C18.2 indicate the presence of a vegetable oil. However, the detection of isopropyl palmitate (IPP) and isopropyl stearate (IPS) in the Py-GC experiment (not shown) means that there would also be a contribution of their fatty acids to this pyrogram. The methyl ethers of cetyl alcohol (C16-OME) and stearyl alcohol (C18-OME) suggest the presence of the respective free fatty alcohols, and this is confirmed by an inspection of the

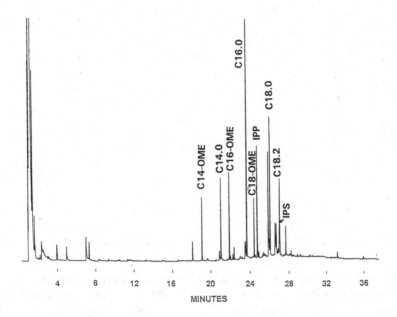

FIGURE 8.14 THM-GC profile of a body lotion.

pyrogram from the Py-GC experiment (not shown). Myristyl alcohol methyl ether (C14-OME) and myristic acid methyl ester (C14.O) originate from myristyl myristate used in the formulation.

X PERSONAL LUBRICANTS

Condom evidence and, more specifically, condom lubricants, can provide valuable evidence in sexual assault and rape cases. It is usually used as associative evidence in cases where the way it happened—and not the sexual act per se—is being questioned [42].

Modern condoms primarily consist of a latex or polyurethane body, mainly 'covered with lubricants and solid particles, but spermicide or fragrances can also be added [42,43]. More than 90% of the condoms found on the market are lubricated with a silicon-based lubricant, more commonly known as polydimethylsiloxanes (PDMS); the remaining are either polyethylene glycol (PEG)-lubricated or non-lubricated condoms [42–45]. While spermicides, fragrances, and PEG lubricants can be detected using techniques such as GC or LC [46,47], Py-GC enables the characterization of silicone lubricants [48–52]. Long chain polydimethylsiloxanes are converted into smaller cyclic oligomers that can easily be detected and are characteristic of PDMS degradation. Figure 8.15 illustrates the typical chemical profile obtained from silicon-based condom lubricant. Py-GC allows the detection of up to ten cyclic oligomers and over 40 minor degradation residues originating from PDMS degradation. Small cyclic oligomers D3 to D6 were the major peaks in the chromatograms (peaks 1–3), although oligomers up to D13 were observed. The 40 minor degradation products are as-yet unidentified products that are attributed to the full degradation of PDMS.

Other sexual lubricants and personal hygiene products may also contain PDMS in addition to water and several other components such as glycerol, oils, and other chemicals. A range of various products belonging to each category were analyzed and characterized using Py-GC, and it was found possible to differentiate these products based on their chemical profiles, which were very distinct [53]. Figures 8.16–8.18 illustrate the three different chemical profiles that were identified from the different products investigated.

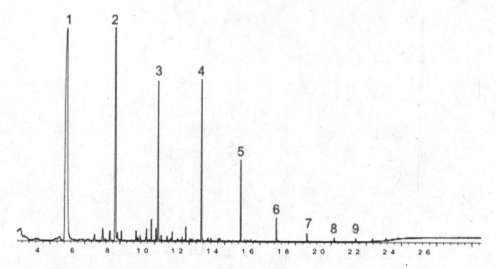

FIGURE 8.15 Py-GC-MS profile of a silicone-based condom 1 = cyclotrisiloxane (D3), 2 = cyclotetrasiloxane (D4), 3 = cyclopentasiloxane (D5), 4 = cyclohexasiloxane (D6), 5 = cycloheptasiloxane (D7), 6 = cyclooctasiloxane (D8), 7 = cyclononasiloxane (D9), 8 = cyclodecasiloxane (D10), 9 = cycloundecasiloxane (D11).

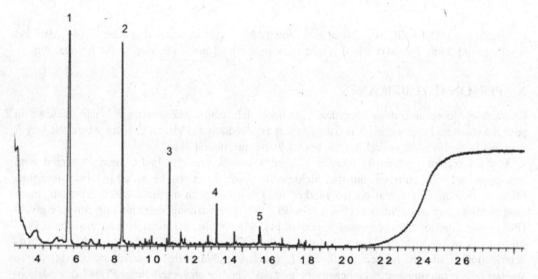

FIGURE 8.16 Py-GC-MS profile for silicone containing lubricant, Ansell LifeStyles Luxe Silicone-based lubricant 1 = cyclotrisiloxane (D3), 2 = cyclotetrasiloxane (D4), 3 = cyclopentasiloxane (D5), 4 = cyclohexasiloxane (D6), 5 = cycloheptasiloxane (D7).

XI NATURAL RESINS – ANCIENT AND MODERN APPLICATIONS

To establish authenticity of oil paintings and works of art, an examination of the surface coating resins and pigments may be required. In older objects, resins may be oleoresins, wood rosin, amber, dammar, shellac, oriental lacquers, or other natural resins.

Oleoresinous varnishes contain vegetable drying oils such as linseed and tung oils. Wood rosins comprise a mixture of rosin acids. Shellac is the flaked form of purified lac, the natural secretion of the insect *laccifer lacca kerr;* it contains 46% aleuritic acid (a trihydroxy substituted hexadecanoic acid), and 27% shelloic acid (a dihydroxy polyacyclic aliphatic dicarboxylic acid), together with

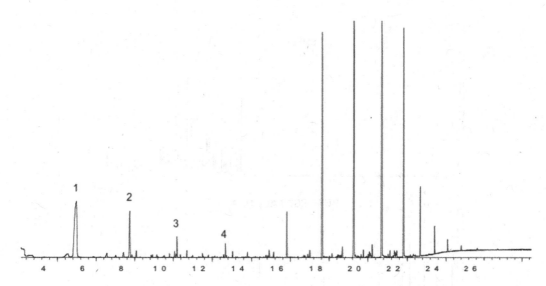

FIGURE 8.17 Py-GC-MS profile for silicone containing lubricant, Astroglide diamond silicone gel personal lubricant 1 = cyclotrisiloxane (D3), 2 = cyclotetrasiloxane (D4), 3 = cyclopentasiloxane (D5), 4 = cyclohexasiloxane (D6).

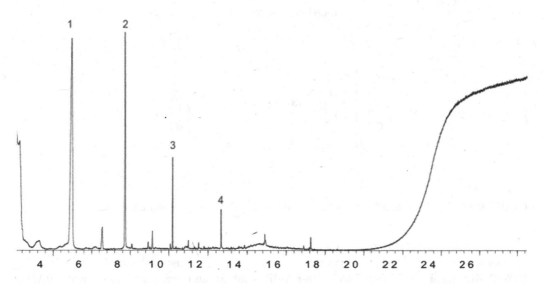

FIGURE 8.18 Py-GC-MS profile for silicone containing lubricant, Astroglide Waterproof Silicone Liquid 1 = cyclotrisiloxane (D3), 2 = cyclotetrasiloxane (D4), 3 = cyclopentasiloxane (D5), 4 = cyclohexasiloxane (D6).

other components, many of which are undetermined. The main components of oriental lacquer are urushiol, laccol, and thitsiol—which are C15- and C17- alkyl and alkenyl substituted benzene-1,2 diol compounds. Printing inks on questioned document letterheads may also contain natural resins in combination with contemporary synthetic modifiers including phenol formaldehyde resins.

Considering the polar nature of these constituents, it is not surprising that THM-GC would be an appropriate technique to characterize these resins [41,54]. THM-GC pyrolysis profiles of shellac, a Burmese lacquer and a printing ink resin, are shown in Figure 8.19.

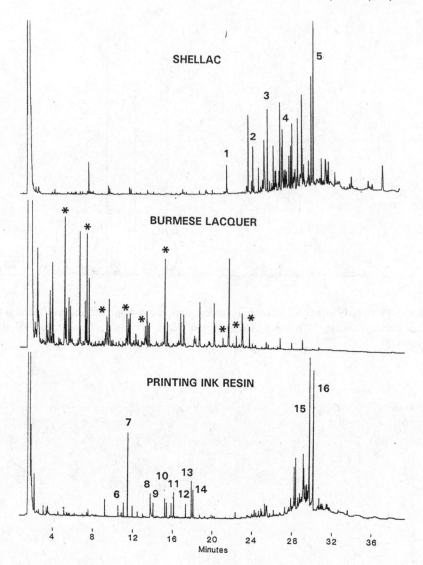

FIGURE 8.19   THM-GC profiles of shellac, a Burmese lacquer and a printing ink resin.

Shellac displays a THM profile distinctly different to the Py-GC pyrogram (not shown) in-dicating a largely hydrolysable component in the resin. There is some evidence of C14-, C16-, and C18-FAMES (peaks 1–2), 6-hydroxy C18-FAME (peak 3), an oxirane group containing FAME (peak 4), and a polymethoxy substituted C18 FAME (peak 5) in the products.

THM-GC analysis of the cured Burmese lacquer results in a series of straight chain and branched chain alkyl benzenes ranging from toluene to dodecyl benzene, and C6- to C17-alkanes and alkenes. These products are similar in composition to the products obtained by conventional pyrolysis. Compounds corresponding to peaks marked with an asterisk are FAMES that are pro-ducts of the THM reaction and become apparent as the lacquer cures. These two natural resins are, therefore, clearly distinguished by this method.

In the THM-GC profile of the printing ink resin [54], dimethyl fumarate (peak 6) is detected, suggesting the presence of maleic or fumaric acid. Pentaerythritol is indicated by the presence of the tetra- and tri-methyl ethers (peaks 7 and 8, respectively). Tertiary butyl phenol, methyl, and dimethyl tertiary butyl phenols are detected as their methyl ethers (peaks 9, 10 and 11, respectively). Rosin acid methyl esters are detected with dehydroabietic acid and abietic acid methyl esters predominating (peaks 15 and 16, respectively). This resin is, therefore, diagnosed as a tertiary butyl phenol—formaldehyde condensate modified pentaerythritol rosin maleic ester type.

XII INKS

Forensic document examiners are asked to identify altered ballpoint ink writing and, in some cases, determine when such an entry has been made. Ballpoint inks are composed of an intimate blend of resin, pigment or dye, and solvent.

In principle, the age of a ballpoint ink entry may be determined by its dye component composition. With a sufficiently wide historical database of such information, it has been possible to "date" a ballpoint entry. An alternative approach to determine the composition and relative age is by Py-GC methodology.

Py-GC, using the THM technique, has shown some potential both in identifying solvent composition and resin type [54]. A THM-GC profile of a typical ballpoint ink is shown in Figure 8.20.

From these data, the solvents are identified as phenoxy ethanol (PE), benzyl alcohol (BA), and propylene glycol (PG). The methyl ethers that are also detected are suffixed by ME. It would appear that the resin is a polyvinyl pyrrolidone type, as indicated by the detection of the dimer (PVP2) and trimer (PVP3). Mass spectral data suggests that peak marked DIAZ is derived from a diazo dye.

The relative age of the ink on the document can be assessed by monitoring the reduction in concentration of individual solvent components; however, these solvents evaporate rapidly from the document entry, so the method has limited application. In alkyd resin binder inks, it may be possible to determine the age of the ink entry by monitoring changes in the fatty acid composition due to curing, as discussed previously.

XIII MISCELLANEOUS

A discussion of the applications of Py-GC to forensic chemistry would not be complete without including the role of the technique to the characterization of proprietary drugs and other organic substances used illegally. Irwin has adequately described this topic [3]. Despite the advantages,

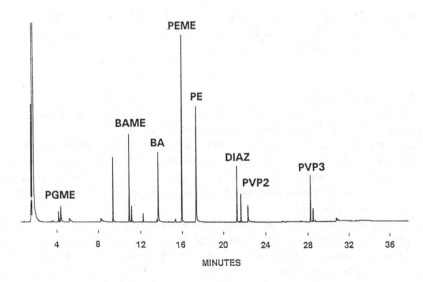

FIGURE 8.20 THM-GC profile of a typical ballpoint ink.

Py-GC has not been widely adopted for this purpose in forensic laboratories. The pyrolysis derivatization techniques previously described have potential for identifying drugs in the form of esters and carboxylic acids (such as Ibuprofen) and their excipients including vegetable oils, sugars, and preservatives [31].

Soils containing high levels of organic matter may be characterized by Py-GC techniques. The presence of humic acids in soils infer that there should be potential for soil differentiation by THM techniques [55,56].

The identification and comparison of wood is an area of forensic interest. Microscopy methods based on morphology are generally used for this purpose. However, developments in the mid-1990s indicate that there is potential for the chemical taxonomy of wood species by THM [57] and by the pyrolysis of TMAH extractives [58].

XIV CASE STUDY

During the police investigation of the scene of a multiple murder at the home of the victims, a greasy substance was found on a bedroom door and a fingerprint was visible within the grease.

The forensic laboratory was asked to identify the substance so that appropriate fingerprint development techniques could be used to visualize the print. Some of the greasy substance was carefully removed from the door and analyzed by THM-GC. The substance was a cosmetic body lotion (as shown previously in Figure 8.14). After development, the print matched a person known to have been in the house several days previously and was, therefore, not necessarily incriminatory.

However, this was suspicious, and the laboratory asked for swabs from three of the victims to be forwarded for comparison, as there was evidence of sexual activity. Methylene chloride extracts of the swabs were evaporated on pyrolysis wires and the experiments were repeated. The results were compared to the body lotion identified on the door. There was a high degree of correspondence between the composition of the body lotion on the door, and the composition of the components in the swabs, considering that some adsorption of some of the components could have taken place on the skin surfaces. The link between the suspect's fingerprint and the victims was established. As a result, the suspect was questioned again and confessed to all aspects of the crime.

XV CONCLUSIONS

It is evident that Py-GC may be used to characterize different types of material likely to be associated with a forensic investigation. It is a versatile and discriminatory technique that may be used for the forensic examination of materials by:

- High temperature pyrolysis to determine the structure of macromolecular materials. This may lead to products that do not directly relate to the parent polymer as a result of dehydration and decarboxylation mechanisms, for example. This is particularly where the polymer is derived from polar monomers such as polyesters.
- Elevated temperature reaction pyrolysis where the analyte is reacted with a chemical reagent. The method is appropriate for materials that contain hydrolysable bonding or free reactive groups—for example acids, alcohols, phenols. Usually, the "pyrograms" more clearly reflect the composition of the parent material. A review of the technique describes applications to non-forensic materials [59].
- A system for thermally desorbing chemical compounds from the sample. This is useful for introduction by volatilization of the total sample to the GC system and compounds adsorbed on an adsorbent, such as activated charcoal or porous polymer. The latter method can be used in the analysis of hydrocarbon accelerants from fire scenes in cases of arson or environmental or toxicological monitoring.

An effective interpretation of data, and a diagnosis of the composition of the original polymer, cannot be made without knowledge of the identity of the pyrolysis products. Therefore, every effort should be made to identify and interpret the significance of—at least—the major products by mass spectrometry, relative retention times, or infrared spectroscopy (in the case of GC-FTIR).

While this discussion has been limited to Py-GC, other pyrolysis techniques should not be forgotten. Py-MS, in its simplest form (in contrast to other more sophisticated techniques, such as field ionization-MS, laser ablation-MS and fast atom bombardment), provides data that can give useful information about chemical structure.

Future directions include the application of other separation techniques, such as supercritical fluid chromatography [60] and the development of novel high temperature chemolytic methods and high-resolution mass spectrometers, which may yield even more information about the structure of macromolecules. Additionally, interpretation may be enhanced using chemometric analysis of resulting pyrograms, increasing the objectivity of associations, especially in conjunction with larger scale data bases.

REFERENCES

1. B. B. Wheals, *J. Anal. Appl. Pyrol.*, 2: 277. (1980).
2. B. B. Wheals, *J. Anal. Appl. Pyrol.*, 8: 503. (1985).
3. W. J. Irwin, *Analytical Pyrolysis, A Comprehensive Guide*, Chromatographic Series Vol. 22, Marcel Dekker, Inc., New York, N. Y. (1982).
4. P. K. Clausen and W. F. Rowe, *J. For. Sci.*, 25, 4: 765 (1980).
5. R. O. Keto, *J. For. Sci.*, 34, 1: 74 (1989).
6. T. O. Munson and J. Vick, *J. Anal. Appl. Pyrol.*, 8: 493 (1985).
7. J. B. F. Lloyd, Hadley and Roberts, *J. Chrom.*, 101: 417 (1974).
8. L. Ortiz-Herrero, M. E. Blanco, C. Garcia-Ruiz, and L. Bartolomé, *J. Anal. Appl. Pyrol.*, 131: 9 (2018).
9. C. S. Lee, T. M. Sung, H. S. Kim, and C. H. Jeon, *J. Anal. Appl. Pyrol.*, 96: 33 (2012).
10. T. P. Wampler, *J. Anal. Appl. Pyrol.*, 16: 291 (1989).
11. T. P. Wampler, *J. Anal. Appl. Pyrol.*, 71: 1 (2004).
12. P. R. De Forest, *J. For. Sci.*, 19: 113–120 (1974).
13. R. W. May, E. F. Pearson, and D. Scothern, (1977) *Pyrolysis Gas Chromatography*, Analytical Science Monographs No. 3, The Chemical Society.
14. T. P. Wampler and E. J. Levy, *Crime Lab. Digest*, 12: 25 (1985).
15. A. Parabyk, The characterization of plastic automobile bumper bars using FTIR, PGC and SEM-EDX, Master of Science in Forensic Science Thesis, George Washington University (1988).
16. R. Saferstein, *Polymer and GC Analysis* (S. A. Liebman and E. J. Levy eds.) Marcel Dekker, New York, p. 339 (1985).
17. B. A. Stankiewicz, P. F. van Bergen, M. B. Smith, J. F. Carter, D. E. G. Briggs, and R. P. Evershed, *J. Anal. Appl. Pyrol.*, 45: 133 (1998).
18. K. Vèkey, J. Tamàs, A. Somogyi, A. Bertazzo, C. Costa, G. Allegri, R. Seraglia, and P. Traldi, *Org Mass Spectrom.*, 27: 1216–1219 (1992).
19. J. M. Challinor, *For. Sci. Int.*, 21: 269 (1983).
20. S. Tsuge, Y. Sugimura, and T. Nagaya, *J. Anal. Appl. Pyrol.*, 1: 221 (1980).
21. S. Tsuge, H. Ohtani, 1997. Polym. Degrad. Stab., 58, 1–2: 109-130. ISSN 0141-3910, https://doi.org/10.1016/S0141-3910(97)00031-1.
22. J. M. Challinor, *J. Anal. Appl. Pyrol.*, 16: 323 (1989).
23. J. I. Thornton, *Forensic Science Handbook* (R. Saferstein. ed.) Prentice Hall. Inc., New Jersey (1982).
24. J. M. Challinor, *Expert Evidence, Advocacy and Practice* (I. Freckleton and H. Selby, eds.) Law Book Company, Melbourne, Australia.
25. P. J. Cardosi, *J. For. Sci.*, 27: 695 (1982).
26. K. Fukuda, *For. Sci. Int.*, 29: 227 (1985).
27. D. McMinn, T. L. Carlson, and T. O. Munson, *J. For. Sci.30*, 4: 1064 (1985).
28. J. M. Challinor, *J. Anal. Appl. Pyrol.*, 18: 233 (1991).
29. J. M. Challinor, Proceedings of the 13th Meeting of the International Association of Forensic Sciences, Oxford (1984).

30. J. W. Bates, T. Allinson, and T. S. Bal, *For. Sci. Int.*, *40*: 25 (1989).

31. J. M. Challinor, *J.Anal. Appl. Pyrol.*, *20*: 15 (1991).

32. I. Skeist, *Handbook of Adhesives*, 2nd Edition, Van Nostrand Reinhold Company, New York, N. Y. (1977).

33. N. L. Bakowski, E. C. Bender, and T. O. Munson, *J. Anal. Appl. Pyrol.*, 8: 483 (1985).

34. J. C. Hughes, B. B. Wheals, and M. J. Whitehouse, *For. Sci.*, *10*: 217 (1977).

35. R. D. Blackledge, *J. For. Sci.*, *26*: 557 (1981).

36. J. Chi-an Hu, *J. Chrom. Sci.*, *19*: 634 (1991).

37. L. Gueissaz, G. Massonnet, *J. Anal. Appl. Pyrol.*, *124*: 704 (2017).

38. S. Tsuge and H. Ohtani, *Analytical Pyrolysis, Techniques and Applications* (K. J. Vorhees, ed.) Butterworths (1984).

39. J. M. Home, J. D. Twibell, and K. W. Smalldon, *Med. Sci. Law*, *20*, 3: 163 (1980).

40. C. Roux, R. Morison, and P. Maynard, *Forensic Examination of Fibres,* 3rd Edition (J. Robertson, C. Roux and K. G. Wiggins, eds.) CRC Press (2017).

41. J. M. Challinor, *J. Anal. Appl. Pyrol.*, *25*: 349 (1993).

42. C. Burnier and G. Massonnet, *For. Sci. Int.*, *302* (2019).

43. P. Maynard, K. Allwell, C. Roux, M. Dawson, and D. Royds, *For. Sci. Int.,124*: 140 (2001).

44. R. D. Blackledge and M. Vincenti, *J. For. Sci. Soc.*, *34*: 245 (1994).

45. R. D. Blackledge, *J. For. Sci. 40*: 467 (1995).

46. R. A. Musah, A. L. Vuong, C. Henck, and J. R. E. Shepard, *J. Am. Soc. Mass Spectrom.*, *23*: 996 (2012).

47. A. O. Onigbinde, G. Nicol, and B. Munson, *Eur. J. Mass Spectrom.*, 7: 279 (2001).

48. S. Fujimoto, H. Ohtani, and S. Tsuge, *Fresenius Z Anal. Chem.*, *331*: 342 (1988).

49. J. C. Kleinert and C. J. Weschler, *Anal. Chem.*, 52: 1245 (1980).

50. L. S. Tottey, S. A. Coulson, G. E. Wevers, L. Fabian, H. McClelland, and M. Dustin, *J. For. Sci.*, *64*: 207 (2019).

51. G. P. Campbell and A. L. Gordon, *J. For. Sci. 52*: 630 (2007).

52. C. Burnier, G. Massonnet, 2020. *For. Sci. Int.*, 310, 110255. ISSN 0379-0738, https://doi.org/10.1016/j.forsciint.2020.110255.

53. C. Burnier, G. Massonnet, D. DeTata, and K. Pitts, unpublished work (2019).

54. J. M. Challinor, Proceedings of the 12th International Association of Forensic Sciences, Adelaide, 1990.

55. C. Lee, T. Sung, H. Kim, and C. Jeon, *J. Anal. Appl. Pyrol.*, *96* (2012).

56. A. Lara-Gonzalo, M. Kruge, I. Lores, B. Gutiérrez, and J. Gallego, *Org. Geochem.* (2015).

57. J. M. Challinor, *J. Anal. Appl. Pyrol.*, *35*: 93 (1995).

58. J. M. Challinor, *J. Anal. Appl. Pyrol.*, *37*: 1 (1996).

59. J. M. Challinor, *J. Anal. Appl. Pyrol.*, *61*: 3 (2001).

60. P. R. DeForest, et al., *Gas Chromatography in Forensic Science* (I. Tebbett, Ed.) Ellis Horwood Limited (1992).

9 Characterization of Microorganisms by Pyrolysis-GC, Pyrolysis-GC/ MS, and Pyrolysis-MS

Stephen L. Morgan, Bruce E. Watt, and Randolph C. Galipo
The University of South Carolina Columbia, SC

I INTRODUCTION

Detecting, identifying, and characterizing microorganisms is vital in solving important environmental, biological, and medical problems. In monitoring a food production process for pathogenic microorganisms or surveying a battlefield for biological warfare agents, detection may require recognizing that microbial species other than those expected in the normal environment are present. A diagnosis of pneumonia for a patient may require classifying a microbial sample as one of several closely related species. Microorganisms play a central role in soil biodegradation processes; analysis and characterization of environmental microbial communities can contribute to the design and control of waste bioremediation processes. Characterization of different organisms might also provide supporting evidence for or against relatedness of species for taxonomic purposes.

Bergey's *Manual of Determinative Bacteriology* [1] catalogs over 250 bacterial genera. Traditional microbiological identification of bacterial species is based on appearance under a microscope (shape, size, presence of particular structures), response to staining (e.g., the classic Gram stain), or indirect characteristics (growth under aerobic or anaerobic conditions, generation of specific enzymes or biochemical products, etc.). Morphology is closely related to taxonomy but its use for microbial characterization is not definitive [2,3]. Structural characteristics of "Gram type" (e.g., thick peptidoglycan in Gram-positive organisms versus lipopolysaccharide in Gram-negative organisms) do not always correlate with the results of the Gram stain. Other cell types (such as acid-fast species) also exist [4,5]. The presence of specific enzymes, the response of an organism to growth substrates, or susceptibility of an organism to specific antibiotics is often due to metabolism rather than structure. Although valuable for identification, the use of these largely phenotypic characteristics requires culturing and isolation of viable cells, which can be time-consuming and may even fail for fastidious organisms [6].

The complex mixture of microbial cellular and extracellular components is an appealing target for chemotaxonomic analysis by gas chromatography (GC), mass spectrometry (MS), or the combination of the two techniques together (GC/MS). Such instrumental methods focus on chemical structures present in a sample and are not dependent—as are biological tests—on the microorganism being viable. Additionally, since only microgram amounts of sample are required, instrumental methods offer enhanced sensitivity.

Chemical component characteristics of specific microorganisms or microbial groups are usually enclosed in or part of a polymeric cellular matrix. Such nonvolatile and intractable biological samples present difficulties for their direct characterization by GC, MS, or GC/MS. For these techniques to be useful, chemical components characteristic of the microbial sample must be released intact or a related compound must be generated before GC or MS analysis. Depolymerization is usually performed off-line by acid hydrolysis, methanolysis, saponification, or other reactions. One or more chemical derivatization steps might then be employed to produce volatile and thermally stable derivatives suitable for GC or MS.

When analytical pyrolysis is applied to a microbial sample, depolymerization and volatilization of bacterial components is accomplished simultaneously. The volatile thermal products may be separated on-line by capillary GC with flame ionization detection (Py-GC-FID), separated by GC and detected by MS (Py-GC/MS), or detected directly by MS (Py-MS). In contrast to derivatization-based methods—and like other direct approaches—microbial characterization by analytical pyrolysis requires minimal sample pretreatment and short total analysis times.

The use of analytical pyrolysis for biomedical taxonomy was proposed by Zemany [7] in 1952. A single strain of an unidentified microorganism was used in fundamental studies of pyrolysis temperatures and residence times by Oyama [8] in 1963. The first comprehensive work was that of Reiner [9], who used pyrolysis-GC in 1965 to identify different strains of *E. coli*, *Shigella sp.*, *Streptococcus pyogenes*, and *Mycobacteria*. Although most early work was performed with packed columns and non-selective flame ionization detection, the potential of analytical pyrolysis for making sensitive discriminations between various microorganism samples was well established by the late 1970s.

In 1979, Gutteridge and Norris [10] reviewed the application of pyrolysis methods in the identification of microorganisms. Significantly, this review appeared in a bacteriology journal and listed 148 references. Meuzelaar, Haverkamp, and Hileman [11] discussed pyrolysis-MS methods for biomaterials in their 1982 text, describing nearly 30 applications in clinical microbiology, quality control, and other biological areas. Irwin [12] included a chapter on microbial taxonomy in his comprehensive 1982 guide to analytical pyrolysis and listed 124 references. The use of analytical pyrolysis in clinical and pharmaceutical microbiology was described by Wieten et al. [13] in 1984 and over 100 references were cited. Bayer and Morgan [14] also reviewed the analysis of biopolymers by pyrolysis-GC in 1984 and listed 80 literature citations on microorganisms. In 1985, Fox and Morgan [6] discussed a chemotaxonomic approach in characterizing microorganisms based on analysis of chemical signatures detected by analytical pyrolysis and derivatization GC/MS. Wampler's 1989 bibliography of analytical pyrolysis references [15] listed 29 microbial applications along with other examples of biopolymer analysis. Analytical methods for microorganisms, including pyrolysis, were also surveyed by Eiceman, Windig, and Snyder [16]. More recently, Morgan et al. [17,18] reviewed their work on identification of chemical markers for microbial differentiation produced by pyrolysis.

Improvements in analytical capability for the analysis of complex pyrolysate mixtures have appeared during the last decade—high resolution capillary GC with more polar and selective stationary phases coated on inert fused silica columns; coupling of capillary GC with sensitive, selective, and lower cost mass spectrometric detectors; enhanced pyrolysis-MS techniques; hyphenated analysis methods, including GC-Fourier transform infrared spectroscopy (GC/FTIR) and tandem MS; and better strategies for handling complex multidimensional pyrolysis data. The present chapter reviews the known chemotaxonomy of microorganisms, summarizes practical considerations for the use of pyrolysis in microbial characterization, and critically discusses selected applications of analytical pyrolysis to microbial characterization.

II MICROBIAL CHEMOTAXONOMY

Chemotaxonomy is the study of chemical differences in organisms and the use of these differences to identify and classify them. As suggested in the last section, direct chemical analysis provides a powerful supplement to the more traditional methods of classifying microorganisms by morphological and biochemical characteristics. Although variations in growth, treatment, and sampling conditions can influence the structure and composition of, microorganisms also exhibit many invariant structures that can be employed for differentiation.

Bacteria and other prokaryotes lack the complex intracellular structures (endoplasmic reticulum, Golgi apparatus, lysosomes, mitochondria, etc.) found in plant and animal cells (eukaryotes). Bacterial morphology (shape and size) is limited to a few simple forms, such as rods, cocci, chains, and spirals [5,19]. The chemical composition of the cell walls, outer membranes, and capsules of bacteria is sufficiently diverse to offer potential for taxonomic discrimination [3,4]. For example, the chemical makeup of a single colon bacillus (*Escherichia coli*) has been estimated to contain 2,000–4,500 different types of small molecules distributed as 120 different amino acids and their precursors and derivatives, 250 different carbohydrates and their precursors, 50 different fatty acids and their precursors, 100 different nucleotides and their precursors and derivatives, and 300 quinones, polyisoprenoids, porphyrins, vitamins, and other small organic molecules [20]. Many of these molecules are common to all organisms and not well suited for discrimination among microorganisms. Some of these molecules are specific to various major groups of organisms and can serve as chemical markers for their detection or identification.

The cell envelope usually consists of a cell membrane, a cell wall, and an outer membrane [21,22]. Figure 9.1 shows schematic representations of the structure of Gram-positive and Gram-negative cell envelopes. The cell wall consists of the peptidoglycan (PG) layer and associated structures. PG is the only substance common to almost all bacteria (except *Mycoplasma* and *Chlamydia*) and absent in non-bacterial matter [23]. PG and its associated chemical components may account for up to 10–40% of the dry weight of the cell [5]. As seen in Figure 9.2, PG consists of a polysaccharide backbone that is a repeating polymer of *N*-acetylglucosamine and *N*-acetylmuramic acid. Attached covalently to the lactyl group of muramic acid are tetra- and pentapeptides (composed of repeating L- and D-amino acids) cross-linked by peptide bridges. The amino sugar muramic acid (3-O--carboxethyl-D-glucosamine) is a fairly definitive marker for bacteria. Other chemical markers in PG include D-alanine, D-glutamic acid, and diaminopimelic acid [24,25]. The D-amino acids are sometimes found in other bacterial components but are not synthesized by mammals. Different bacteria may vary in the sequence of the amino acids in the peptide sidechains and crossbridge.

The PG layer is thicker in Gram-positive bacterial cell envelopes than in Gram-negative bacteria (Figure 9.1A). As a result, chemical markers for PG are found in higher amount in Gram-positive cells. Teichoic acids (with ribitol or glycerol phosphate backbones and sidechains of variable amino acid composition) are covalently bound to the thick peptidoglycan layer in some Gram-positive species [26]. Lipoteichoic acids (LTA) composed of glycerophosphate polymers terminated in glycolipid are not bound to the cell wall but are sometimes linked through the cell wall to muramic acid [5,26]. Teichuronic acids (glucuronic acid polymers with variable sidechains) and neutral polysaccharides are also bound to PG.

Gram-negative cell envelopes usually have a thin PG layer (Figure 9.1B). PG is attached to the outer membrane by lipoprotein containing phospholipids and other hydrophobic substances. A variety of phospholipids and proteins are found on the inner side of the outer membrane; some of these (porins) spanning the outer membrane. The external surface of the outer membrane of Gram-negative bacteria contains its primary endotoxin, a unique lipopolysaccharide (LPS) consisting of an outer O-antigen, a middle core, and an inner lipid A region [27]. The lipid A region contains a glucosamine disaccharide with covalently bound 2- and 3-hydroxy fatty acids. Glucosamine is common, but 3-hydroxy and 2-hydroxy fatty acids are unusual. Although the fatty acid composition of LPS varies among Gram-negative bacteria, -hydroxymyristic acid is a chemical marker for LPS.

Fatty acid profiling by GC is routine in some clinical reference laboratories, particularly in identification of anaerobic bacteria [28,29]. Fatty acids and lipids are bonded to proteins, carbohydrates, or other chemical entities in microbial cell walls and membranes. Fatty acids of chain length from C_9 to C_{20} are useful in identifying Gram-negative organisms at the species and genus level.

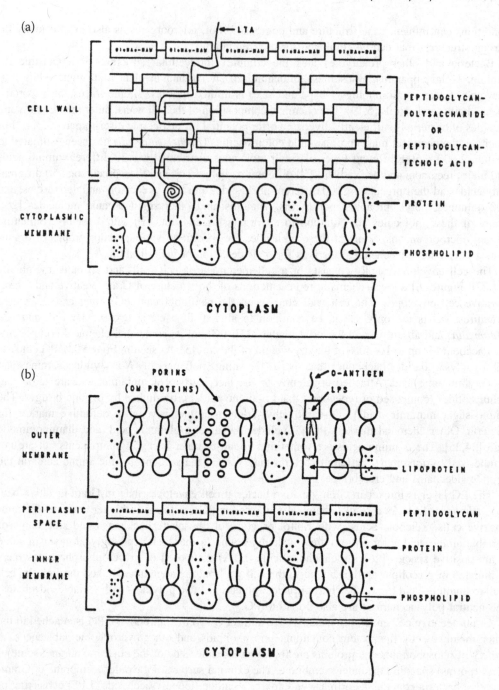

FIGURE 9.1 Generalized structure of bacterial cell envelope. (A) Gram positive organism. (B) Gram negative organism.

Perhaps the only automated GC-based microbial characterization system that is commercially available is a microbial analysis system based on derivatization GC of fatty acid methyl esters (Microbial ID, Inc., Newark, NJ) [30].

The core region in the LPS of Gram-negative bacteria contains two unusual sugars, keto-deoxyoctonic acid (KDO) and L-D-glycero-mannoheptose. KDO and heptoses are not commonly

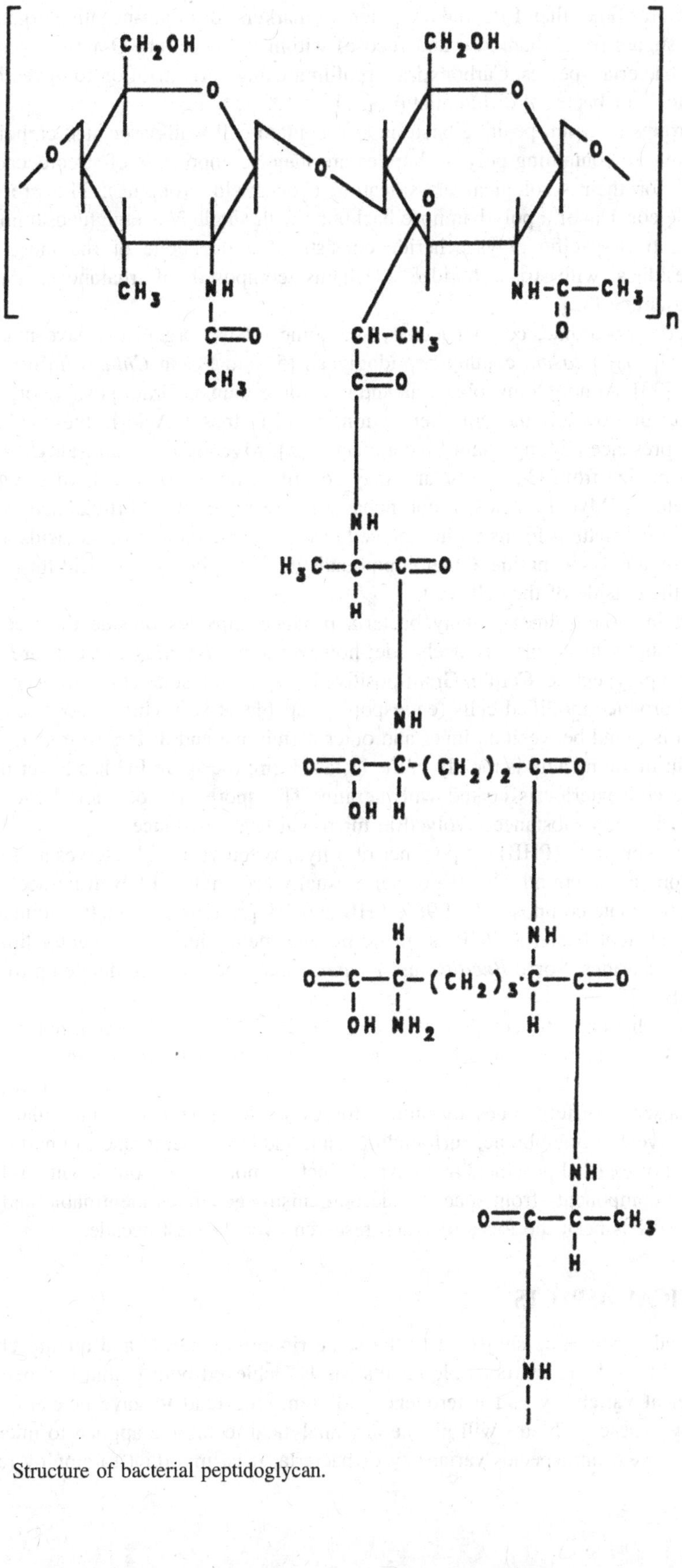

FIGURE 9.2 Structure of bacterial peptidoglycan.

found in structures other than LPS and are potential markers for Gram-negative organisms. Levels and types of sugars (e.g., rhamnose and fucose) within the core and O-antigen regions of LPS differ among bacterial species. Carbohydrate profiling using derivatization followed by GC/MS is an excellent tool for bacterial differentiation [31].

Various groups of Gram-positive bacteria also contain cell wall-associated carbohydrates. For example, rhamnose-containing polysaccharides are major components of streptococcal cell walls and are used for their serological classification (Lancefield grouping). The group A-specific polysaccharide consists of a polyrhamnose backbone with single N-acetylglucosamine sidechains, while the group B-specific polysaccharide consists of a backbone of rhamnose and glucitol phosphate residues with trisaccharide sidechains composed of rhamnose, galactose, and N-acetylglucosamine [32].

Mycobacteria, nocardiae, corynebacteria, and some related organisms have unusual cell envelopes. Not all *Mycoplasma* contain peptidoglycan [5] and PG in *Chlamydia* does not contain muramic acid [33]. Although mycobacteria and nocardiae stain as Gram-positive organisms, some resist decolorization by acid alcohol after staining (acid-fastness). Acid-fastness is believed to be related to the presence of long chain mycolic acids [5]. Mycolic acids are -alkyl, -hydroxy fatty acids ranging in size from C_{30} to C_{86} and may constitute up to 60% of the dry weight of these bacterial envelopes. Mycolic acids in corynebacteria are relatively short-chained. Mycolic acids have a variety of structural forms within the cell envelope and some mycolic acids are covalently bound to PG by a polysaccharide. Other compounds that contain mycolic acid form a thick waxy layer around the outside of the cell wall.

Under certain circumstances, many bacteria produce capsules outside their cell envelopes. Capsules are usually made of polysaccharide; however some *Bacillus* capsules are composed of D-glutamic acid polypeptide. Certain Gram-positive bacteria—most notably, strains of *Bacillus* and *Clostridium*—produce modified cells (endospores) capable of surviving in adverse environments [5]. Spore PG is found between an inner and outer membrane and differs from that in the normal vegetative cell; muramic acid is mostly in the lactam form; the spore PG has fewer peptide cross-links; and the cell exterior is coated with keratin. The spore also contains large quantities of calcium dipicolinate, a substance involved in microbial heat resistance.

Poly--hydroxybutyrate (PHB)—a polymer of 3-hydroxybutyric acid—serves as a major energy and carbon storage compound. Diverse bacteria usually accumulate PHB in intracellular granules having an approximate composition of 98% PHB and 2% proteins and small amounts of lipids. In genetically competent bacteria, PHB is in the plasma membrane or covalently linked to extracellular polysaccharides. Some *Bacillus* and *Pseudomonas* species accumulate up to 30% of their weight as PHB [5].

Many of the characteristic components mentioned in this section are common to diverse bacterial species. For example, chemical markers for peptidoglycan (muramic acid, etc.) are ubiquitous in bacteria. Other chemical components have the potential to be chemical markers whose presence or absence differentiates dissimilar (or even closely related) microbial species. The bacterial cell envelope (membrane, surrounding wall, and outer membrane), in particular, offers a rich assortment of unusual polymers that have distinctive monomeric constituents. Whether any of these chemical components from specific microorganisms generates identifiable and unique pyrolysis signatures has been the focus of much research over the past decade.

III PRACTICAL ASPECTS

When analytical pyrolysis is subjected to the same rigorous controls and quality checks as any other analytical technique, reproducible results can be achieved both within and between laboratories. Biological variability and heterogeneity of sampling tend to have deleterious effects on reproducibility—these problems will plague any analytical technique applied to microorganisms. Because of the great intraspecies variability of bacteria, it is important to employ representative samples of more than one strain of each organism if possible. Ruggedness, simplicity, and ease of interpretation of results are other performance characteristics relevant in developing pyrolysis methods for microbial characterization.

A   MICROBIAL SAMPLE HANDLING

The first step in the handling of any microbial sample, whether it is taken from a clinical specimen or from the environment, is usually to culture the organism in growth media under controlled conditions. This step serves several purposes: the number of cells of the organisms is increased, thereby increasing the detectability, and the organism(s) present may be isolated as nearly pure strains, thus increasing the chances for identification. Such cleanup or isolation of strains can be important. Cultures containing mixed populations of microorganisms can present difficulties for identification schemes, even those based on chemotaxonomic markers.

The formulation of growth media, inoculation, incubation times and temperatures, and sampling of a microorganism should be carefully standardized and controlled to not introduce undesired variability into later attempts at characterization. Media should be formulated in a standardized manner and of a known composition. Inoculation of the microbial cells into the culture media should be done with the same quantity of cells in the same growth stage. All strains should then be cultured for the same period and at the same temperature to provide uniformity. Ideally, organisms should be grown under the same conditions (media, temperature, stage of growth). Unfortunately, this is not always practical. More often, the isolation of an unknown microbiological sample might require a battery of different media appropriate for different growth types. Related strains can be grown under identical conditions to the same growth stage. Organisms having very different physiological requirements for growth clearly cannot be grown under the same conditions (e.g., obligate anaerobes and aerobes). In any case, standardization and quality control are required.

Changes in the type or composition of growth media may influence the production of extra-cellular components (enzymes, metabolic products, etc.) as well as some cellular components. For example, Gutteridge and Norris [34] studied the effects of different growth conditions on the discrimination of three different bacteria by Py-GC. Pyrograms of the same organism grown on different media were more different from one another than pyrograms of different organisms grown on the same media. The effect of culture time and incubation temperature was less obvious, producing variations in two organisms but not for the other organism. Engman et al. [35] also examined the effect of growth media, incubation time, and sample storage time on the pyrograms of five bacterial species of the genus *Flavobacterium*. Pyrograms of organisms grown on two different but similar media such as *Flavobacterium* medium and Trypticase Soy Agar (both consisting of a protein hydrolysate agar with no added carbohydrate) showed few differences. When glucose was added to the growth medium, however, several new peaks appeared in the pyrograms. Samples incubated for different times also showed few differences except for two peaks that increased in height upon longer incubation. It is desirable, of course, for differences in the amounts of discriminating pyrolysis products to be representative of relatively stable genetic or structural differences between organisms that do not vary appreciably with culture conditions.

Depending on its origin, incubation time may affect the levels of a particular pyrolysis product. Watt et al. [36] profiled the production of (*trans*) 2-butenoic acid—a pyrolysis product of poly--hydroxybutyrate (PHB)—as a function of incubation time for *Legionella pneumophila*. The amount of 2-butenoic in the pyrograms followed the expected growth curve for PHB. PHB is found in the plasma membrane of *E. coli* and only present in excessive amounts when growth is limited and cells are genetically competent. Pyrolysis products of PHB were not found in *E. coli* pyro-grams, perhaps because the growth conditions were not optimal to force production of PHB.

After harvesting bacteria, it is usually desirable to wash the samples several times with sterile distilled water to remove contamination from the media. Samples are often autoclaved or heat-killed (at 60°–120 °C in air, depending on the susceptibility of the organism) for safety reasons, then lyophilized (freeze-dried) to a dry powder. Heating may be performed under an inert gas to avoid the possibility of oxidation. Since the sample is exposed to much more energetic degradation conditions during pyrolysis, most researchers have assumed that heat-killing and lyophilization do not appreciably change the chemical structure of the microbial sample. Organisms should be prepared and killed in a similar fashion whenever possible. Preparing samples in widely different fashions may not permit hypotheses concerning observed pyrolysis patterns to be tested. The sample may then be refrigerated for storage before analysis.

In addition to being cultured in a well-defined manner, strains should also be systematically characterized by traditional microbiological tests. Only in this manner can quality control be assured. Traditional microbiological characterization should be performed before killing, as most of these tests require viable cells. In a pyrolysis study of streptococci, a sample labeled as Group F strain was found to contain a chemical marker (due to glucitol phosphate) previously detected only in Group B organisms [37]. Routine tests (hippurate hydrolysis and serological grouping) identified the organism as a group B streptococcus. It was later confirmed that this sample had been incorrectly identified by the originating laboratory. Bacteria are not simple pure chemical substances; incorrect assignments and sample contamination are common when routine microbiological control is not carried out. This is not a criticism of interlaboratory studies as much as a comment on the complexity of adequate sample and information transmittal between laboratories.

Knowledge of chemical composition with respect to relevant carbohydrate, amino acid, or fatty acid chemical markers is often critical for proper interpretation of the pyrolysis of a microbial sample. Sometimes, this information is already available in the literature regarding specific bacteria. More often, however, complementary analyses for that class of components by an independent GC, GC/MS, or spectroscopic method may be needed.

Finally, researchers working with potentially pathogenic microorganisms should be aware of local and government regulations pertaining to operating a safe and healthy workplace [38]. In the United States, this involves compliance with the Occupational Safety and Health Act of 1970 and regulations of the Department of Labor including OSHA document 29 CFR 1910 [39]. Employees are required to inform and train employees regarding the specific hazards associated with their work. Safety techniques and devices should be used to isolate personnel from biohazards. Specific examples include use of protective laboratory clothing, aseptic technique to avoid environmental spread of the pathogenic agent, and safety cabinets to minimize spread of aerosols. Containment considerations also include use of non-porous surfaces for laboratory bench tops; type of door used to separate the laboratory from public areas; the type of ventilation system in use; and accessibility of autoclaves for sterilization. Biohazard labels should be affixed to waste containers, refrigerators, freezers, and transport containers holding potentially infectious materials. Specific procedures for storage, decontamination, and disposal of biohazardous agents should also be documented.

B ANALYTICAL PYROLYSIS TECHNIQUES FOR MICROORGANISMS

Ease of sample handling is often cited as an advantage in pyrolysis-based methods. This advantage refers to the fact that lengthy sample pretreatment, derivatization, or clean-up steps are not required. As pointed out by Windig [40], biological samples are often heterogeneous and the preparation of a representative suspension involving microgram amounts of bacteria can be tedious. Additionally, Montaudo [41] points out that the assumption that *any* sample pretreatment should be avoided is naive. Indeed, if the characteristic discriminating information for a microbial application is known *a priori* to reside in a certain cellular component, a carefully designed pretreatment step may be able to provide a suitable isolated cell fraction. This approach will give better reproducibility because the background of extraneous chemical components irrelevant to decision-making is eliminated.

To prepare a microbial sample for analysis, an amount of the lyophilized powder is weighed and added to a measured volume of water to give a solution of known concentration (e.g., 10 g/L).

The solution should be vortexed or sonicated to produce a homogeneous suspension before sampling onto the pyrolysis sampling device.

Alternatively, a microbial sample may be transferred by a plastic inoculation loop directly from the culture media into a quartz sampling tube or onto a metal wire. Care should be taken to avoid carrying media with the sample; blank pyrolysis runs on media alone should also be done to check for background contamination. Sample homogeneity (composition, shape, and amount) is desirable for reproducible results and this may be difficult to achieve. In some instances, however, analytical pyrolysis may be conducted directly on original samples with little or no sample pretreatment. Gilbart et al. [42] discriminated group B *Streptococci* from groups A, C, F, and G following swabbing of the samples direct from sheep blood agar plates and suspension of the sample in distilled water to a consistent opacity.

When a resistively heated filament pyrolyzer (e.g., the Pyroprobe from CDS Analytical, Inc., Oxford, PA) is used, the sample may be placed directly on a platinum filament or may be placed in a quartz tube or boat inside a platinum coil. In either case, the placement of sample in the sampling tube or the ribbon should be the same for all samples. For liquid sample suspensions placed on a ribbon or in a coil, the solvent is evaporated prior to pyrolysis. Solid microbial samples can be sandwiched between quartz wool plugs inside the quartz sampling tube to reduce extraneous nonvolatile material from leaving the sampling tube during pyrolysis. With quartz, the sample never comes into direct contact with the pyrolyzer filament, as it does when the sample is coated directly on a thin ribbon filament. Ribbon filaments sometimes exhibit a memory effect (particularly with polar components), are harder to clean, and typically have a shorter lifetime. Quartz tubes may be reused after cleaning.

The temperature actually experienced by the sample in a resistively heated probe may be different from the pyrolysis temperature setting. Factors influencing the sample temperature include sample amount, positioning of sample, and whether a quartz sampling tube is employed or not. To achieve reproducibility, sample sizes for analytical pyrolysis should be in the lower g range. Large sample sizes (more than 100 g) make controlled heat transfer and uniform heating difficult to achieve; secondary reactions are promoted, resulting in more complicated and less reproducible pyrograms. Small sample sizes allow formation of thin films and permit effective and rapid heat transfer from the heat source [43]. If analytical profiling is the goal, sample amounts applied to the tube are typically in the range of 10–100 g. Acquisition of quality mass spectra for all peaks (including minor pyrolysis products) may require samples of several hundred micrograms.

With Curie-point pyrolysis, the active heating element is a ferromagnetic metal wire (0.5 mm in diameter). The sample (10–20 g) may be coated on the wire by depositing drops (5 L each) of suspension (2–4 g/L) and then dried while rotating the wire [11]. As mentioned already, placement of sample should be done reproducibly to reduce variability in pyrolysis results. Carbon disulfide has been recommended as a solvent for sampling onto Curie-point wires as it evaporates easily and leaves no detectable residue [44]. Sampling factors influencing Curie-point techniques in Py-MS have been extensively studied [45–47]. The deposition of unpyrolyzed sample or nonvolatile residue onto the walls of the pyrolysis interface increases with the amount of sample loaded on Curie-point wires, with optimal loading between 1 and 20 g [11,45].

Following loading of sample, the filament probe or Curie-point wire is inserted into a heated interface connected to the GC injection port or Py-MS vacuum. In the original design of filament pyrolysis systems from CDS Analytical, the ribbon or coil is an element in a Wheatstone bridge circuit. The bridge is set balanced at the setpoint temperature and a capacitor is discharged to rapidly heat the filament (up to 20 C/ms and 1000 °C). Although resistively heated filament pyrolyzers are generally calibrated by the manufacturer, the actual temperature of the pyrolyzer element can be determined by appropriate calibration [48].

Curie-point pyrolysis involves placing the sample wire into a radio frequency field that induces eddy currents in the ferromagnetic material and causes a temperature rise. When the wire reaches the Curie-point temperature, it becomes paramagnetic and stops inducting power. The temperature at which the wire stabilizes (the Curie-point) is a function of the type of metal. For example, the Curie-points of cobalt, iron, and nickel are 1128 °C, 770 °C, and 358 °C, respectively. Wires made from alloys of these metals produce intermediate temperatures. For example, the commonly used nickel-iron wire has a Curie-point of 510 °C [11]. Differences between filament and Curie-point pyrolyzers depend on the pyrolysates examined and may be obscured by other instrumental differences, including the design of the transmission system to the detector.

Long term reproducibility is affected by eventual deterioration of resistive filaments or sample wires. All components exposed to sample during pyrolysis (GC injection port liners and quartz sample tubes if used) often require acid cleaning, solvent washing, and oven drying. Active pyrolyzer elements (coils and ribbons in filament pyrolyzers) can be heated without sample to remove contamination (1000 °C for 2 s is usually adequate). Curie-point wires are inexpensive enough to be discarded after use.

Preliminary experiments are usually required to choose pyrolysis conditions. The pyrolysis literature may guide the researcher, but optimal conditions for a particular sample must be determined empirically. The question of appropriate heating rates and final temperatures has been addressed by other authors [11,12]. Pyrolysis heating conditions interact with sample size and effects are usually system-dependent (i.e. resistive heating *versus* Curie-point, wires *versus* ribbons *versus* quartz tubes). With microorganisms, temperatures that produce pyrolysates characteristic of the parent sample may range from 400 °C to 800 °C or even 1000 °C. Generally, lower pyrolysis temperatures induce less fragmentation and decrease the total amount of pyrolysates produced. Higher temperatures ensure more complete fragmentation and, thus, might reduce the structural information obtained. Extremely low pyrolysis temperatures may not produce significant amounts of volatile fragments from intractable biomaterials and may lead to the accumulation of nonvolatile residues contaminating the analytical system. Pyrolysis at a variety of different temperatures may provide selective information about degradation patterns and thereby reveal distinct aspects of structure. A time profile of volatile products generated at different temperatures by linear programmed thermal degradation may also provide characteristic structural information [49].

C CHROMATOGRAPHIC SEPARATION AND MASS SPECTROMETRIC DETECTION OF PYROLYSIS PRODUCTS FROM MICROORGANISMS

Once the sample has been pyrolyzed, volatile fragments are swept from the heated pyrolysis/injection port by carrier gas into the GC or GC/MS system. In Py-MS, it is likewise desired to transfer pyrolysis products to the ionization source of the MS without appreciable degradation, condensation loss, or recombination. Designs of Curie-point Py-MS systems have incorporated glass reaction tubes, expansion chambers, heated walls, and positioning of the pyrolysis reactor directly in front of the ion source [11].

In Py-GC, if the column is kept relatively cold at the start of the run (e.g., as in a programmed temperature run), pyrolysates will be focused on the head of the column until eluted by the increasing temperature. The amount of pyrolysis products transferred to the chromatographic column is dependent on whether a split injection or a direct injection without splitting is employed. For highest sensitivity, direct flow without splitting is preferred; good chromatography may dictate using a reasonable split ratio (e.g., 10–20 parts split flow, 1-part column flow).

In their profiling of oral streptococci, French and his co-workers [50,51] used 1.5 m × 4 mm columns packed with a porous polymer (Chromosorb 104) combined with isothermal chromatographic conditions for fast repetitive analysis. Porous polymer columns are suitable for the analysis of low molecular weight volatile pyrolysates but provide only limited higher molecular weight information. Although some conventional packed columns have been used in pyrolysis studies of microorganisms, over the past decade, GC applications have converted completely to the use of fused silica open tubular capillary columns. Because microbial pyrolysates are often labile polar species, the column material and chromatographic column should be as inert as possible. Fused silica capillary columns coated with "bonded" phases satisfies these needs. Fused silica columns offer improved resolution, increased inertness, and better analytical precision than packed columns. Superior resolution per unit time available with capillary columns means that adequate resolution can often be achieved using short (5–10 m) columns. With optimized conditions, analysis times are usually less than 10–30 min. Engman et al. [35], in studying the pyrolysis of *flavobacteria*, employed fast temperature programs to shorten analysis times when using selected ion monitoring (SIM). Despite the resultant loss of resolution, capillary pyrograms were found to retain discriminating information. Another advantageous approach is to directly interface short (1–5 m) capillary columns to low pressure MS ion sources. The decrease in analysis time, maintenance of resolution, and elution of labile components at reduced temperatures have been described by Trehy et al. [52] and applied to microbial pyrolysis by Snyder et al. [53].

Choice of chromatographic stationary phases for fused silica capillary columns can range from non-polar (e.g., 100% methyl silicone) to polar (e.g., polyethylene glycol) phases. A particular stationary phase is employed, of course, to affect elution order of components being chromatographed. Figure 9.3A and 9.3B shows replicate pyrolysis-GC/MS analyses of *B. anthracis* strain VNR-1-D1. These pyrograms were obtained using a fused silica capillary column coated with 1.0 film thickness of DB-1701 (moderate polarity, 14% cyanopropylphenyl–86% methyl silicone). Figure 9.3C and 9.3D shows replicate pyrograms *of* the same strain on a different fused silica capillary column coated with a 0.25 film thickness of free fatty acid phase (FFAP, polar, polyethylene glycol-acid modified). Correlation of peaks between the two sets of chromatograms on different columns is extremely difficult without an additional dimension of information (e.g., mass spectra for all peaks). Nevertheless, replicate chromatograms on the same phase are reproducibly identical. The different stationary phases employed here "process" the mixture of pyrolysis products differently. Both stationary phases provide a characteristic and reproducible signature for the microorganism. In the same way, chromatographic systems (instrument plus column) with differing active metal or other sorption sites may preferentially filter polar or reactive components from the chromatogram.

Contamination in GC systems is a continual problem in analytical pyrolysis. Early reports [54,55] recognized that production of a residue from a sample during pyrolysis may be unavoidable, particularly with biological samples. Extraneous peaks that mysteriously appear in some pyrograms have determinate causes that, if understood, can be eliminated [56]. Contamination in pyrolysis is often due to carry-over of higher molecular weight, polar, or less volatile material produced during pyrolysis that remains behind the interface, connecting tubing, or other parts of the chromatographic system. Tarry products or smoke may be produced by many samples and build-up inside the system after a period of use, confounding long term reproducibility. During temperature programming, tiny amounts of the contaminating materials are flushed from the system. Possible trouble-shooting solutions include investigating different pyrolysis temperatures and times to pyrolyze the sample more effectively and evaluating the pyrolysis elements for possible contamination. Wiping the inner surfaces of the pyrolysis interface with a cotton swab will reveal if residue accumulation is the problem. If residue is present, the pyrolysis interface and GC interface should be cleaned using various solvents including dilute nitric acid, hexane, chloroform, and acetone—in that order. Irwin [12] recommends overnight heating (600°–700 °C) of solvent washed pyrolysis elements, heating in a water-saturated stream of hydrogen at 550 °C, or chemical polishing followed by washing with acetone and heating. Direct flame heating should be avoided as it may oxidize metal, contaminating surfaces and changing heat transfer characteristics. Quartz sample tubes should be cleaned in dilute nitric acid at 100 °C overnight, rinsed with methanol and acetone, then dried in an oven.

Sample sizes that are too large (approaching 1 mg) may contaminate the column. If residue from pyrolysis cannot be avoided, using less sample may reduce contamination and extend column life. However, contamination becomes even more obvious when small samples are involved.

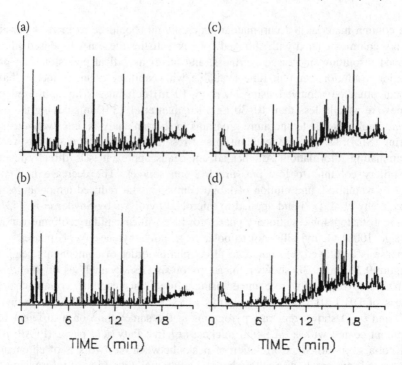

FIGURE 9.3 Pyrolysis of *B. anthracis* strain VNR-1-D1 at two different temperatures and analyzed by two different columns. (A, B) Replicate pyrograms produced by pyrolysis at 650 °C with pyrolysates separated on a 1.0 film thickness DB-1701 fused silica column (J & W Scientific). (C, D) Replicate pyrograms produced by pyrolysis at 800 °C with pyrolysates separated on a 0.25 film thickness FFAP fused silica column (Quadrex).

Occasionally with increased usage, contamination may be visibly detected on the first few cm of a column. A rather drastic but appropriate solution is to clip off the initial section of column. Removal of a section of contaminated column does not have a profound effect on resolution and may restore performance to a degraded system. Crosslinked or bonded phase columns can sometimes be rejuvenated by rinsing with organic solvent (such as methylene chloride) and then reconditioning. Mitchell and Needleman [57] described a disposable pre-column with backflushing capability to remove contaminants from a pyrolysis-GC system. Sugimura and Tsuge [58] designed a splitter system for pyrolysis with capillary GC that may be used to reduce column contamination without decreasing chromatographic performance. More recently, Halket and Schulten [59] modified a Curie-point inlet system for capillary columns to permit better identification of all products, including fewer volatile products left behind in a quartz pyrolysis chamber.

In Py-GC or Py-MS systems, contamination may be caused by a cold spot that is trapping less volatile residue. To see if any temperature zones are set too low, temperature settings should be increased by small increments (without exceeding temperature limits of the column or instrument) and the results checked. Wrapping connecting tubing with heating tape or insulation may eliminate cold spots. Condensation losses due to cold spots can also severely limit the ability to observe higher molecular weight and less volatile components. Flexible fused silica columns can be inserted directly into the pyrolysis interface near the pyrolyzer element at one end and threaded straight through to the FID flame tip or the MS ion source at the other end. Contact with active surfaces is minimized and more complete transfer of the pyrolysis products from the pyrolysis interface to detector is achieved.

Similar considerations are important in Py-MS: cold spots and wall contact in the Py-MS vacuum systems can also affect transmission of pyrolysates [11]. Py-MS produces a pyrogram of all compounds in the pyrolysis product mixture superimposed in a single mass spectrum. For that reason, interpretation of Py-MS results from a complex sample can be more difficult than interpretation of a Py-GC/MS pyrogram, in which pyrolysis products are chromatographically separated before MS detection. As pointed out by Snyder et al. [53], Py-MS with relatively cold pyrolysis interface walls or with expansion chambers tend to provide only low mass range analysis (under m/z 200). When direct Py-MS is performed with the pyrolysis reactor close to the ion source to detect larger mass pyrolysis products (m/z 200–1,000), the ion source tends to contaminate quickly, jeopardizing reproducibility.

Long term reproducibility of pyrograms in Py-GC or Py-GC/MS depends on the lifetime of chromatographic columns. Fused silica columns coated with polar bonded phases (such as the DB-1701 and FFAP columns mentioned above) have been used in our laboratory for periods longer than one year without appreciable degradation.

In any pyrolysis work, a valuable habit to acquire is the analysis of a standard sample at regular intervals between other samples. When "ghost peaks" appear in a pyrogram, analysis of the standard sample as a quality check—coupled with one or more of the ideas described above—may help resolve the contamination problem.

IV SELECTED APPLICATIONS

A summary of experimental conditions for selected analytical pyrolysis studies of microorganisms is given in Table 9.1. The literature review included here is not intended to be comprehensive but is designed to summarize selected recent applications of analytical pyrolysis to the analysis of microorganisms.

A DISCRIMINATION OF GRAM-POSITIVE AND GRAM-NEGATIVE BACTERIA

Perhaps the simplest and most basic distinction made of microbial cells is that of Gram type and numerous studies have assessed the value of analytical pyrolysis for this task. Early systematic studies included those of Simmonds et al. [60,61], who correlated the formation of specific pyrolysis products with their site of origin in the microorganism. Pyrolysis products were assigned to protein, carbohydrate, nucleic acid, lipid, and porphyrin sources in the cell. Acetamide was identified in pyrograms of whole cells and it was speculated that acetamide was produced from the N-acetyl groups on the glycan backbone of peptidoglycan (see Figure 9.2). Model compounds with chemical structures similar to substructures of PG were pyrolyzed by Hudson et al. [62] and Eudy et al. [63]. Only exceptionally low amounts of acetamide were found in pyrograms of glucosamine, muramic acid, and trialanine. Acetylated compounds such as N-acetyl glucosamine, N-acetyl muramic acid, and muramyl dipeptide produced much larger amounts of acetamide when pyrolyzed. It was not suggested that PG is the only source of acetamide in microbes, only that it is a major source. Levels of acetamide were greater in pyrograms of Gram-positive bacteria than in Gram-negative bacteria [18,62,63]. These results correlate well with the higher levels of PG found in Gram-positive cells [18,64]. A second pyrolysis product—propionamide—was also seen to originate in part from the lactyl-peptide bridge region of PG [63]. This pyrolysis product is not produced in sufficient amount, and probably has other sources of origin in the microbial cell, to not be valuable in differentiating Gram type [18].

Furfuryl alcohol has been suggested to derive in minor amounts from carbohydrates and in major amounts from RNA and DNA in microorganisms [60,61,63]. As Gram-negative bacteria have relatively thin cell walls compared to Gram-positive bacteria, the higher levels of furfuryl alcohol in pyrograms of Gram-negative bacteria may be due to intracellular DNA being more readily released from Gram-negative bacteria [18]. Although not directly related to Gram typing, a

TABLE 9.1

Summary of Selected Experimental Conditions for Pyrolysis-GC and Pyrolysis-GC/MS of Microorganisms

Organism/Components	Pyrolysis Conditions	Stationary Phase	Column and Material	Detector	Reference
Bacillus species	Curie-point 510 C	-	-	VG Pyromass 8-80 Py-MS	[80]
Bacilli	CDS Pyroprobe 650 C, 10 s fast ramp	DB-1701	30 m × 0.329 mm id, 1 film fused silica	Hewlett-Packard MSD EI GC/MS	[83]
	CDS Pyroprobe 800 C, 5 s fast ramp	FFAP	25 m × 0.25 mm id, 0.25 film fused silica	Hewlett-Packard MSD EI GC/MS	[36]
B. Cereus	Curie-point	-	-	Extrel ELQ400	[79]
B. globigii sporesB. subtilis, MS-2aldolase, E. coli, Secale cereale coliphage, Fog oil, diesel and woodsmoke, pollen, Type X, dry yeast	610C	-	-	Extrel EL-400triple quadrupole MS, EI	[82]
B. cereus, B. licheniformis B. Subtilis, E. coli, S. aureus,	Curie-point 610C	-	-	Extrel EL-400triple quadrupoleMS, EI	[97]
Organism/components	Pyrolysis conditions	Stationary phase	Column and material	Detector	Reference
B. subtilis, E. coli, E. coli, M. tuberculosis	Curie-point 358, 510, and 610C	BP-1	12 m × 0.22 mm id, 0.15 m film fused silica	Perkin Elmer8500 GC,Finnigan 700 MATion trap MS	[93]
B. cereus,B. licheniformis,B. subtilis, B. thuringiensis, E. coli, S. aureus,	Curie-point 610C	007-1Quadrexmethylsilicone	15 m × 0.250 mm id, 100% film fused silica	Extrel EL-400triple quadrupoleMS, EIVarian 4000 GC	[98]
B. anthracis, B. cereus, B. licheniformis, B. thuringiensis, B. subtilis, E. Coli, P. fluorescens, S. Albus, S. aureus,	CDS Pyroprobe 1000C, 20 s fast ramp	BP-1	5 m × 0.25 m film, capillary	HP 5890 GC and Finnigan-MAT 700 ion trapMS	[95]
B. cereus, B. subtilis, E. coli	Curie-point 510C	methyl silicone	4 m × 0.25 mm id, fused silica	Extrel triplequadrupole MS,CI	[81]
Bacteroidesgingivalis	Curie-point 358C, 510C	-	-	FOM Institutequadrupole MS -	[100]
Organism/Components	Pyrolysis Conditions	Stationary Phase	Column and Material	Detector	Reference
Biomaterials	Curie-point 500C	DB-1	-		[59]

(Continued)

TABLE 9.1 (Continued)

E. coli HB101, HB101(pSLM204)	Curie-point 530C	—	30 m × 0.32 mmid, 0.25 m film, fused silica	Varian 3700 GCFinnigan MAT 212MS, EI/FID, FID	[102]
Flavobacterium	CDS Pyroprobe 900C, 10 sfast ramp	methyl-silicone or SE-52	—	HorizonInstrumentsPYMS-200X quadrupole MS, EI	[35]
Gram positive, Gram negative, bacteria, and cell walls	CDS Pyroprobe 800C fast ramp	Carbowax 20M or Superox-4	20 m × 0.21 mm id fused silica or 8 m glass capillary	Hewlett-Packard5985 EI GC/MS	
Capsular polysaccharide from *Klebsiella*	CDS Pyroprobe 650C, 10 s	DB-1701	20 m × 0.25 mm id glass capillary or25 m × 0.2 mm idfused silica	Hewlett-Packard 5592EI GC/MS and Finnigan EI GC/MS	[62–64]
Lipids, lipopoly-saccharides, fatty acids,	direct CI 570C	—	30 m × 0.329 mm id 1 film fused silica	Finnigan MATGC/MS, EI/CI	[87]
Listeria monocytogenes	Curie-point 358C	—		Scientific Res.Instruments CI MS	[99]
*Mycobacteria*Lipids	Curie-point 610C	SE-30 BP-1	5 m × 0.32 mm id fused silica 5 m, 0.25 l film	Spectrel quadrupole MS	[88]
Mycobacteria/methyl mycolates	Curie-point 358C	OV-1	3% (w/w) packed column 1 m × 3 mm id	Finnigan ion trap GC/MS	[94]
Methylated Microbial fatty acids	Curie-point 510C	DB-5	7.2 m × 0.2 mm id fused silica 0.11 film	Hitachi EIGC/MS	[90]
Penicillium italicum/polysaccharides	DCI pyrolysis	—		Hewlett-Packard MSD,EI GC/MS, FID	[92]
Pseudomonas and *Xanthomonas* bacteria	Curie-point 510C	—		Finnigan MAT HSQ-30	[105]
Streptococci/glucitol phosphate	CDS Pyroprobe 800C fast ramp	FFAP	25 m × 0.25 mm id fused silica EI GC/MS	Extrel quadrupole MS	[101]
				Hewlett-Packard MSD	[18,37,42,75,76]

pyrolysis-GC/MS method for measuring DNA content of cultured cells was recently compared to results from an established colorimetric method based on reaction with diphenylamine [65]. The single-step procedure used cell samples containing 0.1–25 g of DNA and had a detection limit of about 100 ng DNA. Furfuryl alcohol from DNA was also found to be the source peak in pyrograms of virally transformed mammalian cells that could discriminate them from their normal counterparts [66,67]. This was not surprising as DNA levels are known to be higher in cancer cells.

Figure 9.4 shows relative amounts of acetamide and furfuryl alcohol measured in the GC/MS pyrograms of a group of microorganisms [18]. Replicate pyrolysis results are connected by lines to indicate their reproducibility. Fungal samples included in this study as a control produced low levels of both pyrolysis products. Gram-positive bacteria cluster towards higher levels of acetamide and moderate levels of furfuryl alcohol. Gram-negative bacteria produced higher levels of furfuryl alcohol and moderate levels of acetamide. Diagonal lines (with no statistical significance) can be drawn separating the Gram-positive and Gram-negative groups. The data from some organisms are close to the dividing lines. Obviously, acetamide and furfuryl alcohol have other sources within cells and are not generally useful as specific chemical markers for bacterial PG. Morgan et al. [18] concluded that it remains to be seen if individual bacteria can be Gram-typed by analytical pyrolysis.

Adkins et al. [49] reported ions at m/z 166 in negative ion mass spectra following pyrolysis of N-acetylglucosamine, chitin, and B. subtilis peptidoglycan. It has been speculated that larger pyrolysis products more specific for bacterial PG might be generated by pyrolysis, but a comprehensive study with diverse Gram-positive and Gram-negative bacteria has not validated this possibility. However, recent pyrolysis results from Voorhees et al. [68] indicate that Gram typing may be more readily demonstrated with the higher selectivity of the triple quadrupole MS.

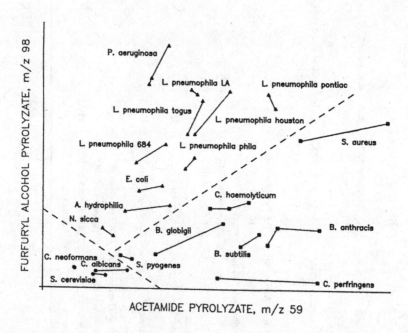

FIGURE 9.4 Plot of relative amounts of acetamide and furfuryl alcohol in a diverse group of microorganisms. Amounts are measured as the integrated reconstructed ion intensity (m/z 59 for acetamide, m/z 98 for furfuryl alcohol) at the appropriate retention time as a percentage of the total ion intensity. Gram-positive bacteria are represented by squares, Gram-negative bacteria by triangles, and fungi by circles. Reproduced with permission from Ref. 18. Copyright 1990 Plenum Publishing Corporation.

B   ANALYSIS OF GRAM-POSITIVE ORGANISMS

Streptococci have been popular targets for analytical pyrolysis [18,37,42,69–76], perhaps because their chemical composition has been well-characterized by other techniques. One of the first papers describing the additivity of pyrograms of individual components in a complex microbial pyrogram was published by Huis In't Veld et al. [69]. The pyrogram of a streptococcal strain containing a polysaccharide cell wall antigen was the sum of the pyrogram of its streptococcal mutant lacking this antigen and the pyrogram of the isolated antigen.

In a study involving 14 different strains, Smith et al. [75] found that group B streptococci (*Streptococcus agalactiae*) could be differentiated from group A streptococci by a single peak. Electron impact (EI) mass spectra of model compounds, including group B polysaccharide antigen and glucitol, suggested that the distinctive peak in the pyrogram of group B organisms was derived from glucitol phosphate residues in the group-specific polysaccharide. This carbohydrate-derived chemical marker for group B streptococci was identified by EI and methane CI MS as a dehydration product of glucitol-6-phosphate, dianhydroglucitol [18]. Figure 9.5 shows this differentiation using selected ion monitoring to focus on the characteristic m/z 86 ion of the dianhydroglucitol marker. Later studies using 29 different strains differentiated groups A, B, C, F, and G streptococci from one another [37,42,76]. Group B pyrograms were clearly distinguishable from the other groups, and samples from groups A and F were usually classified correctly [37]. Group C and G strains being indistinguishable was not surprising because no commercially available system was capable of distinguishing these groups. The discrimination by pyrolysis was judged achievable by any technique except serogrouping.

Identification of *Staphylococci* tends to be slow and irreproducible and a commercially available system (APIStaph) only reduces identification time to 24 h [77]. Hindmarch and Magee [78]

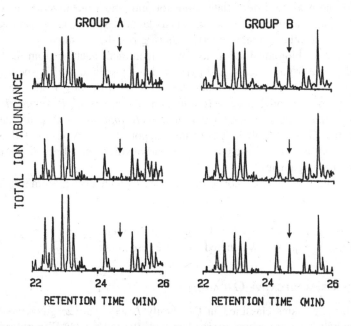

FIGURE 9.5   Discrimination of group A and Group B streptococci shown in an expanded view of total ion abundance pyrograms in the region from 22–26 min. Pyrograms from three different samples of group A and group B isolates are shown. The peak marked with an arrow is dianhydroglucitol. Pyrolysis conditions: CDS model 120 Pyroprobe, 80 g sample, pyrolysis settings at 800 °C and 75 °C/ms. Chromatographic conditions: 25 m × 0.24 mm FFAP fused silica capillary, Hewlett-Packard model 5880A GC coupled to a model 5970 mass selective detector. Reproduced with permission from Ref. 75. Copyright 1987 American Chemical Society.

analyzed a total of 451 strains of staphylococci from a clinical investigation by Curie-point pyrolysis-GC. Samples were pyrolyzed at 610 °C and packed GC columns were used to acquire pyrograms. After discarding some results from anomalous strains, discrimination by pyrolysis for only 60 of 415 strains disagreed with biochemical testing. At least a third of the misidentified strains were atypical strains.

Samples of a single *S. aureus* strain were pyrolyzed in an air atmosphere tube furnace and in a vacuum Curie-point pyrolyzer connected to a triple quadrupole mass spectrometer [79]. Pyrograms of *S. Aureus* were distinguishable from those of *E. coli* and *bacillus* strains. Peaks related to fatty acids were identified in the mass pyrograms.

Curie-point pyrolysis MS was applied to the identification of selected *Bacillus* species by Shute et al. [80]. 53 strains of *B. subtilis, B. pumilus, B. licheniformis,* and *B. amyloliquefaciens* in both a sporulating and non-sporulating state were pyrolyzed at 510 °C using a PyroMass 8–80 instrument (VG Gas Analysis, Middlewich, Cheshire, UK). The four strains could be differentiated by Py-MS from non-sporulated cultures. With sporulated cultures, the pyrograms were more similar and only *B. Licheniformis* could be completely differentiated.

Adkins et al. [49] presented temperature-resolved pyrolysis-mass spectra of cell fractions from Gram-positive bacteria of the genus *Bacillus*. Culture conditions were found to play a significant role in determining exact pyrolysis profiles. Figure 9.6 shows two pyrolysis mass spectra of *B. subtilis* samples grown under different conditions. Pyrolysis products were identified from amino sugars in peptidoglycan as well as from teichoic acids present in the *Bacilli*. In one of the first pyrolysis-triple quadrupole MS analyses of bacteria, Voorhees et al. [81] compared Curie-point pyrograms of *B. cereus, B. subtilis,* and *E. coli*. Parent ion scans of daughter ions selected by pattern recognition were found to give the best identification of the three species. Morgan et al. [17] reported that dipicolinic acid from sporulating organisms produces pyridine upon pyrolysis and found a pyridine peak to differentiate spore-forming and vegetative strains of *B. anthracis*. Voorhees et al. [82] have also recently used pyrolysis-tandem MS to analyze samples of *B. subtilis, B. globigii, E. coli,* and other possible interferents (fog oil, diesel smoke, dry yeast, and pollen). Four useful markers for biological substances were identified: adenine from nucleic acids; diketopiperazine and indole from proteins; and pyridine from picolinic acid in sporulating bacteria.

A pyrolysis product of galactose was identified in *B. anthracis* by Watt et al. [83] The anhydrosugar 1,6-anhydro-galactopyranose was found in pyrograms of *B. anthracis* and absent in pyrograms of *B. cereus*. The presence of galactose in *B. anthracis* was confirmed by an alditol acetate derivatization GC/MS method. The results support the assertions of Helleur et al. [84–87] that analytical pyrolysis can be used for rapid carbohydrate profiling of complex biological samples such as microorganisms.

Samples of the Gram-positive coccobacillus *Listeria monocytogenes* from infected patients and food sources were analyzed by Py-MS [88]. Samples within the serogroup could be differentiated to the extent that the strain obtained from the food was ruled out as a possible source for human infections. The independence of Py-MS from the labor-intensive use of reagent probes for the classification of these organisms was noted by these researchers.

C ANALYSIS OF GRAM-NEGATIVE ORGANISMS

During the 1980s, organisms classified in the family *Legionellaceae* grew rapidly as new serogroups of Legionella-like organisms were identified. Pyrolysis-GC/MS with selected ion monitoring was used to differentiate 21 strains of *Legionella* organisms [6]. Replicate samples (100 g) were pyrolyzed at 800 °C using a model 120 Pyroprobe (CDS Analytical) and 15 pyrolyzates were monitored by SIM GC/MS. The pyrograms of *Legionella pneumophila, Tatlockia micdadei,* and *Fluoribacter* could be distinguished from one another. In a Py-MS study of *Legionella* species, Kajioka and Tang [89] suggested that differences in ion profiles could be used to distinguish the various species from one another. Samples were pyrolyzed in a Curie-point pyrolyzer at 358 °C

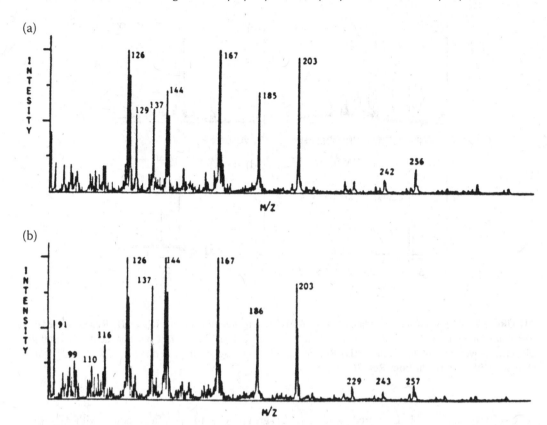

FIGURE 9.6 Pyrolysis mass spectra of *B. subtilis* grown (A) on minimal media, and (B) in penassay broth. Reproduced with permission from Ref. 49. Copyright 1984 Elsevier Science Publishers.

and analyzed by a quadrupole mass spectrometer. Specific and reproducible pyrograms providing distinctive fingerprints for *Legionella* strains were obtained by Py-MS.

A pyrolysis fragment, 2-butenoic acid, from the lipid substance poly--hydroxybutyrate present in microbial samples was identified by Watt et al. [36]. The use of this pyrolysis product as a chemical marker was validated by correlating results from an extraction, hydrolysis, and derivatization GC/MS method, by growth trials that profiled PHB content as a function of culture age, and by comparative pyrolysis studies of a diverse group of organisms. PHB was shown to be common at moderate to high levels in the family *Legionellaceae*. Background to low levels of PHB was only in organisms of the family *Enterobacteriaceae*. Figure 9.7 compares total ion abundance GC/MS pyrograms and reconstructed ion *m/z* 86 pyrograms for two organisms, *L. pneumophila* and *P. vulagaris*. In common with many other bacilli, *B. anthracis* was also found to contain PHB.

The bacterium *Klebsiella* is known to possess capsular polysaccharides. Helleur [87] studied isolated capsular polysaccharides from *Klebsiella* by pyrolysis-GC/MS. The polysaccharide is composed of repeating three glucopyranosyl units, one galactopyranosyl unit, one galactofuranosyl unit, one rhamnopyranosyl unit, and one glucuronosyl unit. Specific anhydrosugars produced upon pyrolysis for each type of saccharide unit were identified and quantified (Figure 9.8).

Pyrolysis-GC/MS of 32 species of *mycobacteria* for classification based upon the methyl ester fatty acid profiles was performed by Kusaka et al. [90]. The mycobacteria were successfully classified into four groups: 22-group, C_{20} to C_{24} fatty acids with C_{22} predominating; 24-group, C_{22}

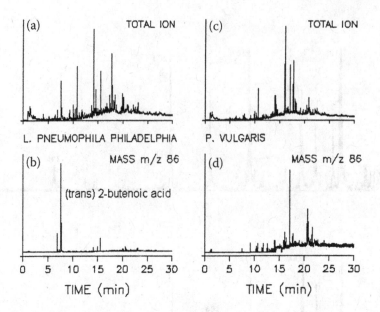

FIGURE 9.7 (A) Total ion abundance pyrogram of *L. pneumophila* Philadelphia. (B) Reconstructed ion pyrogram for ion mass *m/z* 86 of *L.* pneumophila Philadelphia. Peak 3 is *trans*-2-butenoic acid. (C) Total ion abundance pyrogram of *P. vulgaris*. (D) Reconstructed ion pyrogram for ion mass *m/z* 86 of *P. Vulgaris*. Adapted with permission from Ref. 36.

to C_{24} fatty acids with C_{24} predominating; 24'-group, C_{22} to C_{26} fatty acids with C_{24} predominating; and 26-group, C_{22} to C_{26} fatty acids with C_{26} predominating.

Curie-point pyrolysis coupled with short-column GC interfaced to an ion trap mass spectrometer was employed by Snyder et al. [53,91] to identify lipid components of microorganisms. A portion of the lipid component was further distinguished as dehydrated mono- and diacylglycerides. Discrimination of diverse microorganisms (four different *bacilli, S. aureus, E. coli. L. pneumophila*) was performed by visual comparison of total ion chromatograms and selected reconstructed ion chromatograms.

Holzer et al. [92] employed Curie-point pyrolysis to analyze microbial fatty acid by *in situ* methylation with trimethylanilinium hydroxide (TMAH). The fatty acid methyl ester profiles produced during pyrolysis of the whole cell agreed well with the analysis of lipid extracts from the same microorganism. Dworzanski et al. [93] also used pyrolytic methylation-GC to profile fatty acids in whole cells of *E. coli, Mycobacterium tuberculosis*, and *B. subtilis*. Curie-point pyrolysis coupled to GC/ion trap MS was used by Snyder et al. [94] to characterize 14 strains representing nine bacterial species for their lipid biomarker content. A quartz-tube Pyroprobe from CDS Analytical was used with GC/ion trap MS by Smith and Snyder [95] to analyze lipid and fatty acid components in a variety of bacteria with a total analysis time under 10 min. Snyder et al. [96] also used 13 strains of eight bacterial species to demonstrate the use of Curie-point wire and quartz tube pyrolysis coupled to short column GC/MS for the detection of biological warfare agents. Lipid and nucleic acid components were the primary target for discrimination purposes. Two articles by DeLuca et al. [97,98] studied Curie-point pyrolysis-tandem MS for direct analysis of bacterial fatty acids. Py-MS results correlated well with previous Py-GC/MS data on the same bacteria. Interestingly, better classification results were obtained when just Py-MS fatty acid profiles were employed rather than total ion spectra.

Adkins et al. [99] also produced temperature resolved pyrolysis results for cell envelope fractions and whole cells of *Salmonella typhimurium* and for several *Salmonella* LPS samples.

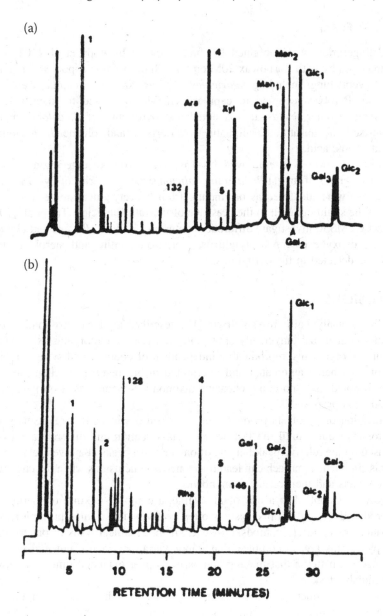

FIGURE 9.8 Capsular polysaccharides identified by pyrolysis GC/MS from Klebsiella K41 strains. Reproduced with permission from Ref. 87. Copyright 1987 Elsevier Science Publishers.

Model compounds (fatty acids, phospholipids, cholines, etc.) for LPS structures were also pyrolyzed using linear-programmed thermal degradation strategies. Common ions were found in the Py-MS fingerprints of cell wall components and the model compounds. Boon et al. [100] applied Py-MS to the analysis of samples from *Bacteroides gingivalis* isolated from dental patients. Pyrolysis mass spectra were dominated by volatile components, but clear differences were detected between related B*acteroides* strains. A rule-building expert system was shown to classify pyrolysis mass spectra from Gram-negative *Pseudomonas* samples versus samples containing *Xanthomonas*, a plant pathogen [101]. Goodacre et al. [102] demonstrated the use of Py-MS to detect the fimbrial adhesive antigen F41 from *E. Coli* and discriminate between bacteria that differ genotypically in that regard.

D ANALYSIS OF FUNGI

Pyrolysis-GC fingerprints were obtained for fungal spores by Papa et al. [103] using packed column chromatography with Carbowax 20M on 80–100 mesh Supelcoport. Papa et al. [104] also characterized several fungi including *Agaricales, Boletus, Russula, Amanita, Lepiota, Agaricus,* and *Lycoperdon.* Pyrolysis coupled to capillary GC/MS was used to identify discriminating components. Among other results, it was noted that pyrograms of *B. calopus* and *B. bovinus* contained saturated and unsaturated aliphatic hydrocarbons and toluene. *B. bovinus* pyrograms contained esadecanoic acid.

Pyrolysis-direct CI MS was employed by Tas et al. [105] to detect ions related to poly-saccharides in pyrograms of fungal *Penicillium italicum* strains. A series of Py-MS ions at *m/z* 167, 185, 187, and 210 were identified as originating from *N*-acetylglucosamine residues, present at higher levels in the strain cultured in the presence of imazalil fungicide. Tas et al. [106] also used pyrolysis-direct CI MS to investigate differences between *Candida albicans* and *Ophiostoma ulmi* fungi. Hexoses, deoxyhexose, *N*-acetylglucosamine from chitin, and sterols in treated yeast samples could be detected in the pyrograms.

V CONCLUSIONS

In his 1982 text on analytical pyrolysis, Irwin [12] described the ideal taxonomic method for the analysis of microorganisms: universally applicable to diverse microorganisms, capable of differentiating groups of organisms, reproducible independent of operator and system, rapid, sensitive, and capable of both being automated and interpreted in biochemical or chemical terms. At the time, Irwin also noted that "no comprehensive taxonomic scheme based on pyrograms of mass pyrograms had been proposed."

The identification of pyrolysis products and their sources in intact organisms was not widespread in pyrolysis studies until the last decade. Classification and differentiation of organisms were often based on the relative peak heights of one or more peaks at a given retention time in the pyrogram. This simplistic approach can lead to erroneous conclusions when the chemical identities of pyrolysis products and their origin is unknown.

When pyrolysis is applied to a microbial sample, a complex mixture of thermal degradation products is produced. Analytical pyrolysis is often performed in a fingerprinting mode using sophisticated pattern recognition methods. However, if the chemical basis of pattern differences is not defined, pyrograms are so complex that minor variations in instrumental conditions cause sufficient changes in patterns that making comparisons over extended time and between laboratories is a formidable task.

Ideally, pyrolysis products are preferred that retain as much of the structural integrity and chemical uniqueness of the original monomer so that their origin can be identified. Simple scission fragments may not retain as much chemical uniqueness of the parent structure as products, such as those from dehydration, simple rearrangements, or generated by newer soft ionization methods and detected directly by MS. Many pyrolysis products are common organic compounds that could potentially be derived from multiple sources. Some pyrolysis products, although their origin may be well-defined, may not be useful as chemical markers for bacterial discrimination.

By identifying specific chemicals in pyrograms as originating from taxonomically relevant microbial structures, discrimination may be based on well-defined features simplifying this process dramatically. Invariant chemical features of the organism will provide effective discrimination if distinctive pyrolysis products can be generated under a wide range of experimental conditions. Furthermore, the discrimination thus achieved should be reproducible between different instruments and laboratories. While quantitative amounts of pyrolysates generated may vary, the absence or presence of pyrolysis products due to a true chemotaxonomic marker should not depend dramatically on the choice of the pyrolysis system.

Appropriate groups of organisms for chemotaxonomic studies involving pyrolysis can often be chosen based on *a priori* information concerning microbial chemical structures. Interpretation is simplified by selecting groups of organisms, or cell fractions, that differ in defined structural characteristics. Analysis of bacteria possessing unrelated differences may not permit the significance of a particular pyrolysis product to be evaluated. Organisms not containing known chemical structures act as a blank indicating background levels produced from other sources. Analysis of multiple strains also confirms the consistency of correlations established among results from different organisms.

Automation of sampling is more readily accomplished using Py-MS than Py-GC/MS; however, better and automated sampling methods are needed for all forms of analytical pyrolysis. The absence of standard reference materials for many microbial components makes peak validation difficult and is a hindrance to systematic studies. Improved high-resolution capillary GC with inert polar phases capable of separating the complex mixture of pyrolysis products from microbial samples are needed. A new generation of MS techniques are evolving and being applied to microbiological problems: laser and plasma desorption, fast-atom bombardment, new chemical ionization approaches, and tandem MS. Many of these advances in MS detection will find their way to pyrolysis applications. Multiple detectors coupled in "hyphenated" modes will allow a more thorough characterization of the microbial pyrolysis product mixture. Finally, advances in small computers are also driving data analysis techniques closer to the analytical instrument. The future may bring us closer to Irwin's ideal taxonomic method for the analysis of microorganisms.

REFERENCES

1. R. E. Buchanan and N. E. Gibbons, Eds., *Bergey's Manual of Determinative Bacteriology*, 8th edition, Williams and Wilkins, Baltimore (1974).
2. C. S. Cummins, Chapter 3 in *Analytical Microbiology Methods: Chromatography and Mass Spectrometry* (A. Fox, S. L. Morgan, L. Larsson, and G. Odham, Eds.), Plenum Press, New York, NY, p. 53–57 (1990).
3. S. L. Morgan, A. Fox, and J. Gilbart, *J. Microbiol. Methods*, *9*: 57–69 (1989).
4. A. Fox, J. Gilbart, and S. L. Morgan, *Analytical Microbiology Methods: Chromatography and Mass Spectrometry* (A. Fox, S. L. Morgan, L. Larsson, and G. Odham, Eds.), Plenum Press, New York, Chapter 1, pp. 1–17 (1990).
5. W. K. Joklik, H. P. Willet, and D. B. Amos, Eds., *Zinsser Microbiology*, 17th edition, Appleton-Century-Crofts, New York, NY (1980).
6. A. Fox and S. L. Morgan, *Instrumental Methods for Rapid Microbiological Analysis*, (W. H. Nelson, Ed.), VCH Publishers Inc., Deerfield Beach, FL, Chapter 5, pp. 135–164 (1985).
7. P. D. Zemany, *Anal. Chem.*, *24*: 709 (1952).
8. V. I. Oyama, *Nature (London)*, *200*: 1058 (1963).
9. E. Reiner, *Nature*, *206*: 1272 (1965).
10. C. S. Gutteridge and J. R. Norris, *J. Appl. Bacteriol.*, *47*: 5 (1979).
11. H. L. C. Meuzelaar, J. Haverkamp, and F. D. Hileman, *Pyrolysis Mass Spectrometry of Recent and Fossil Biomaterials*, Elsevier, Amsterdam (1982).
12. W. J. Irwin, *Analytical Pyrolysis: A Comprehensive Guide*, Marcel Dekker, New York, NY (1982).
13. G. Wieten, H. L. C. Meuzelaar, and J. Haverkamp, *Gas Chromatography/Mass Spectrometry Applications in Microbiology*, (G. Odham, L. Larsson, P.-A. Maardh, Eds.) Plenum, New York, NY, Chapter 10, pp. 335–380 (1984).
14. F. L. Bayer and S. L. Morgan, *Pyrolysis and GC in Polymer Analysis* (E. J. Levy and S. Liebman, Eds.), Marcel Dekker, New York, NY, Chapter 6, pp. 277–337 (1985).
15. T. P. Wampler, *J. Anal. Appl. Pyrol.*, *16*: 291–322 (1989).
16. G. A. Eiceman, W. Windig, and A. P. Snyder, *Gas Chromatography: Biochemical, Biomedical, and Clinical Applications* (R. E. Clement, Ed.), John Wiley & Sons, New York, NY, Chapter 12, pp. 327–347 (1990).

17. S. L. Morgan, A. Fox, J. C. Rogers, and B. E. Watt, *Modern Techniques for Rapid Microbiological Analysis* (W. H. Nelson, Ed.), VCH Publishers, Inc., New York, NY, Chapter 1, pp. 1–18 (1991).

18. S. L. Morgan, B. E. Watt, K. Ueda, and A. Fox, *Analytical Microbiology Methods: Chromatography and Mass Spectrometry* (A. Fox, S. L. Morgan, L. Larsson, and G. Odham, Eds.), Plenum Press, New York, Chapter 12, pp. 179–200 (1990).

19. J. Postgate, *Microbes and Man*, 3rd edition, Cambridge University Press, Cambridge (1992).

20. J. D. Watson, N. H. Hopkins, J. W. Roberts, J. A. Steitz, and A. M. Weiner, *Molecular Biology of the Gene*, 4th edition, The Benjamin/Cummings Publishing Company, Inc., Menlo Park, CA, pp. 101 (1987).

21. H. J. Rogers, *Bacterial Cell Structure*, Van Nostrand Reinhold Co. Ltd., Wokingham, Berkshire (1983).

22. H. J. Rogers, H. R. Perkins, and J. B. Ward, *Microbial Cell Walls and Membranes*, Chapman and hall, London (1980).

23. K. H. Schliefer and O. Kandler, *Bacteriol. Rev.*, *36*: 407 (1972).

24. K. Ueda, S. L. Morgan, A. Fox, A. Sonesson, L. Larsson, and G. Odham, *Anal. Chem.*, *61*: 265–270 (1989).

25. A. Fox, K. Ueda, and S. L. Morgan, *Analytical Microbiology Methods: Chromatography and Mass Spectrometry* (A. Fox, S. L. Morgan, L. Larsson, and G. Odham, Eds.), Plenum Press, New York, Chapter 6, pp. 89–99 (1990).

26. J. Baddiley, *Essays in Biochemistry*, Vol. 8 (P. N. Campbell and F. Dickens, Eds.), Academic Press, London, pp. 35–78 (1972).

27. O. Luderitz, O. Westphal, A. M. Staub, and H. Nikaido, *Microbial Endotoxins*, Vol. 4 (G. Weingbaum, S. Kadis, and S. J. Ayl, Eds.), Academic Press, London, pp. 145–223 (1971).

28. C. W. Moss, *Analytical Microbiology Methods: Chromatography and Mass Spectrometry* (A. Fox, S. L. Morgan, L. Larsson, and G. Odham, Eds.), Plenum Press, New York, NY, Chapter 4, pp. 59–69 (1990).

29. V. Holdeman, E. P. Cato, and W. C. Moore, *Anaerobe Laboratory Manual*, 4th edition, Virginia Polytechnic Institute and State University, Anaerobe Laboratory, Blacksburg (1977).

30. L. Miller and T. Berger, Bacteria identification by gas chromatography of whole cell fatty acids, Hewlett-Packard Application Note *228*: 41 (1985).

31. A. Fox, S. L. Morgan, and J. Gilbart, *Analysis of Carbohydrates by GLC and MS* (C. J. Bierman and G. McGinnis, Eds.), CRC Press, Boca Raton, FL (1989).

32. D. G. Pritchard, B. M. Gray, and H. C. Dillon, *Arch. Biochem. Biophys.*, *235*: 385 (1984).

33. A. Fox, J. C. Rogers, J. Gilbart, S. L. Morgan, C. H. Davis, S. Knight, and P. B. Wyrick, *Infect. Immun.* *58*: 835–837 (1990).

34. C. S. Gutteridge and J. R. Norris, *Appl. Environm. Microbiol.*, *40*: 462 (1980).

35. H. Engman, H. T. Mayfield, T. Mar, and W. Bertsch, *J. Anal. Appl. Pyrol.*, *6*: 137 (1984).

36. B. E. Watt, S. L. Morgan, and A. Fox, *J. Anal. Appl. Pyrol.*, *20*: 237–250 (1991).

37. C. S. Smith, S. L. Morgan, C. D. Parks, A. Fox, *J. Anal. Appl. Pyrolysis*, *18*: 97–115 (1990).

38. NIH Biohazards Safety Guide, U. S. Department of Health, Education, and Welfare, Washington, DC (1974).

39. Occupational Safety and Health Administration, United States Department of Labor, Occupational Safety and Health Standards for General Industry (29 CFR Part 1910), Commerce Clearing House, Inc., Chicago, IL.

40. W. Windig, *J. Anal. Appl. Pyrol.*, *17*: 283–289 (1990).

41. G. Montaudo, *J. Anal. Appl. Pyrol.*, *13*: 1–7 (1988).

42. J. Gilbart, A. Fox, and S. L. Morgan, *Eur. J. Clin. Microbiol.*, *6*: 715–723 (1987).

43. T. P. Wampler and E. J. Levy, *J. Anal. Appl. Pyrol.*, *12*: 75–82 (1987).

44. H. L. C. Meuzelaar and R. A. in't Veld, *J. Chromatogr. Sci.*, *10*: 213–216 (1972).

45. W. Windig, P. G. Kistemaker, J. Haverkamp, and H. L. C. Meuzelaar, *J. Anal. Appl. Pyrol.*, *1*: 39–52 (1979).

46. W. Windig, P. G. Kistemaker, J. Haverkamp, and H. L. C. Meuzelaar, *J. Anal. Appl. Pyrol.*, *2*: 7–18 (1980).

47. A. van der Kaaden, R. Hoogerbrugge, and P. G. Kistemaker, *J. Anal. Appl. Pyrol.*, *9*: 267 (1986).

48. G. Wells, K. J. Voorhees, and J. H. Futrell, *Anal. Chem.*, *52*: 1782 (1980).

49. J. A. Adkins, T. H. Risby, J. J. Scocca, R. E. Yasbin, and J. W. Ezzell, *J. Anal. Appl. Pyrol.*, *7*: 15–33 (1984).

50. G. L. French, I. Phillips, S. Chin, *J. Gen. Microbiol.*, *125*: 347 (1981).

51. G. L. French, H. Talsania, and I. Philips, *Med. Microbiol.*, *29*: 19 (1989).

52. M. L. Trehy, R. A. Yost, and J. G. Dorsey, *Anal. Chem.*, *58*: 14 (1986).

53. A. P. Snyder, W. H. McClennen, and H. L. C. Meuzelaar, *Analytical Microbiology Methods: Chromatography and Mass Spectrometry* (A. Fox, S. L. Morgan, L. Larsson, and G. Odham, Eds.), Plenum Press, New York, Chapter 13, pp. 201–217 (1990).

54. P. A. Quinn, *J. Chromatogr. Sci.*, *12*: 796–806 (1974).

55. H. L. C. Meuzelaar, H. G. Ficke, and H. C. den Harink, *J, Chromatogr. Sci.*, *13*: 12–17 (1975).

56. J. E. Purcell, H. D. Downs, and L. S. Ettre, *Chromatographia*, 8: 605–616 (1975).

57. A. Mitchell and M. Needleman, *Anal. Chem.* 50: 668–669 (1978).

58. Y. Sugimura and S. Tsuge, *Anal. Chem.* 50: 1968–1972 (1978).

59. J. M. Halket and H.-R. Schulten, *J. High Resolut. Chromatogr. & Chromatogr. Commun.*, *9*: 596–597 (1986).

60. P. G. Simmonds, *Appl. Microbiol.*, *20*: 567 (1970).

61. E. E. Medley, P. G. Simmonds, and S. L. Mannatt, *Biomed. Mass Spectrom.*, 2: 261, (1975).

62. J. R. Hudson, S. L. Morgan, and A. Fox, *Anal. Biochem.*, *120*: 59–65 (1982).

63. L. W. Eudy, M. D. Walla, J. R. Hudson, S. L. Morgan, and A. Fox, *J. Anal. Appl. Pyrol.*, *7*: 231–247 (1985).

64. L. W. Eudy, M. D. Walla, S. L. Morgan, and A. Fox, *Analyst*, *110*: 381–385 (1985).

65. R. S. Sahota, S. L. Morgan, and K. E. Creek, *J. Anal. Appl. Pyrol.*, *24*: 107–122 (1992).

66. R. S. Sahota and S. L. Morgan, *Anal. Chem.*, *64*: 2383–2392 (1992).

67. R. S. Sahota, S. L. Morgan, K. E. Creek, and L. Pirisi, unpublished manuscript (1994).

68. K. J. Voorhees, Department of Chemistry & Geochemistry, Colorado School of Mines, Golden, CO, personal communication (1994).

69. J. H. J. Huis In't Veld, H. L. C. Meuzelaar, and A. Tom, *Appl. Microbiol.*, *26*: 92–97 (1973).

70. M. V. Stack, H. D. Donoghue, J. E. Tyler, and M. Marshall, *Analytical Pyrolysis* (C. E. R. Jones, C. A. Cramers, Eds.), Elsevier, Amsterdam, pp. 57 (1977).

71. M. V. Stack, H. D. Donoghue, and J. E. Tyler, *Appl. Environ. Microbiol.*, *35*: 45, (1980).

72. G. L. French, I. Phillips, and S. Chin, *J. Gen. Microbiol.*, *125*: 347 (1981).

73. M. V. Stack, H. D. Donoghue, and J. E. Tyler, *J. Anal. Appl. Pyrol.*, *3*: 221 (1981/1982).

74. G. L. French, H. Talsania, and I. Phillips, *Med. Microbiol.*, *29*: 19 (1989).

75. C. S. Smith, S. L. Morgan, C. D. Parks, A. Fox, and D. G. Pritchard, *Anal. Chem.*, *59*: 1410–1413 (1987).

76. A. Fox, J. Gilbart, B. Christensson, and S. L. Morgan, *Rapid Methods and Automation in Microbiology and Immunology* (A. Balows, R. C. Titon, and A. Turano, Eds.), Brixia Academic Press, Brescia, Italy, pp. 379–388 (1989).

77. Y. Brun, J. Fleurette, and F. Forey, *J. Clin. Microbiol.*, 8: 503–508 (1978).

78. J. M. Hindmarch and J. T. Magee, *J. Anal. Appl. Pyrol.*, *11*: 527–538 (1987).

79. S. J. DeLuca and K. J. Voorhees, *J. Anal. Appl. Pyrol.*, *24*: 211–225 (1993).

80. L. A. Shute, C. S. Gutteridge, J. R. Norris, and R. C. W. Berkeley, *J. Gen. Microbiol.*, *130*: 343–355 (1984).

81. K. J. Voorhees, S. L. Durfee, J. R. Holtzclaw, C. G. Enke, and M. R. Bauer, *J. Anal Appl. Pyrol.*, *14*: 7 (1988).

82. K. J. Voorhees, S. J. DeLuca, and A. Noguerola, *J. Anal. Appl. Pyrol.*, *24*: 1–21 (1992).

83. B. E. Watt, S. L. Morgan, K. Fox, and A. Fox, *Anal. Chem.*, submitted (1994).

84. D. R. Budgell, E. R. Hayes, and R. J. Helleur, *Anal. Chim. Acta.* *192*: 243 (1987).

85. R. J. Helleur, E. R. Hayes, J. S. Craigie, and J. L. MacLachlan, *J. Anal. Appl. Pyrol.*, 8: 349 (1985).

86. R. J. Helleur, D. R. Budgell, and E. R. Hayes, *Anal. Chim. Acta.*, *192*: 367 (1987).

87. R. J. Helleur, *J. Anal. Appl. Pyrol.*, *11*: 297–311 (1987).

88. R. Kajioka and M. A. Noble, *J. Anal. Appl. Pyrol.*, *22*: 29–38 (1991).

89. R. Kajioka and P. W. Tang, *J. Anal. Appl. Pyrol.*, *6*: 59–68 (1984).

90. T. Kusaka and T. Mori, *J. Gen. Microbiol.*, *132*: 3403 (1986).

91. A. P. Snyder, W. H. McClennen, J. P. Dworzanski, and H. L. C. Meuzelaar, *Anal. Chem.*, *62*: 2565–2573 (1990).

92. G. Holzer, T. F. Bourne, and W. Bertsch, *J. Chromatogr.*, *468*: 181 (1989).

93. J. P. Dworzanski, L. Berwald, and H. L. C. Meuzelaar, *Appl. Environ. Microbiol.*, *56*: 1717–1724 (1990).

94. A. P. Snyder, W. H. McClennen, J. P. Dworzanski, and H. L. C. Meuzelaar, *Anal. Chem.*, *62*: 2565–2573 (1990).

95. P. B. Smith and A. P. Snyder, *J. Anal. Appl. Pyrol.*, *24*: 23–38 (1992).

96. A. P. Snyder, P. B. W. Smith, J. P. Dworzanski, and H. L. C. Meuzelaar, *Mass Spectrometry for the Characterization of Microorganisms* (C. Fenselau, Ed.), ACS Symposium Series 541, American Chemical Society, Washington, DC, Chapter 5, pp. 63–84 (1994).

97. S. J. Deluca, E. W. Sarver, P. De B. Harrington, and K. J. Voorhees, *Anal. Chem. 62*: 1465–1472 (1990).

98. S. J. DeLuca, E. W. Sarver, and K. J. Voorhees, *J. Anal. Appl. Pyrol., 23*: 1–14 (1992).

99. J. A. Adkins, T. H. Risby, J. J. Scocca, R. E. Yasbin, and J. E. Ezzell, *J. Anal. Appl. Pyrol., 7*: 35–51 (1984).

100. J. J. Boon, A. Tom, B. Brandt, G. B. Eijkel, P. G. Kistemaker, F. J. W. Notten, and F. H. M. Mikx, *Anal. Chim. Acta, 163*: 193–205 (1984).

101. P. B. Harrington, T. E. Street, K. J. Voorhees, F. R. Brozolo, and R. W. Odom, *Anal. Chem., 61*: 715–719 (1989).

102. R. Goodacre, R. C. W. Berkeley, and J. E. Beringer, *J. Anal. Appl. Pyrol., 22*: 19–28 (1991).

103. G. Papa, P. Balbi, and G. Audisio, *J. Anal. Appl. Pyrol., 11*: 539–548 (1987).

104. G. Papa, P. Balbi, and G. Audisio, *J. Anal. Appl. Pyrol., 15*: 137 (1989).

105. A. C. Tas, A. Kerkenaar, G. F. LaVos, and J. Van der Greef, *J. Anal. Appl. Pyrol., 15*: 55–70 (1989).

106. A. C. Tas, H. B. Bastiaanse, J. Van der Greef, and A. Kerkenaar, *J. Anal. Appl. Pyrol., 14*: 309–321 (1989).

10 Analytical Pyrolysis of Polar Macromolecules*

Charles Zawodny and Karen D. Sam
CDS Analytical, Oxford, Pennsylvania

I INTRODUCTION

There is a wide range of polymers that are polar in their chemical make-up; frequently, they are macromolecules formed by condensation reactions or through other catalyzed means. Analytical pyrolysis can be a useful method for the characterization of such polymers, both from a qualitative and quantitative standpoint.

When heated sufficiently to cause bond dissociation, these polymers degrade to produce a pyrolysate that is individually indicative of the particular polymer. The result in pyrolysis-gas chromatography (Py-GC) is a fingerprint chromatogram that qualitatively identifies the macromolecule.

Quantitative information can also be obtained in the analysis of copolymeric systems. Many polar macromolecular systems produce high yields of monomers that may be used in quantitatively determining monomer composition in the polymer. This type of information can be valuable not only in terms of characterizing the polymer but also in performing degradation studies.

The macromolecules discussed in this chapter are synthetic polymers such as poly(acrylates), poly(esters), poly(amides), and poly(urethanes), as well as surfactants and natural polymers like cellulose, chitin, chitosan, and proteins.

II SYNTHETIC POLYMERS

A ACRYLICS

Poly(methacrylates) fit into a class of polymers that tend to undergo depolymerization under pyrolysis conditions. Poly(methyl methacrylate) — or PMMA, when pyrolyzed — yields primarily methyl methacylate monomer. This process begins by initial cleavage of the skeletal backbone of the polymer, forming two free radical ends. Subsequent beta scissions produce an unwinding effect as sequential monomer units are formed. Once initiated, this process proceeds down the entire length of the polymer. This process is also known as "unzipping." Figure 10.1 shows a typical pyrogram of the poly(methyl methacrylate) at 750°C. From this pyrogram, it is evident how extensively the depolymerization process works.

Figure 10.2 shows the mechanistic description of depolymerization initiation and product formation for methacrylate polymers. From the literature concerning monomer yields for poly (methyl methacrylate), typical values are 92–98% recovery of methyl methacrylate monomer.

These recovery values are fairly consistent regardless of pyrolysis temperature and heating rates. In the case of PMMA, as long as there is sufficient energy delivered to the sample to break a C–C bond, then depolymerization occurs readily.

Pyrolysis of methacrylate polymers tend to get more complex as the length of the R group increases. Due to the increasing length and increased steric interactions, the amount of depolymerization may be reduced. This increases the probability that products other than the monomer will be produced. For

*This chapter was based on the original work of the late John W. Washall

Abundance

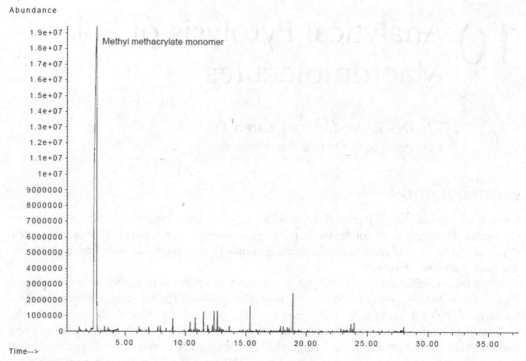

FIGURE 10.1 Pyrolysis of polymethyl methacrylate at 750°C.

instance, the pyrogram of poly(butyl methacrylate) — PBMA — is shown in Figure 10.3. The monomer yield for PBMA is in the 93–95% range. Minor products include 1-butanol, butyl acrylate, 1-propene, 2-methyl-l-propene, and butanal. Among these minor products, 1-butanol is the most abundant and is formed as a result of cleavage of the ethereal C–O bond of the ester group.

The last methacrylate polymer example is poly(lauryl methacrylate), or PLMA. This polymer is interesting because it contains an extremely long R group. Having such a long hydrocarbon chain

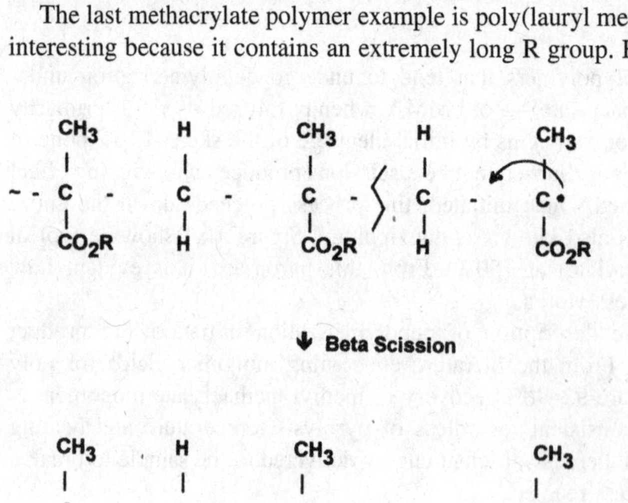

FIGURE 10.2 Depolymerization mechanism for poly(methacrylates).

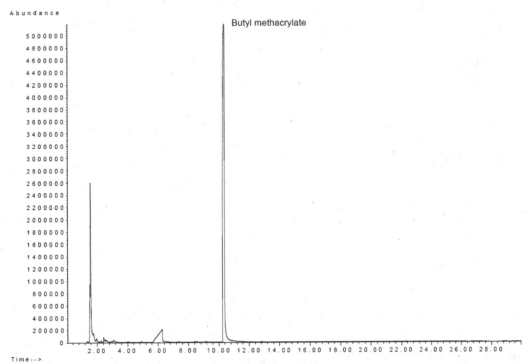

FIGURE 10.3 Pyrogram of polybutyl methacrylate pyrolyzed at 750°C. The small peak at 6 min is methacrylic acid.

dangling from the polymer backbone makes it much easier to cleave. The result is a pyrolyzate composed not only of lauryl methacrylate monomer, but also of many unsaturated hydrocarbons as a result of random scission reactions in the chain of the C_{12} group. Figure 10.4 shows an example of PLMA pyrolyzed at 750°C. There is quite a noticeable repeating series of peaks between 10 and 16 min in the pyrogram, corresponding to the various mono-unsaturated hydrocarbons. In this particular example, the yield for lauryl methacrylate was 68%. This value is much lower than those obtained for PMMA and PBMA, which may be attributed to the length of the R group.

The qualitative aspect of methacrylate polymer pyrolysis as well as the quantitative determination of recovered monomer have been described. What remains to be discussed is a quantitative assessment of monomeric composition in a copolymer sample. It is often helpful for quality purposes to know the composition of a copolymer product. One copolymer system that has been studied extensively using pyrolysis is poly(methyl methacrylate/butyl acrylate). A pyrogram of this copolymer is shown in Figure 10.5.

The two primary peaks in this pyrogram correspond to the respective monomers. A series of such copolymers with varying concentrations of monomers can be pyrolyzed to obtain a calibration curve by plotting the peak area ratios for compounds generated from each of the monomers versus concentration. A typical calibration curve is shown in Figure 10.6, plotting the area ratio of the methyl methacrylate peak to the butanol peak against the concentration of methyl methacrylate. Proper calibration of the pyrolyzer makes it possible to obtain linear quantitative relationships for most polymeric systems.

As opposed to Poly(methacrylates) — which unzip or depolymerize into monomer — poly-acrylates break apart using random scission to produce monomer, dimer, trimer, and tetramer. Because hydrogens are available for a 1–5 hydrogen shift, the trimer peak is larger (Figure 10.7).

Structural determination often hinges on identification of higher-molecular-weight pyrolysis fragments. For this reason, it may be necessary to choose a pyrolysis temperature that gives a

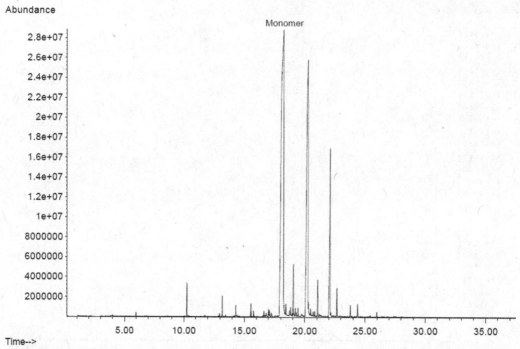

FIGURE 10.4   Pyrolysis of polylauryl methacrylate at 750°C.

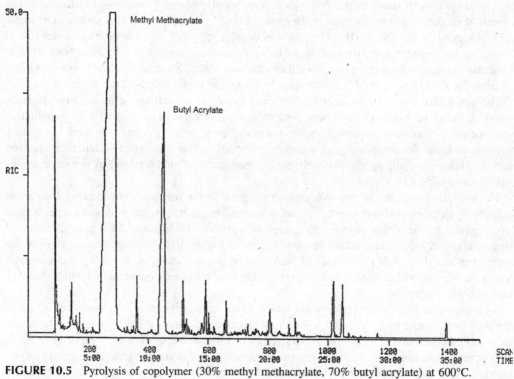

FIGURE 10.5   Pyrolysis of copolymer (30% methyl methacrylate, 70% butyl acrylate) at 600°C.

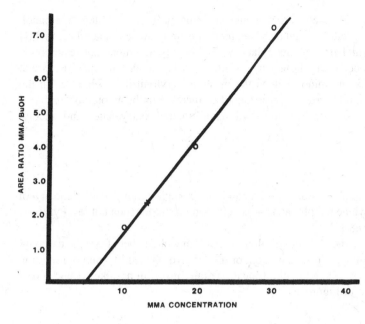

FIGURE 10.6 Graph of peak area ration of methyl methacrylate monomer (from PMMA) to butanol (from PBA) vs. the concentration of methyl methacrylate in the copolymer.

product distribution that favors formation of dimeric, trimeric, and higher units. This provides information on the arrangement of the various monomer units and will also provide an insight as to whether a copolmer is random or block. For example, a styrene-glycidyl methacrylate copolymer shows monomer, dimer, and codimer fragments [1].

Many acrylic copolymers are currently used in the textile industry as binders for non-woven fabrics, pigment printing, or finishing. The purpose of these fibers is to stabilize the material. In many instances, these copolymers are used in conjunction with amino resins. Casanovas and Rovira [2] have done a study of methyl methacrylate-ethyl acrylate-*N*-methylol-acrylamide by

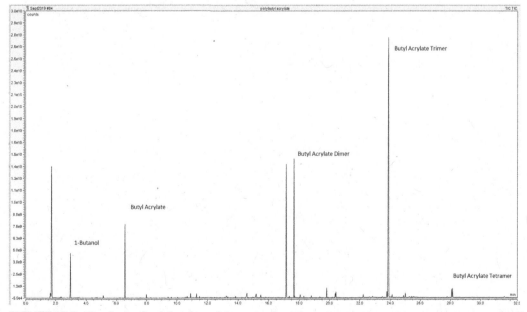

FIGURE 10.7 Pyrolysis of polybutyl acrylate.

PY/GC-MS. Among the products identified were methane, ethylene, propene, isobutene, methanol, propionaldehyde, ethanol, ethyl acetate, methyl acrylate, methyl isobutyrate, ethyl acrylate, methyl methacrylate, *n*-propyl acrylate, and ethyl methacrylate. With this polymer, monomer reversion is the primary degradation process occurring; however, several other degradation mechanisms are at work. When the sample contains an amino resin in the mixture, acrylonitrile is observed in the pyrogram. Another effect of the amino presence was a marked increase in the amount of methanol detected. Other products detected were methoxyhydrazine, methyl isocyanate, and methyl isocyanide.

B POLYESTERS

Polyesters can be made from a variety of monomers, a common example being polyethylene glycol terephthalate also called polyethylene terephthalate — a common plastic for soda bottles. Pyrolysis products are shown in Figure 10.8 [3].

Weak linkages of polyesters are the C–O bonds along the polymer chain, beta from the carbonyl bond. Scission occurs there, involving Cis-elimination, or the 1,5-hydrogen shift. (random scission of ester linkages, through a six-membered cyclic transition state, result in an alkenyl and a carboxyl end group). This is shown in Figure 10.9.

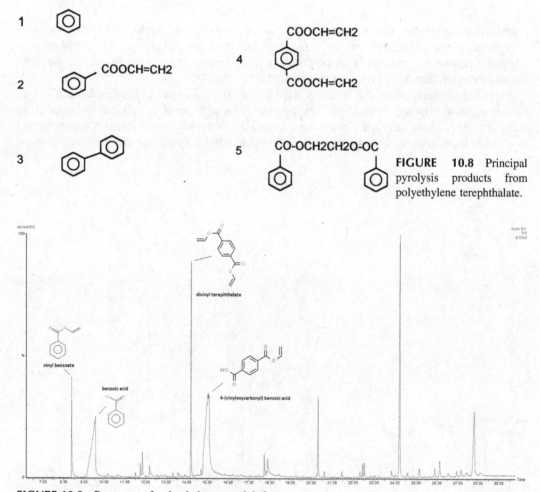

FIGURE 10.8 Principal pyrolysis products from polyethylene terephthalate.

FIGURE 10.9 Pyrogram of polyethylene terephthalate.

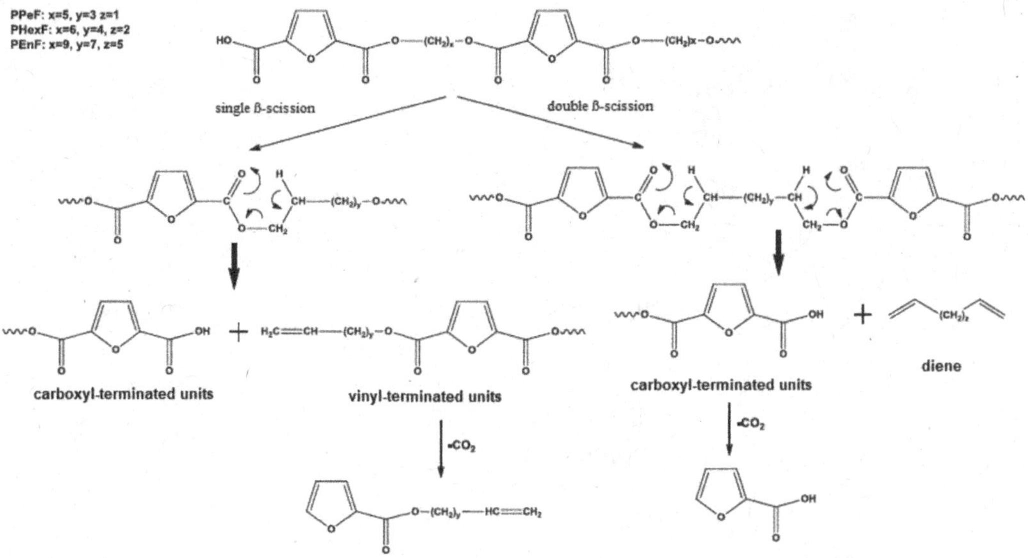

FIGURE 10.10 Proposed decomposition mechanism for poly(hexylene 2,5 furan dicarboxylate) (Ref. 4).

Due to the concern over the use of petroleum resources, there has been an increased interest to create polymers from renewable resources such as biomass. Furanoate-based polyesters have been studied as possible replacement for petrochemical derived polyesters like poly(ethylene terephthalate). Bikiaris et al. used Py GC/MS to study the thermal degradation products of furanoate based polyesters. As with other polyesters, they concluded that beta-hydrogen bond scission was the major pathway for decomposition. Decomposition of PHF (Poly(hexylene 2,5 furan dicarboxylate)) proceeded as described Figure 10.10 [4].

Some of the pyrolysis products are found in Figure 10.11.

Polylactic acid and polyhydroxybutyrate are other polyesters considered to be biodegradable as well as biobased and have been studied using pyrolysis to understand their decomposition [5,6].

Cis-elimination (or random scission of ester linkages through a six-membered cyclic transition state to result in an alkenyl and a carboxyl end group) is the basic pyrolysis of polyhydroxybutyrate and other polyesters, producing monomers, dimers, and trimers [5].

Thin film polylactide underwent sequential pyrolysis at a setpoint of 600°C for 0.5 s, produce lactide, mesolactide, dimer, three trimer isomers and tetramer, as shown in Figure 10.12 [6]. Degradation with sequential pyrolysis occurred in a sigmoidal fashion.

In the pyrolysis of polar macromolecules, stability of the final products plays a crucial role in determining the relative abundances. With polyesters, the polymer will produce an original monomer unit. However, when ester copolymers are pyrolyzed, several other reactions are likely. For instance, to better understand the degradation behavior of chloro-norbornene-dicarboxylic acid containing polyesters used to impart flame retardancy, an unsaturated polyester was prepared by the reaction of chlorinated norobornene dicarboxylic acid, maleic anhydride, and 1,2-propanediol under vacuum. Pyrograms of this unsaturated polyester [7] show the following major pyrolysis products: 1,2,3,4,5-pentachlorocyclopenta-1,3-diene and 1,2,3,4,5,5-hexachlorocyclopenta-1,3-diene. The monomer yield in this case was very small. Product formation takes place via retro-Diels Alder reactions due to the presence of various polychloro-5-norbornene-dicarboxylic acid units in the backbone of the polymer.

Styrene-cured unsaturated polyesters made from chlorinated norbornene dicarboxylic acids, 1,2-propanediol, and maleic anhydride showed increased quantities of toluene, ethylbenzene, and alpha-methylstyrene in their pyrograms. Interestingly, dimer formation was not seen under these

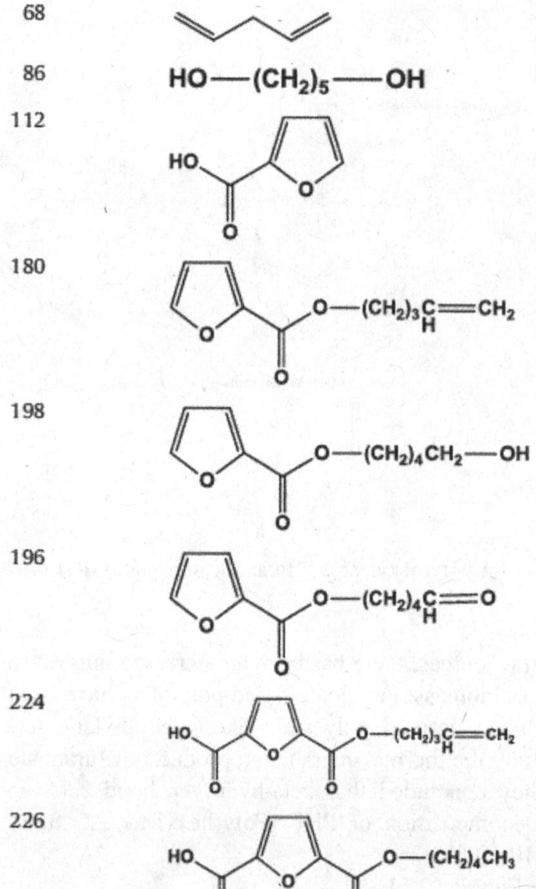

68

86   HO—(CH$_2$)$_5$—OH

112

180

198

196

224

226

292

336

404

FIGURE 10.11   Pyrolysis products of poly(hexylene 2,5 furan dicarboxylate) (PHF) (Ref. 4).

circumstances. It is believed that chlorine radicals participate to prohibit dimer formation through transfer reactions. The result of this reaction is the formation of phenylalkyl chlorides [8].

More information on polyesters can be found in Chapter 5, which details degradation mechanisms of condensation polymers.

C   WATER-SOLUBLE POLYMERS

Water-Soluble Polymers are particularly useful in medical and pharmaceutical industries. One of these is Polyvinyl Pyrrolidone (PVP). During pyrolysis, PVP unzips to form a monomer (Figure 10.13).

Polyvinyl Alcohol is another water-soluble polymer. It is made by first polymerizing polyvinyl acetate, then hydrolyzing it to form polyvinyl alcohol. When completely hydrolyzed, PVA

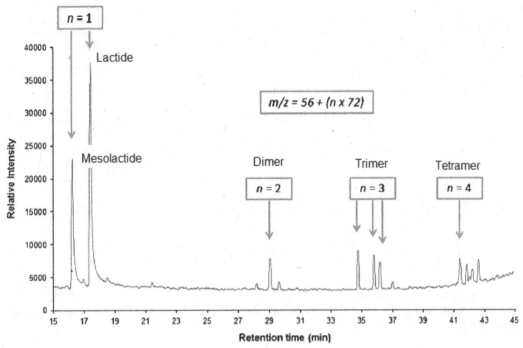

FIGURE 10.12   Pyrogram of thin film polylactide at 600°C (Ref. 6).

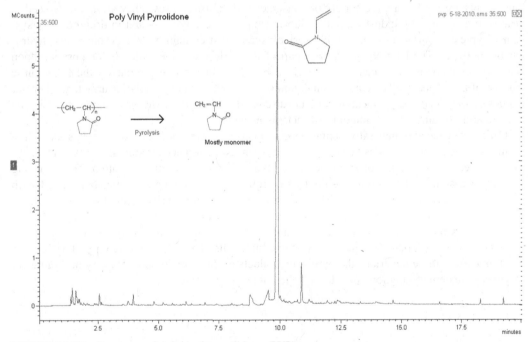

FIGURE 10.13   Pyrogram of polyvinyl pyrrolidone (PVP).

pyrolyzes; OH group is removed, attracting an H from the neighboring alcohol group, creating butenal, 2,4 hexadienal as oligomers — the partially hydrolyzed PVA also produce acetic acid, as the acetic acid group undergoes side-group elimination. Benzaldehyde is also more prevalent in partially hydrolyzed PVA (Figure 10.14).

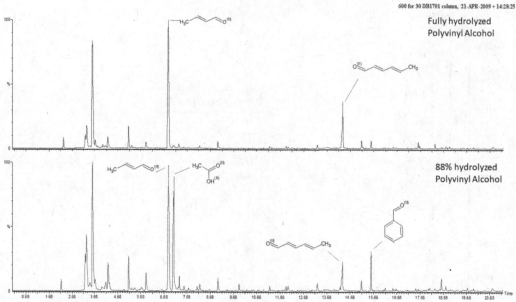

FIGURE 10.14 Pyrograms of fully hydrolyzed (top) and 88% hydrolyzed (bottom) Polyvinyl alcohol, 600°C.

III SURFACTANTS

Surfactants are used in a wide variety of consumer and industrial uses. They are not only used in detergent and cosmetic industries, but in lubrication — as catalysts — and in drug delivery systems. Typically, surfactants are soluble surface agents that contain both hydrophilic and hydrophobic portions. The long chain aliphatic part of the molecule is the hydrophobic portion. Upon aqueous dissociation, they can be categorized based on the charge present on the hydrophilic portion of the molecule. There are anionic, nonionic, cationic, and ampholytic surfactants. For our purposes, only cationic surfactants will be discussed, although pyrolysis is appropriate in the investigation of anionic and nonionic surfactants as well.

Cationic surfactants generally contain a central heteroatom with four side groups extending from the central atom. A class of surfactants known as quaternary ammonium salts contain typically three small aliphatic or aromatic groups bonded to a central nitrogen atom. The cleansing properties arise from the presence of the fourth group, which is a long aliphatic chain. The length of this chain varies from C_{12} to C_{20}. The aliphatic chain, which is derived from the reaction of the amine with a fatty acid, is crucial to the formation of the micelle. Micelle formation is the mechanism by which surfactants get their cleansing properties. Commercially available cleaning products contain a mixture of quaternary ammonium salts. To understand how pyrolysis can be used to analyze these materials, the pyrolysis products of the pure components may be examined. A generic quaternary ammonium salt with the chemical structure:

$$R_1 - \overset{\overset{\displaystyle R_3^+}{\displaystyle |}}{\underset{\underset{\displaystyle R_2}{\displaystyle |}}{N}} - R_4 \quad Cl^-$$

could undergo homolytic cleavage because of bond strength [9]. An R–N bond has a dissociation energy of 78 kcal/mol whereas a C–C bond has one of 83 kcal/mol. A look at the

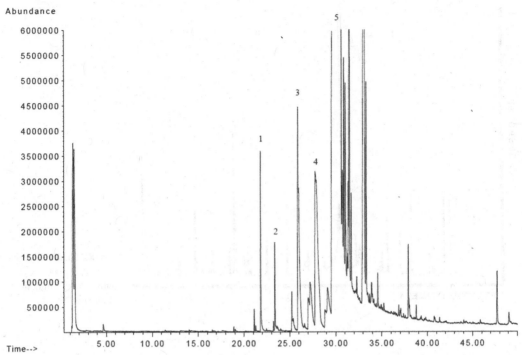

FIGURE 10.15   Pyrogram of hexadecyltrimethyl ammonium chloride at 600°C. Peak 1 = *N,N*-dimethyl dodecanamine, 2 = hexadecane, 3 = *N,N*-dimethyl tetradecanamine, 4 = *N,N*-dimethyl pentadecanamine, 5 = *N,N*-dimethyl hexadecanamine.

pyrogram in Figure 10.15 shows that bond cleavage occurred at the C–N bond of the long aliphatic chain. The characterization of the peak at 23 min shows it to be hexadecene. A hydrogen transfer from the aliphatic free radical to the amine free radical forms the product methyl chloride CH_3Cl. A series of homologous tertiary amines from C_{12} to C_{15} was also observed due to random cleavage of the C_{16} aliphatic chain.

Similarly, when octadecyl dimethylbenzyl ammonium chloride is pyrolyzed (Figure 10.16), the major products are octadecene and dimethylbenzyl ammonium chloride. Other products include benzene, toluene, dimethyl amine, and unsaturated hydrocarbons with carbon numbers less than C_{18}.

The utility of pyrolysis in the analysis of surfactants can be shown by looking at aqueous solutions of these surfactants in dilute quantities. Analytical pyrolysis is frequently thought of as a technique that does not work well with trace quantities. However, dilute solutions of surfactants can be analyzed by applying the aqueous solution to the platinum ribbon of a filament pyrolyzer and allowing the water to evaporate. This procedure can be repeated to allow for the sensitivity needed. In Figure 10.17, the analysis of a commercial surfactant at the low ppm level was analyzed by pyrolysis at 700°C. Among the products seen are benzene, toluene (the peaks at 5 and 8 min, respectively), and many olefinic hydrocarbons. This exercise brings about the prospect of using pyrolysis as a tool for analyzing sediments, high molecular-weight organics, pesticides, and semivolatiles by pyrolysis-gas chromatography. Such work is currently underway by several research groups.

IV SULFUR-CONTAINING MACROMOLECULES

Sulfonamides are known for their ability to form antibacterial drugs, and the derivatives of 4-aminobenzenesulfonamide serve several medical applications. Analytical pyrolysis of these compounds proceeds in a similar manner to that of amino acids.

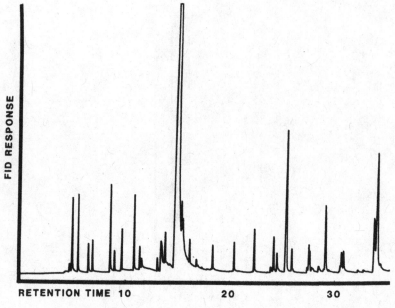

FIGURE10.16 Pyrolysis of an octadecyldimethylbenzyl ammonium chloride at 750°C.

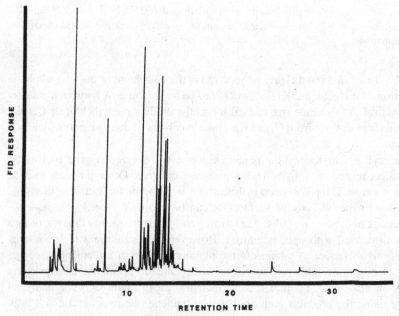

FIGURE 10.17 Pyrolysis of the residue of a 40 ppm solution in water of a quaternary ammonium chloride surfactant (800°C for 10 s).

    Pyrolysis generally occurs at the sulfonamido group, resulting in the liberation of sulfur dioxide. The other products will include aniline and a heterocyclic aromatic amine. The sulfonamide is generally characterized by the heterocyclic amine since aniline is always a product with these compounds [10]. For example, in the pyrolysis of sulfadiazine, pyrolysis occurs at the sulfonamido nitrogen atom to form an unstable SO_2 radical, which readily converts to sulfur dioxide. The other remaining products are aniline and 2-aminopyrimidine. The same process holds for the sulfonamides sulfamerazine and sulfadimidine. The only difference is that the heterocyclic products are 2-amino-4-methylpyrimidine and 2-amino-4,6 dimethylpyrimidine, respectively [11].

These are very simple sulfur compounds in terms of pyrolytic behavior. The problem gets more sophisticated when the macromolecule is a polymer that may be cross-linked and have sulfur bridges. Many attempts have been made to address the topic of sulfur linkages in vulcanized cross-linked rubbers, with varying degrees of success. The problem that arises is that the sulfur bridges make up only a small portion of the polymer. This leads to difficulty because of sensitivity levels and the positive identification of pyrolysis fragments that can be attributed directly to the sulfur bridges. The chromatogram produced is often quite complex not only because of the nature of the rubber itself, but because of the sulfur bridging. The complexity of this situation may make it necessary to use statistical modeling of known cross-linked rubbers to be able to perform adequate quantitative analysis of the degree of cross-linking for these polymers.

V NITROGEN-CONTAINING POLYMERS

Thus far we have looked at several oxygenated polymers and copolymers, and we will look at macromolecules that contain a nitrogen heteroatom.

Before doing so, however, it might be helpful to look at smaller molecules to understand the chemistry surrounding a central nitrogen atom. The first compounds that we will discuss are amino acids. Smaller compounds containing a central nitrogen atom are relatively simple chemically and offer some insight to the pyrolytic behavior surrounding the nitrogen atom. Many alpha-amino acids, when exposed to pyrolysis conditions, undergo initial decarboxylation to form the amine fragment. With aliphatic amino acids, this results in the amine as the primary pyrolysis product in the chromatogram. This is not at all surprising since CO_2 is highly stable and readily available in the amino acid. There are some bimolecular processes that can occur to give the formation of a nitrile upon pyrolysis of amino acids, these have been widely characterized in the literature.

Other products that form from the pyrolysis of amino acids are aldehydes, which contain one less carbon atom than the parent amino acid. This process occurs via an S_Ni deamination mechanism. Another process that can occur is side-chain stripping involving chain homolysis. The result is the production of various saturated and unsaturated compounds. Many alpha amino acids, when pyrolyzed, form an amine fragment, a nitrile fragment, and aldehydes that contain one less carbon atom than the parent amino acid. Also, a side stripping mechanism can result in various saturated and unsaturated compounds [10].

With materials that contain nitrogen heteroatoms, the pyrolysis conditions are of extreme importance. Because of the thermally labile nature of many nitrogen-containing compounds, the thermal stability of the final product is paramount. Sample size can also dramatically affect the degradation mechanism and product formation. The larger the sample size, the greater the corresponding thermal gradient within the sample. Large sample sizes tend to push the product distribution in the direction of secondary degradation processes.

Pyrolysis of phenylalanine, for example, can proceed through several different pathways. Figure 10.18 shows the pyrograms of phenylalanine and tyrosine. In the case of phenylalanine, a primary degradation product is toluene. This product results from the cleavage of the carbon–carbon bond beta to the amine nitrogen atom. There is the need for a hydrogen radical migration to form the final product. Likewise, cleavage of the carbon–nitrogen bond leads to the formation of styrene. Ethylbenzene is another product of the pyrolysis of phenylalanine. Bibenzyl is also a major pyrolysis product of phenylalanine resulting from the reaction of two toluyl radicals [12].

The nature of the pyrolysis products with nitrogen-containing macromolecules can be highly dependent on many analytical conditions, such as sample size, pyrolysis temperature, and inertness of the pyrolysis system. Many nitrogen-containing compounds are thermally labile and, thus, are sensitive to temperature and metal surfaces. Figure 10.18 shows the result of pyrolyzing

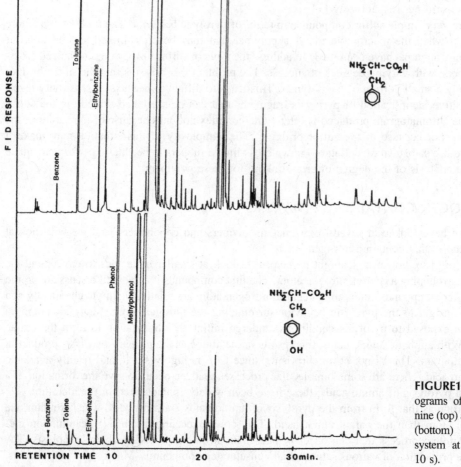

FIGURE10.18 Pyrograms of phenylalanine (top) and tyrosine (bottom) (glass-lined system at 700°C for 10 s).

phenylalanine and tyrosine using a glass-lined analytical system. Irwin (1982) reports on the differences resulting from the presence of stainless steel, quartz, and Pyrex in the same analysis. Although the pyrolysis products remain the same in all three experiments, the product distribution does change. For instance, in a stainless system, the amount of benzene produced is much higher than that of the Pyrex and quartz systems. Quartz, which is the most inert, has the highest level of styrene and ethylbenzene. Pyrolysis at 500°C provides the largest yield of Ph-CH$_2$CH$_2$-CN. As the temperature increases, the amount of nitrile diminishes [12].

Understanding the product formation from the pyrolysis of amino acids is the first step in comprehending pyrolysis behavior for more complex systems, such as proteins and nucleic acids, which comprise a large quantity of the material in biological systems. Later in this chapter, under Natural Polymers, proteins are discussed in more detail.

VI NATURAL POLYMERS

A PROTEINS

Many macromolecules can be found in living organisms. Proteins, large polymeric peptides made up of amino acids, make a particularly important group. Materials like hair, wool, and feathers

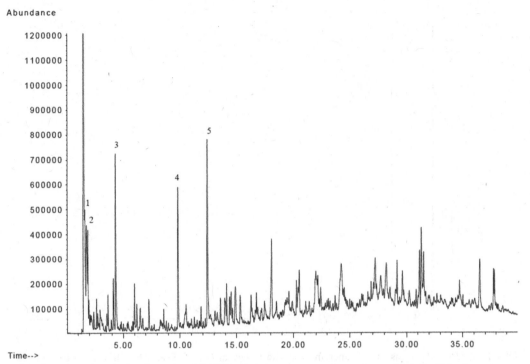

FIGURE 10.19   Pyrolysis of wool at 750°C. Peak 1 = mercaptomethane, 2 = carbon disulfide, 3 = toluene, 4 = phenol, 5 = methyl phenol.

contain proteins called keratins. The chemical structure of wool shows that the cortex consists of at least 100 proteins with polypeptides made up of about 22 different amino acids with different structure and behavior. These amino acids are structurally different as well as chemically — some are acidic, some hydrophobic, and some contain sulfur. The pyrogram shown in Figure 10.19 was produced from wool and shows a peak for mercaptomethane and carbon disulfide indicative of the presence of the amino acid cystine. The pyrogram of a feather (Figure 10.20) shows the presence of sulfur dioxide, indicative of a sulfur containing amino acid in the polypeptide. A homologous series of nitriles was also observed.

Orsini et al. studied ovalbumin (OVA) to learn about protein structure, formation of aggregates, and cross-linked structures using py-GC/MS. Pyrolysis products of ovalbumin include NH_3, HNCO, CO_2, NO, and CO, as aromatics and cyclic dipeptides. The degradation of a protein is related to the reactivity of certain amino acids and their position in the polypeptide chain.

They demonstrated that Py-GC/MS can be used to learn about changes occurring in protein structure. This type of research has applications in all fields of research that involve proteins, like clinical research for the study of neurological disorders, as well as pharmaceutical industries, textiles, cosmetics, food, and antiquities [13].

B   CELLULOSE AND LIGNIN

The most abundant natural organic polymer on earth is cellulose. It is is a major constituent of wood as well as plant leaves and stems. Composed of glucose units, when pyrolyzed, it produces char, CO_2, furan, substituted furans, and levoglucosan. In wood, the phenolic polymer lignin binds the cellulose wood fibers together, helping to give wood structure. The resulting pyrolysis of wood, which contains both cellulose and lignin, yields a large number of organic compounds (see Figure 10.21) plus CO_2, and a considerable amount of non-volatile char. The pyrolysates include several substituted phenols from lignin, as well as furans and levoglucosan from cellulose.

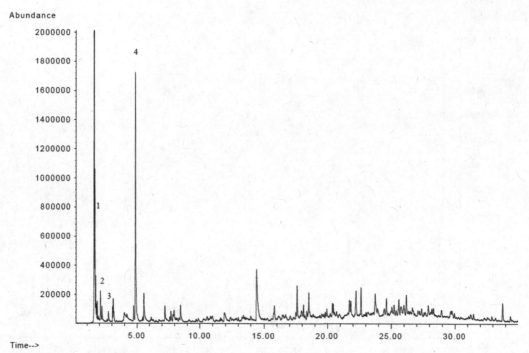

FIGURE 10.20 Pyrolysis of a feather from a wild turkey at 650°C. Peak 1= Sulfur 1 = sulfur dioxide, 2 = Acetonitrile, 3= Propanenitrile, 4 = Toluene.2 = acetonitrile, 3 = propanenitrile, 4 = toluene.

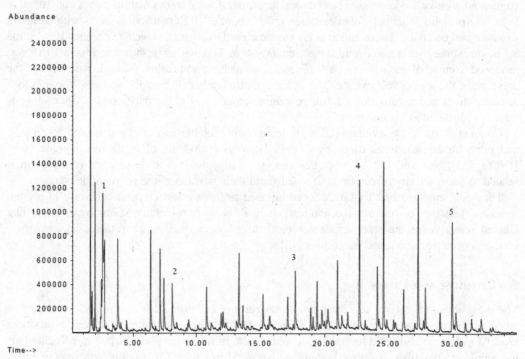

FIGURE 10.21 Pyrogram of oak wood pyrolyzed at 600°C. Peak 1 = Acetic 1 = acetic acid, 2 = Furan 2 = furan carboxaldehyde, 3 = 2-Methoxy-4-methyl 3 = 2-methoxy-4-methyl phenol, 4 = 2,6 Dimethoxy4 = 2,6-dimethoxy phenol, 5 = 2,6 Dimethoxy-4-propenyl5 = 2,6-dimethoxy-4-propenyl phenol.

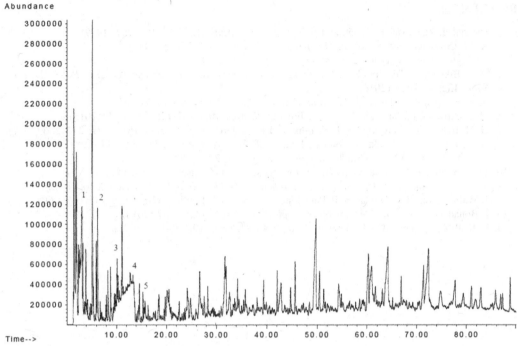

FIGURE 10.22 Pyrogram of the exoskeleton of a salt water saltwater shrimp pyrolyzed at 750°C. Peak 1 = Acetic 1 = acetic acid, 2 = Pyrrole, 3 = Methyl 2 = pyrrole, 4 = Acetamide, 5 = 2,5-Dimethyl 3 = methyl pyrrole, 4 = acetamide, 5 = 2,5-dimethyl pyrrole.

C CHITIN AND CHITOSAN

Chitin is one of the main components in the make-up of insect exoskeletons and those of other arthropods. It is a polysaccharide that is composed of units of acetyl glucosamine (N-acetyl-D-glucos-2-amine) that are linked together by a beta-1,4 bond. They are linked the same way cellulose units are. Chitosan is a linear polysaccharide that contains randomly distributed (deacetylated) beta-(1,4) D-glucosamine units and N-acetyl-D-glucosamine (acetylated) units. Figure 10.22 shows the pyrogram of saltwater shrimp exoskeleton. The typical chemical composition of the exoskeleton is approximately 15% chitin with most of the remaining material made up of protein. The pyrogram of this exoskeleton shows the thermal elimination of acetic acid and acetamide. A number of nitrogenous heterocyclics were also detected, such as pyrrole, methylpyrrole and 2,5-dimethylpyrrole [11,14].

VII SUMMARY

The pyrolysis of polar macromolecules tends to be more complex from a mechanistic standpoint than non-polar materials since many of the products formed go through a cyclic intermediate. However, compounds such as polymethacrylates decompose via a depolymerization mechanism that mostly produces monomer. Analytical pyrolysis can be used to extract a great deal of information concerning nonvolatile samples; for instance, sample identification through fingerprinting, quantitative measurement of copolymers, and structural information related to branching and stereochemistry.

It is clear from this brief overview that analytical pyrolysis can be a useful analytical tool for both qualitative as well as quantitative analysis of large macromolecules. Treatment of non-volatile samples such as synthetic polymers and natural biopolymers using pyrolysis-GC/MS can yield much information about volatiles, structure, and provide a characteristic fingerprint of the macromolecule.

REFERENCES

1. J. Kálal, J. Zachoval, J. Kubát, and F. Svec, *J. Appl. Anal. Pyrol.*, 1: 143–157 (1979).
2. A. M. Casanovas and X. Rovira, *J. Appl. Anal. Pyrol.*, *11*: 227–232 (1987).
3. Y. Sugimura and S. Tsuge, *J. Chrom. Sci.*, *17*: 269–272 (1979).
4. D. N. Bikiaris, Z. Terzopoulou, V. Tsanaktsis, M. Nerantzaki, and G. Z. Papageorgiou, *Polym. Degrad. Stab.*, 132: 126–127 (2016).
5. F. D. Kopinke, K. Mackenzie, *J. Appl. Anal.Pyrol.*, *40-41*: 43–53 (1997).
6. M. P. Arrietta, F. Parres, J. López, A. Jiménez, *J. Appl. Anal.Pyrol.*,101:150–155(2013).
7. G. H. Irzl, C. T. Vijayakumar, J. K. Fink, and K. Lederer, *J. Appl. Anal. Pyrol.*, *11*: 277–286 (1987).
8. G. H. Irzl, C. T. Vijayakumar, and K. Lederer, *J. Appl. Anal.Pyrol.*, *13*: 305–317 (1988).
9. S. J. Abraham and W. J. Criddle, *J. Appl. Anal. Pyrol.*, 7:37–349.
10. W. J. Irwin and J. A. Slack, *Analytical Pyrolysis*, 107–116, Elsevier, Amsterdam (1977).
11. W. J. Irwin, *Analytical Pyrolysis, A Comprehensive Guide*, MarcelDekker, New York (1982).
12. J. M. Patterson, N. F. Haidar, E. P. Papadopoulos, and W. T. Smith, *J. Org. Chem.*, *38*: 663–666 (1973).
13. I. Bonaduce, S. Orsini, E. Bramanti, *J. Appl. Anal. Pyrol.*, 133:59–67 (2018).
14. P. Koll G. Borchers, and J. Metzger, *J. Appl. Anal. Pyrol.*, 19:119–128 (1991).

11 Characterization of Condensation Polymers by Thermally Assisted Hydrolysis and Methylation-GC

Hajime Ohtani[1] and Shin Tsuge[2]
[1]Nagoya Institute of Technology, Nagoya, Japan
[2]Nagoya University, Nagoya, Japan

I INTRODUCTION

Pyrolysis-gas chromatography (Py-GC) has been widely applied in the characterization of various synthetic polymers. However, the characterization of intractable condensation polymers such as aromatic polyesters is often difficult using ordinary Py-GC—even when a high-resolution capillary GC system is employed—because such polymers tend to degrade into a number of polar compounds, along with considerable amounts of solid residues and chars.

Modifications to the pyrolysis process that involve high-temperature chemical reactions other than the conventional thermolysis of macromolecules can often provide additional information regarding the chemical structures of various organic materials that cannot be readily obtained using conventional analytical pyrolysis methods [1]. A representative example is pyrolysis in the presence of an organic base, typically tetramethylammonium hydroxide [$(CH_3)_4NOH$; TMAH]. This procedure is known as thermally assisted hydrolysis and methylation (THM) because the samples are hydrolytically decomposed and most of the products undergo almost simultaneous methylation with TMAH.

The THM reaction has been successfully used in conjunction with GC, GC/mass spectrometry (MS), and MS for the chemical characterization of various synthetic and natural products, including resins, lipids, waxes, wood products, soil sediments, and microorganisms [1]. This technique is also highly effective for the detailed characterization of synthetic polymeric materials, especially condensation polymers such as polyesters and polycarbonates because it typically produces highly simplified pyrograms that consist mainly of peaks corresponding to the methyl derivatives of the constituents of the polymer samples. In this chapter, the instrumental and methodological aspects of Py-GC in the presence of an organic base are briefly described. In the remainder of the chapter, typical applications of this technique to precise compositional analysis and microstructural elucidation are discussed for various condensation-type polymeric materials, including intractable cross-linked structures.

II MEASUREMENT SYSTEM AND PROCEDURE

Figure 11.1 illustrates a THM-GC measurement system using a microfurnace pyrolyzer, and shows typical reaction schemes for ester and carbonate linkages. This system is basically the same as that used for ordinary Py-GC. A polymer sample of known weight is placed in a sample cup and the reagent TMAH is added. The sample cup is then introduced into the pyrolyzer, and the polymer

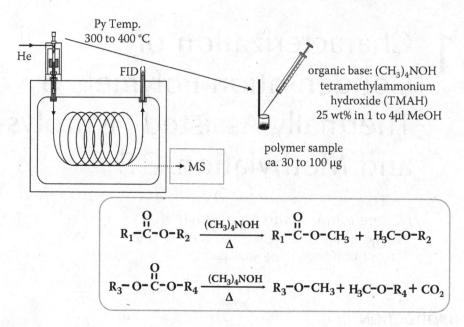

FIGURE 11.1 Typical measurement system for THM-GC.

sample is instantaneously decomposed under a flow of helium carrier gas. The polymer sample should be ground as finely as possible to obtain sufficient reaction efficiency with the organic base.

TMAH is generally added as a solution (typically 25 wt% in methanol or water). The amount of TMAH added should be at least several times higher than that of the hydrolysable linkages in the polymer sample, although excess reagent could contaminate the pyrolysis device. A typical combination is 30–100 μg of the polymer sample and 1–4 μl of TMAH solution. The solvent is typically chosen based on the solvent compatibility of the sample. The solvent methanol has been observed to contribute to the methylation of the products in some cases [2].

Compared with the pyrolysis temperatures typically used in ordinary Py-GC (500°–700 °C), relatively low pyrolysis temperatures (300°–400 °C) are normally chosen in THM-GC to suppress the random thermal cleavage of the polymer chain that occurs at higher temperatures while still enabling instantaneous THM reaction. The resulting products are transferred to the separation column, and the separated species are detected using a flame ionization detector (FID) or a mass spectrometer (MS) to yield a pyrogram.

III APPLICATIONS

A FULLY AROMATIC POLYESTER (COMPOSITIONAL ANALYSIS)

Fully aromatic liquid crystalline polyesters (LCPs) based on *p*-hydroxybenzoic acid (PHB) have been recognized as one of the most promising classes of high-performance engineering plastics, especially in fields where extremely high thermal stability and mechanical strength are required. To modify their processability, copolymer type LCPs are produced industrially, and precise compositional analysis is often required for production management. However, ordinary spectroscopic analysis methods—such as nuclear magnetic resonance (NMR), infrared (IR) spectroscopy, and even ordinary Py-GC—do not necessarily provide sufficient information regarding the LCPs. On the other hand, THM-GC is highly effective for the precise compositional analysis of LCPs [3,4].

Figure 11.2 shows pyrograms of an LCP sample consisting of PHB (50 mol%), terephthalic acid (TA; 25 mol%), and biphenol (BP; 25 mol%); pyrogram (a) was obtained using THM-GC with

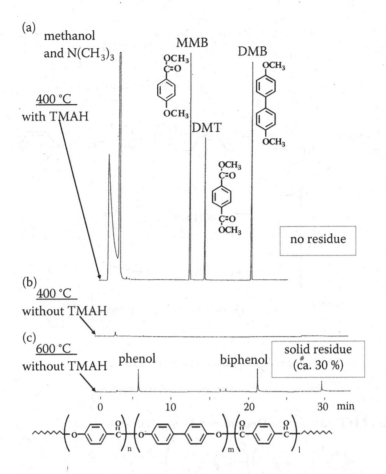

FIGURE 11.2 Pyrograms of LCP obtained (a) at 400°C with TMAH, (b) at 400°C without TMAH, and (c) at 600°C without TMAH [3]. Sample weight: 0.05 mg; Reagent: 1 μl of 25 wt% TMAH solution in methanol; Separation column: fused silica capillary (30 m × 0.25 mm i.d.) coated with 0.25 μm of 5% diphenyl-95% dimethylpolysiloxane; Column temp: 80–280°C at 8°C/min.

TMAH at 400 °C, (b) was obtained using ordinary Py-GC at 400 °C, and (c) at 600 °C [3]. During ordinary pyrolysis at 400 °C (b), hardly any decomposition of the sample into volatile compounds took place due to the high thermal stability of the LCP sample; most of the sample remained as a solid residue in the sample cup. Pyrolysis at 600 °C without the addition of TMAH (c) produced only small amounts of (nonmethylated) phenol and biphenol, and about 30% of the sample remained in the cup. However, in the presence of TMAH at 400 °C, three large peaks corresponding to the dimethyl derivatives of PHB, TA, and BP—that is, methoxy *p*-methylbenzoate (MMB), dimethylterephthalate (DMT), and 4,4′-dimethoxybiphenyl (DMB), respectively—were observed after the elution of the products (methanol and trimethylamine) from TMAH, and no sample residue remained in the cup. This result suggested that TMAH did not methylate the volatile decomposition products, but instead reacted directly with the ester linkages in the polymer chain to yield the methyl derivative of each constituent [5]. Figure 11.3 shows the most probable THM reaction of the ester linkages in LCP.

Table 11.1 summarizes the copolymer composition of various LCP samples estimated using this method, along with the amount of each monomer supplied in the feed. The molar ratios of the products, which were calculated based on the relative peak intensities detected by a flame ionization detector (FID) after correction for molar response, are clearly in particularly good

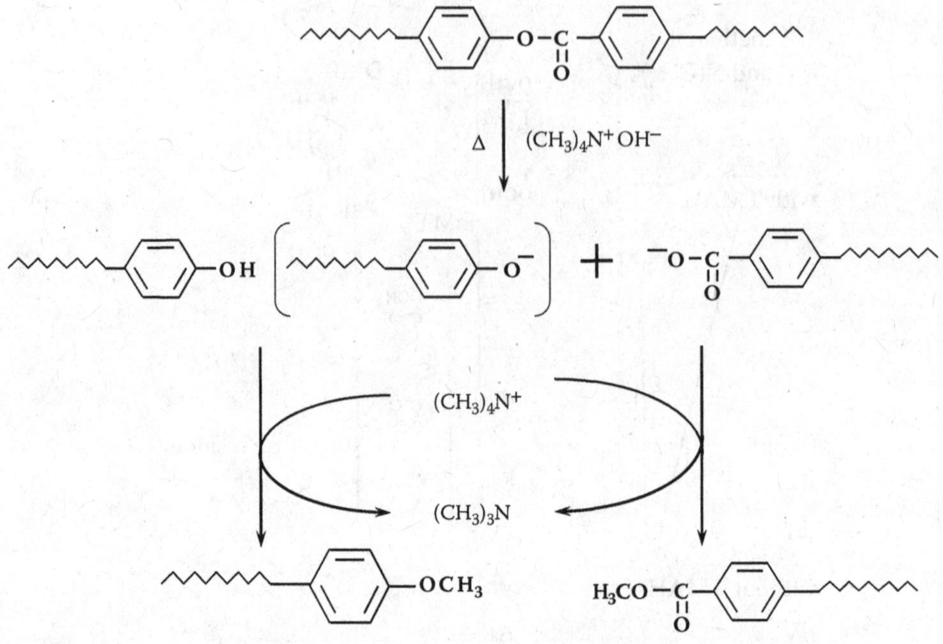

FIGURE 11.3 Typical THM reaction for an ester linkage in LCP.

TABLE 11.1

Composition of LCP Samples as Determined Using THM-GC

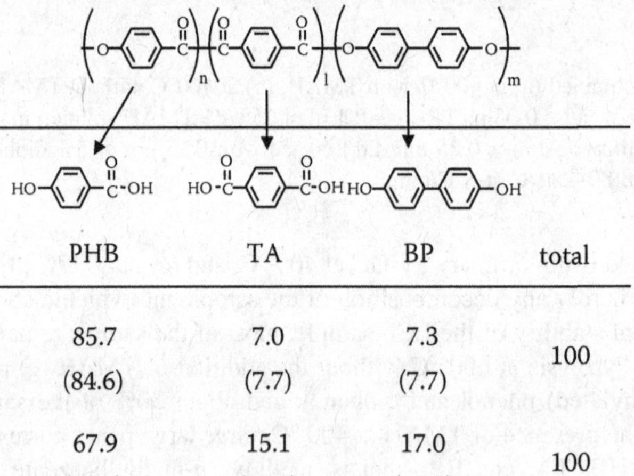

sample	PHB	TA	BP	total
A	85.7	7.0	7.3	100
	(84.6)	(7.7)	(7.7)	
B	67.9	15.1	17.0	100
	(66.7)	(16.7)	(16.7)	
C	50.5	24.0	25.5	100
	(50)	(25)	(25)	
D	34.1	32.0	33.9	100
	(33.3)	(33.3)	(33.3)	

( ): Feed compositionreproducibilities: CV < 1.0%

agreement with the original feed compositions and exhibit superior reproducibility (coefficient of variation [CV] less than 1.0%). This result demonstrates that THM-GC is a powerful method for the rapid and precise compositional analysis of LCPs.

B ALIPHATIC POLYESTERS (COMPOSITIONAL ANALYSIS/BIODEGRADABILITY)

Various aliphatic polyesters are considered to be environmental-friendly materials based on their biodegradable nature and/or nonpetroleum origin. Among these, the poly(hydroxyalkanoate) co-polymer poly(3-hydroxybutyrate-*co*-3-hydroxyvalerate) (P(3HB-*co*-3HV)—which is usually produced via bacterial fermentation by *Alcaligenes eutrophus* — shows good stiffness and toughness for practical applications. The copolymer composition of P(3HB-*co*-3HV) strongly affects not only its physical properties, but also its biodegradability, and can be readily determined using THM-GC [6].

Figure 11.4 shows a typical THM-GC pyrogram of P(3HB-*co*-3HV) with a 3HV content of 12 mol%. As shown in the figure, nine major peaks could be observed and identified [6]. Figure 11.5 shows the reaction scheme by which the 3HB unit could form the observed products. The expected products of the THM reaction of P(3HB-*co*-3HV) are methyl 3-methoxybutanoate (peak 8) from the 3HB unit and methyl 3-methoxypentanoate (peak 9) from the 3HV unit. The other major products in the program—which include methylbutenoates (peaks 1 to 3) and me-thylpentenoates (peaks 4 to 7)—could be formed through thermal cleavage of the ester bond through a six-membered cyclic transition state followed by the THM reaction, as shown in Figure 11.5. Since the nine major products were all attributed to either the 3HB or 3HV unit, the copolymer composition of P(3HB-*co*-3HV) could be accurately determined based on the relative intensities of these peaks [6].

Because of the overly sensitive nature of the THM-GC technique, which requires only trace amounts (ca. 0.1 mg or less) of the sample, the local structural changes in an aliphatic polyester

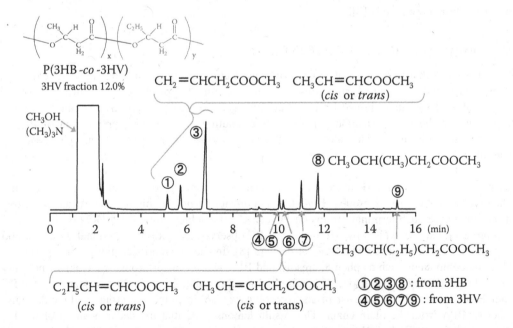

FIGURE 11.4 Pyrogram of P(3HB-*co*-3HV) obtained by THM-GC at 350°C [6]. Sample weight: 0.03 mg; Reagent: 4 µl of 25 wt% TMAH solution in methanol; Separation column: metal capillary (30 m × 0.25 mm i.d.) coated with 0.25 µm of polydimethylsiloxane (Frontier Lab, Ultra ALLOY PY-1); Column temp: 30°C (5 min)-230°C at 5°C/min.

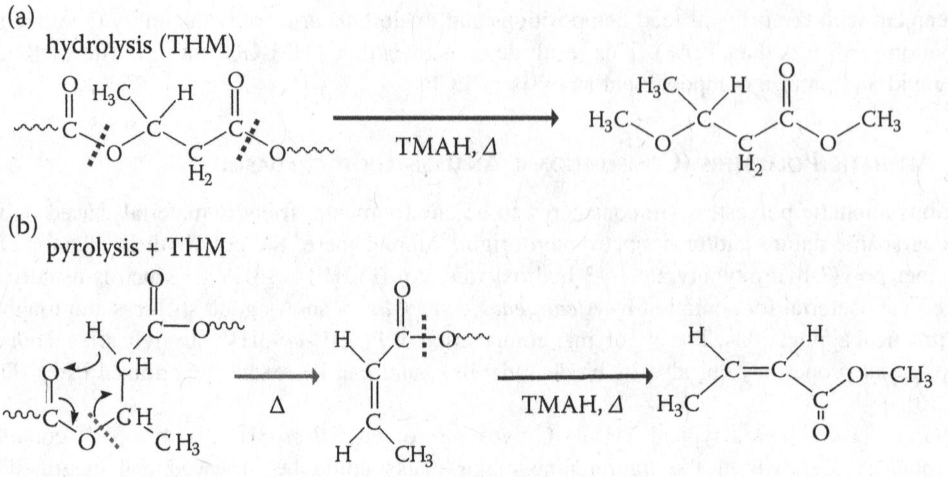

(a)

hydrolysis (THM)

TMAH, Δ

(b)

pyrolysis + THM

Δ

TMAH, Δ

FIGURE 11.5 Scheme of the decomposition of the 3HB unit in the presence of TMAH.

sample during the biodegradation process can be also evaluated. For instance, the enzymatic degradation behavior of poly(ε-caprolactone) (PCL) with an α-benzoyloxy terminal by cholesterol esterase was studied by THM-GC [7]. Figure 11.6 shows pyrograms of the PCL samples obtained (a) before and (b) after an enzymatic degradation test. In both pyrograms, the THM reaction products benzyl methyl ether (B)—from the benzoyloxy terminal group —and the dimethyl derivative of hydroxycaproic acid (dM-CL$_1$)—from the main chain—were observed. Interestingly, the relative intensity of peak B decreased slightly after the enzymatic degradation, while that of peak dM-CL$_1$ showed a relative increase. This result suggested that the enzymatic degradation of the PCL sample proceeded mainly via the exo-cleavage mode from the α-benzoyloxy terminal side under the tested conditions [7].

C POLYCARBONATE (COMPOSITION/END GROUP)

Polycarbonate (PC) is one of the most widely used engineering plastics owing to its excellent transparency and mechanical properties. The main chain of typical PC molecules consists of bisphenol A (BPA) units joined by carbonate linkages; various terminal groups can be formed depending on the polymerization process. In the solution method (SM) preparation of PC, *p-tert*-butylphenol is often added along with the precursors—sodium BPA salt and phosgene—to control the molecular weight of the resulting polymer. This method provides PC molecules with *p-tert*-butylphenoxy endcaps. On the other hand, the melt method (MM) in which BPA and diphenylcarbonate are directly reacted through ester exchange, gives PC molecules with either phenoxy or hydroxyl end groups. Both the identification and the highly precise and sensitive determination of such terminal groups on PC can also be achieved by THM-GC [8].

Figure 11.7 shows typical pyrograms of SM-PC observed using (a) conventional Py-GC at 600 °C without TMAH and (b) THM-GC at 400 °C [8]. In the conventional pyrogram (a), various phenolic compounds such as phenol, cresols, and BPA were formed through cleavage of not only the carbonate linkages, but also C–C bonds. On the other hand, in the THM pyrogram (b), the PC sample produced only two intense peaks: *p-tert*-butyl anisol from the end group, and the dimethyl ether of BPA from the main chain. This result demonstrated that the decomposition of the PC polymer chains through THM reactions occurred selectively and quantitatively at the carbonate linkages to yield the methyl ethers of the components in the PC sample, as shown in Figure 11.8. Therefore, the end group content could be accurately determined from the relative peak intensities of these two peaks after correcting the molar sensitivity of the FID response.

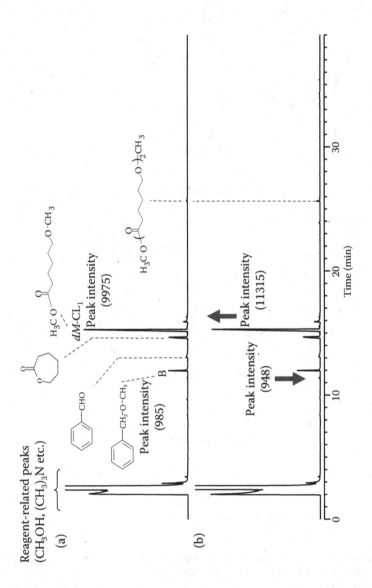

FIGURE 11.6 Pyrograms of PCL samples obtained by THM-GC at 400°C: (a) Control sample, (b) residual sample recovered after a 36 h enzymatic degradation test [7]. Sample weight: 0.05 mg; Reagent: 3 μl of 25 wt% TMAH solution in methanol; Separation column: metal capillary (30 m × 0.25 mm i.d.) coated with 0.25 μm of polydimethylsiloxane (Frontier Lab, Ultra ALLOY PY-1); Column temp: 40°C (5 min)–300°C at 5°C/min.

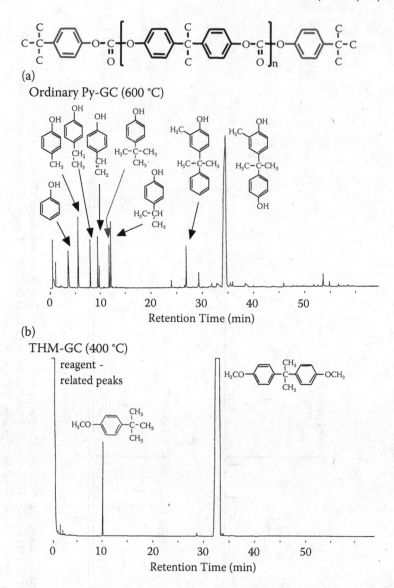

FIGURE 11.7 Pyrograms of a polycarbonate sample (SM-PC) obtained using (a) ordinary Py-GC at 600°C or (b) THM-GC at 400°C in the presence of TMAH [8]. Sample weight: 0.05 mg; Reagent: 1 μl of 25 wt% TMAH solution in methanol; Separation column: fused silica capillary (25 m × 0.25 mm i.d.) coated with 0.25 μm of polydimethylsiloxane; Column temp: 50–300°C at 4°C/min.

The number average molecular weight (M_n) was then estimated based on the relative end group content obtained using THM-GC [8]. Table 11.2 summarizes the M_n values estimated for the fractionated SM-PC samples using various methods. The values determined using THM-GC were in fairly good agreement with those obtained using size exclusion chromatography (SEC) and NMR. Furthermore, highly precise results were obtained using THM-GC—for example, the CV value was 1.8% for S-2 and 0.6% for S-3 based on five repeated measurements. The content of phenoxy end groups in MM-PC samples was also estimated very accurately using this technique [8].

Various PC copolymers with enhanced thermal or photostability have been produced, and characterization of their chemical composition and end groups is often required. However, these

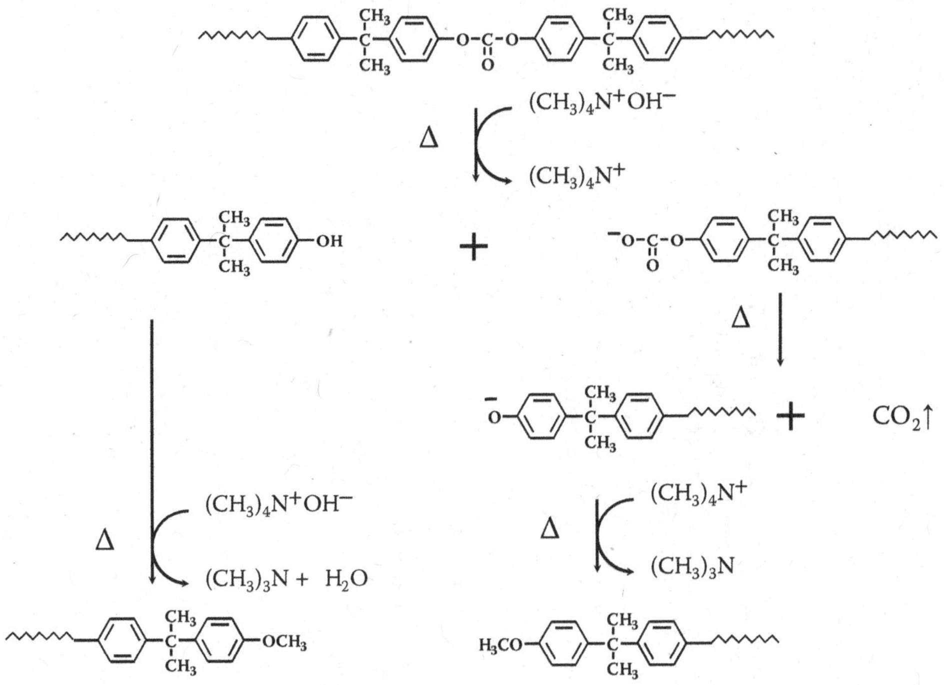

FIGURE 11.8 Typical THM reaction of carbonate linkages in PC [12].

cannot be determined precisely, even using NMR, due to the complexity of the obtained spectra and the insufficient sensitivity toward the end groups, particularly in higher-molecular-weight polymers. THM-GC has also been successfully applied for the compositional analysis and end group determination of such PC copolymers [9].

As a typical example, a thermally stabilized PC copolymer (PC-I) with the comonomer 1,1'-bis-(4-hydroxyphenyl)-3,3,5-trimethylcyclohexane (BHTH, MW = 310), and a light stabilized one (PC-II) with the comonomer 2,2'-methylenebis[6-(2H-benzotriazol-2-yl)-4-(1,1,3,3-tetramethylbutyl)phenol] (MBTP, MW = 656) were analyzed [9]; their molecular structures are shown in Figure 11.9. Figure 11.10 shows typical pyrograms of (a) PC-I and (b) PC-II obtained using THM-GC. In both cases, only three characteristic peaks were observed in the pyrograms. Peaks a, b, and c in pyrogram (a) were assigned to the methyl ether of the *p-tert*-octylphenoxyl terminal group moiety and the dimethyl ethers of the two comonomers (BPA and BHTH)—which

TABLE 11.2

Estimated Number Average Molecular Weight of Fractionated SM-PC Samples [8]

Fraction No.		Estimated *Mn* Values	
	by SEC	by THM-GC*	by ¹H-NMR
S-1	23400	26900	30200
S-2	8300	9500	9600
S-3	4000	4700	-

* CV values: 0.5-2%

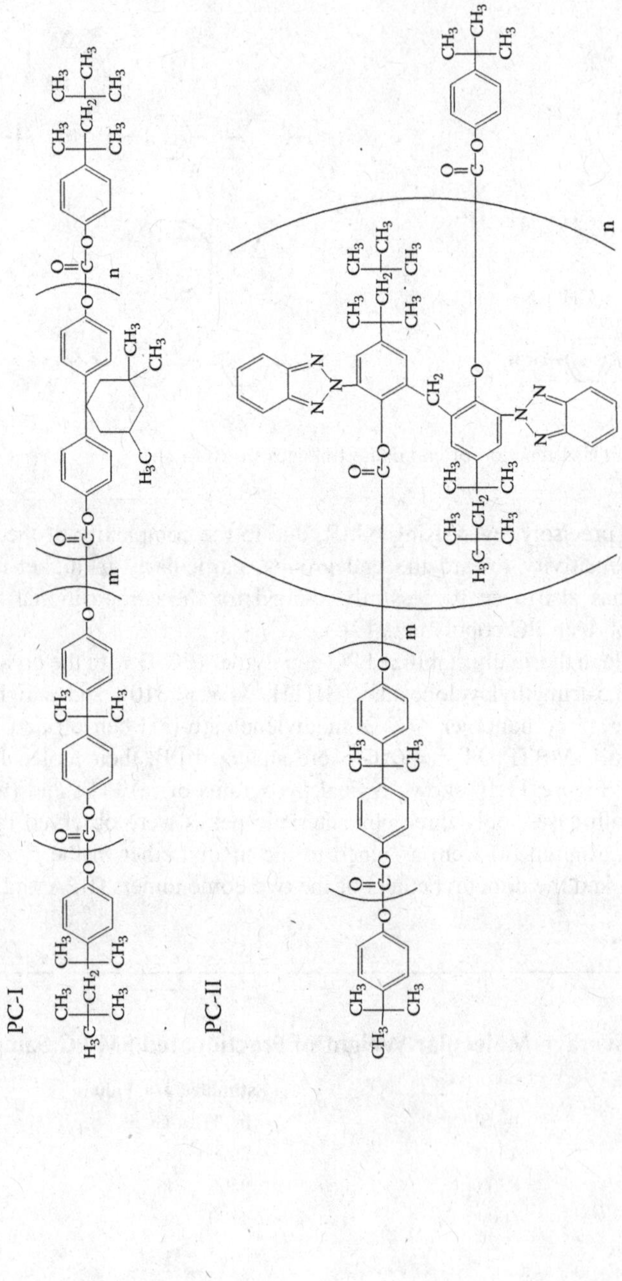

FIGURE 11.9 Molecular structures of thermally stabilized PC-I and light stabilized PC-II.

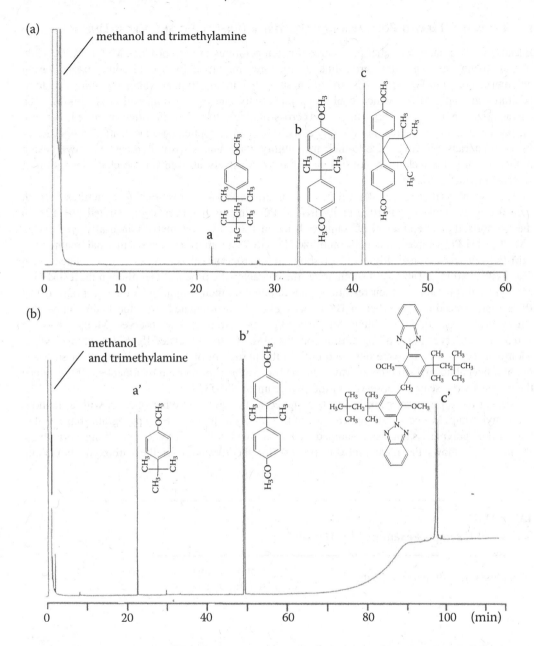

FIGURE 11.10 Pyrograms of PC samples obtained by THM-GC at 400°C: (a) PC-I, (b) PC-II [9]. Sample weight: 0.1 mg; Reagent: 2 μl of 25 wt% TMAH solution in methanol; Separation column: (a) fused silica capillary (30 m × 0.25 mm i.d.) coated with 0.25 μm of polydimethylsiloxane; Column temp: 50–300°C at 5°C/min; (b) metal capillary (30 m × 0.25 mm i.d.) coated with 0.5 μm of polydimethylsiloxane (Frontier Lab, Ultra ALLOY PY-2); Column temp: 50–380°C at 5°C/min.

were formed via the selective hydrolysis of carbonate linkages in the polymer chain followed by simultaneous methylation. Based on the intensities of these peaks, the exact composition of the PC-I sample could be determined, along with its accurate M_n. Pyrogram (b) provides the same information for the PC-II sample. In this case, it should be noted that dimethyl ether of the bulky MBTP (MW = 684) was detected using a thermally stable metal capillary column at a high oven temperature of 380 °C [9].

D   THERMALLY TREATED POLYCARBONATE/POLYESTER (BRANCHING AND CROSS-LINKING)

Because of the relatively higher polymerization temperatures used to produce MM-PC (200°–300 °C), branching and insoluble cross-linked structures might be formed to some extent during polymerization. Furthermore, PCs are often subjected to injection molding operations at temperatures around 300 °C, which could also potentially cause branching and cross-linking. The characterization of these branching and cross-linking structures is often required because the presence of such structures is strongly associated with the optical and mechanical properties of the PC materials. However, due to their insolubility, their analysis is not an easy task even using NMR, which is the technique that has been most extensively utilized for the characterization of soluble polymeric materials.

The use of THM-GC for the verification of branching and cross-linking structures in both industrially available PC and thermally treated PC samples has been demonstrated [10–12]. In these reports, the three kinds of PC samples listed in Table 11.3 – namely, industrially synthesized SM-PC, MM-PC samples, and an insolubilized PC (IS-PC) sample formed via thermal treatment of MM-PC—were examined. Figure 11.11 shows the pyrograms of the PC samples obtained using THM-GC at 400 °C. Table 11.4 summarizes the characteristic products identified in the THM-GC/MS pyrograms in order of their retention times, along with their origin. In all the pyrograms of the PC samples, the dimethyl ether of BPA (peak c)—which originated from the main chain—and ethers (peaks a and b)—originating from the end groups—were clearly observed. Moreover, peaks 1 to 5 were observed in the pyrograms of both the MM-PC (b) and related IS-PC (c) samples, while additional peaks 6 to 10 were observed only in the pyrogram of IS-PC (c). Some of these peaks might reflect abnormal structures formed during the synthesis or thermal treatment at 300 °C, since these peaks were scarcely observed in the pyrogram of SM-PC (a).

Among these products, peak 5 can be assigned to the dimethyl ether of BPA with a methoxycarbonyl group. Moreover, the appearance of this pyrolysis product in the pyrogram suggests that the initial polymer samples contained some amount of carboxylic branching structures. Figure 11.12 shows the most probable process for the formation of the carboxylic branching

TABLE 11.3
Series of PC samples examined by THM-GC

Sample name	Preparation method	Degree of branching/ cross-linking structures	
SM-PC	Solution method	Low	⎫
			⎬ industrial PC
MM-PC	Melt method	Somewhat	⎭
IS-PC	Thermal insolubilization with MM-PC (at 300°C in air for 3 hours)	considerable	

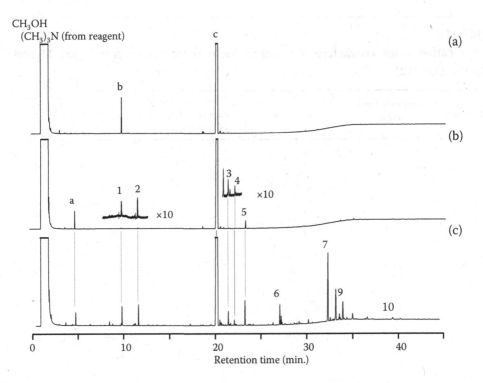

FIGURE 11.11 Pyrograms of PC samples obtained using THM-GC at 400°C: (a) SM-PC, (b) MM-PC, and (c) IS-PC [12]. Peak numbers correspond to those in Table 11.4. Sample weight: 0.1 mg; Reagent: 2 μl of 25 wt% TMAH solution in methanol; Separation column: metal capillary (30 m × 0.25 mm i.d.) coated with 0.25 μm of polydimethylsiloxane (Frontier Lab, Ultra ALLOY PY-1); Column temp: 50–370°C at 8 °C/min.

structure via Kolbe-Schmitt or Fries rearrangement of the carbonate group, along with the formation pathway of the peak 5 component via the THM reaction of the branching moiety [10].

Additionally, since no appreciable cleavage of C–C bonds would occur under the THM conditions used, the appearance of peak 1 (*p*-isopropenyl anisole) in the pyrograms suggests that the original MM-related PC samples contained isopropenyl end groups to some extent, especially the thermally treated PC sample (IS-PC); this compound would be formed through the disproportionation reaction of a C(CH₃)-phenylene bond. In a comparable manner, peak 2 reflected the disproportionation reaction of the C–C bond neighboring the carboxylic branching structure formed during the polymerization or thermal treatment of PC-2.

On the other hand, peaks 6 to 9—which were observed exclusively in the pyrogram of IS-PC (c)—should be derived from other types of branching and cross-linking structures formed during thermal treatment, such as CH_2-O (peaks 6 and 8), biphenyl (peak 7), and phenylene-O (peak 9) bridges[9]. Furthermore, the identification of peak 10 confirmed the existence of xanthone structures in IS-PC [11]. Based on the observed results, a model of the structure of the thermally treated PC sample was produced and is shown in Figure 11.13.

The flame-retardation mechanisms of PC containing silicone additives (FR-PC) were also elucidated using THM-GC [13]. The use of silicone additives as flame retardants has received a great deal of attention because they generate relatively low amounts of toxic substances even in the case of an extreme fire. To elucidate the structural changes induced in the FR-PC system during thermal degradation, FR-PC and a control PC sample were subjected to THM-GC measurement at 400 °C before and after thermal treatment at 380 °C for 2 h.

In the pyrogram of the control PC sample after thermal treatment, peaks corresponding to the abnormal structures 2-(4-methoxyphenyl)-2-phenylpropane and 2-(3-methoxycarbonyl-4-

TABLE 11.4

Identification of the characteristic peaks on the pyrograms shown in Figure 11 and their classification [12].

Peak no[a]	Retention time (min.)	Structure	Classification for the corresponding structure in PC chain[b]
a	5.1	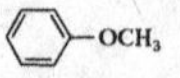 (phenyl)–OCH$_3$	I
1	10.1	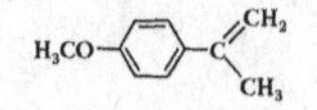 H$_3$CO–(phenyl)–C(=CH$_2$)–CH$_3$	III
b	10.2	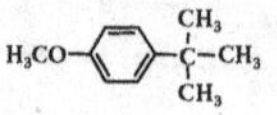 H$_3$CO–(phenyl)–C(CH$_3$)(CH$_3$)–CH$_3$	I
2	11.9	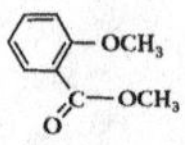 (phenyl)–OCH$_3$; C(=O)–OCH$_3$	III
c	20.5	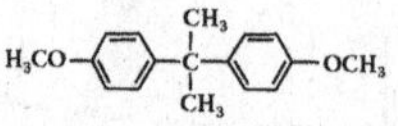 H$_3$CO–(phenyl)–C(CH$_3$)(CH$_3$)–(phenyl)–OCH$_3$	I
3	21.8	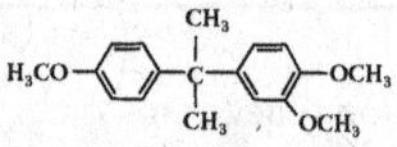 H$_3$CO–(phenyl)–C(CH$_3$)(CH$_3$)–(phenyl)(OCH$_3$)–OCH$_3$	II
4	22.6	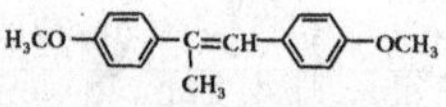 H$_3$CO–(phenyl)–C(CH$_3$)=CH–(phenyl)–OCH$_3$	III
5	23.6	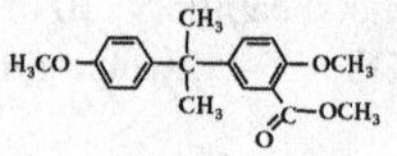 H$_3$CO–(phenyl)–C(CH$_3$)(CH$_3$)–(phenyl)(C(=O)OCH$_3$)–OCH$_3$	II
6	27.5	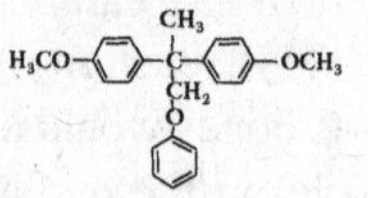 H$_3$CO–(phenyl)–C(CH$_3$)(CH$_2$–O–phenyl)–(phenyl)–OCH$_3$	III
7	32.7	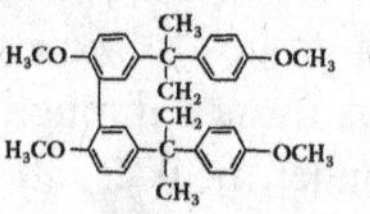 H$_3$CO–(phenyl)–C(CH$_3$)(CH$_2$...)–(phenyl)–OCH$_3$; H$_3$CO–(phenyl)–C(CH$_2$)(CH$_3$)–(phenyl)–OCH$_3$	II
8	33.6	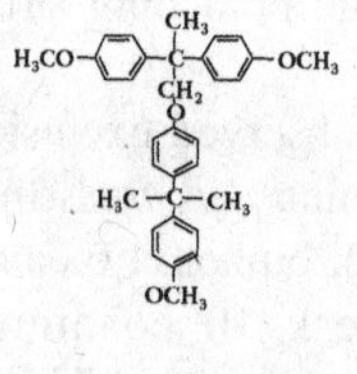 H$_3$CO–(phenyl)–C(CH$_3$)(CH$_2$–O–phenyl–C(CH$_3$)(CH$_3$)–OCH$_3$)–(phenyl)–OCH$_3$	II
9	34.3	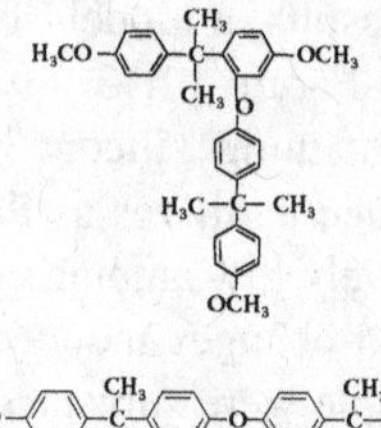 H$_3$CO–(phenyl)–C(CH$_3$)(CH$_3$ O–phenyl–C(CH$_3$)(CH$_3$)–OCH$_3$)–(phenyl)–OCH$_3$	II
10	39.5	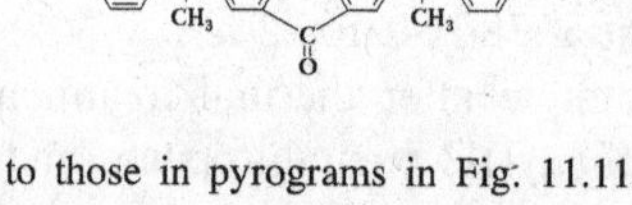 H$_3$CO–(phenyl)–C(CH$_3$)(CH$_3$)–(phenyl)–O–(phenyl)–C(CH$_3$)(CH$_3$)–(phenyl)–OCH$_3$	III

a) Peak numbers corresponding to those in pyrograms in Fig. 11.11.b) I: Main Chain or original terminal groups II: Branching or cross-linking structures formed during the thermal treatments. III: Abnormal structures formed during the thermal treatment except for the branching and cross-linking structures.

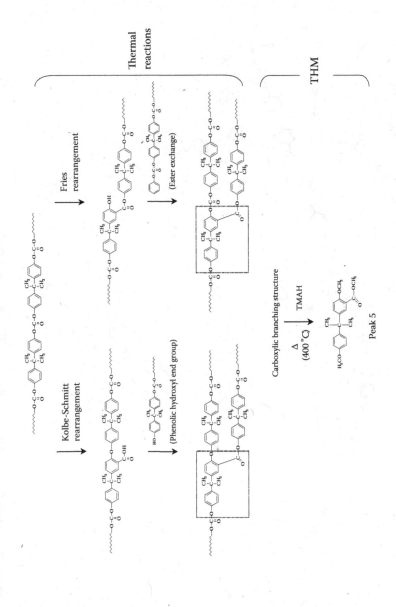

FIGURE 11.12 Possible pathway for the formation of the carboxylic branching structure and its characteristic product (peak 5) [12].

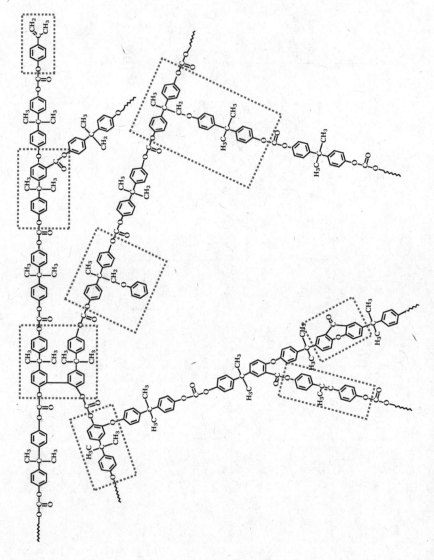

FIGURE 11.13 Model of the speculated structure of IS-PC based on THM-GC measurement.

methoxyphenyl)-2-(4-methoxyphenyl)propane were observed. These peaks were much more prominent in the pyrogram of the thermally treated FR-PC, suggesting that the formation of abnormal structures—including cross-linking structures—might be promoted by the silicone-based flame retardant. The results observed for the FR-PC material suggested that cross-linking structures were formed during an earlier stage of combustion in the presence of the silicone flame retardant and might suppress the thermal decomposition of the PC materials at higher temperatures by restricting the movement of the degradation products in the degrading material into the combustion zone.

In another study, the abnormal structures in LCP—a 4-hydroxybenzoic acid (HBA)/2-hydroxy-6-naphthoic acid (HNA) copolymer—were also characterized using THM-GC [14]. The pyrogram of an LCP sample that had been thermally treated at 500 °C exhibited peaks reflecting branching and condensed structures along with the expected methyl derivatives of HBA and HNA from the main chain and those of the end groups. Generally, the yields of all the products originating from the abnormal structures increased with increasing thermal treatment temperature. Furthermore, the condensation reactions might be preferred over the formation of branching in the LCP sample at higher treatment temperatures of around 500 °C.

E   UV-Cured Resin (Cross-Linking Network)

Ultraviolet (UV)-cured photopolymerization is currently utilized in various industrial fields, including paints, adhesives, dental cements, coatings for optical fibers and disks, photoresists, and three-dimensional stereolithography. To improve the performance of UV-curable resins, both the kinetics of photopolymerization and the chemical structures of the cured resins must be elucidated. However, their structural characterization is not an easy task due to their insolubility. THM-GC has been applied for the characterization of various UV-cured resins including compositional analysis of multi-component cured resins [15], estimation of the original molecular weight of the epoxy acrylate constituents of cured resins [16], and determination of the conversion during UV-curing [17]. Moreover, this method has been extensively utilized to study the network structures in UV-cured acrylic ester resins prepared from poly(ethylene glycol) diacrylate in the presence of a benzoyl type photoinitiator under UV irradiation [18]. In this work, the minor but characteristic peaks of the methyl acrylate oligomers observed in the pyrograms of the cured resin samples were interpreted in terms of the chain distribution of the network junctions in the cured resins.

Figure 11.14 shows a typical pyrogram of the cured resin (expanded for the characteristic region mainly reflecting the network structure) and its model structure, along with possible THM pathways. As shown in this reaction scheme, the ester and ketone linkages in the cross-linked resin structure could be selectively cleaved and methylated by TMAH. Therefore, in addition to the dimethyl ethers of the poly(ethylene glycol)s, methyl acrylate and initiator fragments should be formed from the unreacted acryloyl groups and initiator residues at the chain ends of the cross-linking portion, respectively.

Moreover, based on the THM cleavages in the cross-linking sequences consisting of linked methacryloyl groups, various methyl aclylate oligomers should be produced, reflecting the degree of polymerization. In the observed pyrogram, various minor peaks were observed, along with a series of main peaks corresponding to the dimethyl ethers of poly(ethylene glycol). Among these minor peaks, a series of methyl acrylate oligomers ranging from dimers to at least hexamers were identified, which directly reflected the distribution of the cross-linking sequences in the cured resin [18].

F   Poly(Aryl Ether Sulfone)

In general, ether linkages in polymer chains are less susceptible to THM reactions than ester and carbonate bonds. In some cases, however, ethers can also be hydrolyzed almost quantitatively into

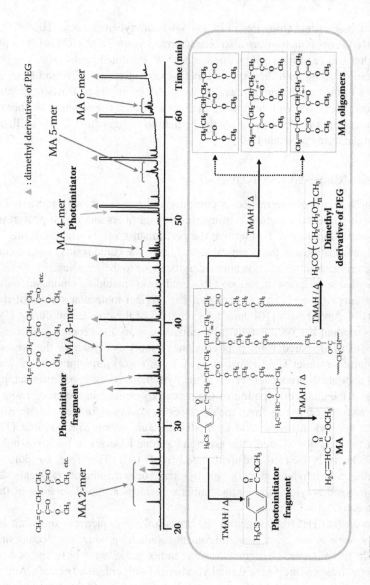

FIGURE 11.14 Pyrogram of UV-cured acrylic ester resin obtained using THM-GC at 400°C, along with its possible cross-linking structure and the formation pathways of the products [18]. Sample weight: 0.1 mg; Reagent: 4 μl of 25 wt% TMAH solution in water; Separation column: metal capillary (30 m × 0.25 mm i.d.) coated with 0.25 μm of 5% diphenyl–95% dimethylpolysiloxane (Frontier Lab, Ultra ALLOY⁺-5); Column temp; 30° (5 min)–230°C at 5°C/min.

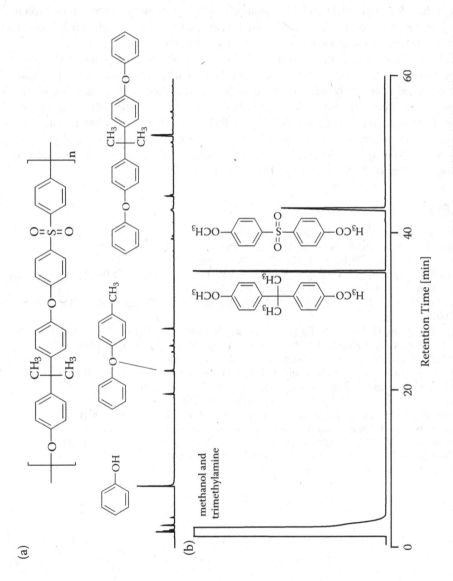

FIGURE 11.15 Pyrograms of polusulfone (a) obtained using ordinary Py-GC at 700°C and (b) THM-GC at 300°C [19]. Sample weight: 0.05 mg; Reagent: 2 µl of 25 wt % TMAH solution in methanol; Separation column: metal capillary (30 m × 0.25 mm i.d.) coated with 0.5 µm of polydimethylsiloxane (Frontier Lab, Ultra ALLOY PY-1); Column temp: 50°–300°C at 5°C/min.

methyl ethers by TMAH, depending on the properties of the units neighboring the ether linkages. The ether bonds in poly(aryl ether sulfone)s (PSU) are a typical example [19].

Figure 11.15 shows typical pyrograms of poly(oxy-*p*-phenylenesulfonyl-*p*-phenylenoxy-*p*-phenyleneisopropylidene-*p*-phenylene)—which is more commonly known as polysulfone (PSF)—that were obtained using (a) ordinary Py-GC and (b) THM-GC [19]. In the conventional pyrogram obtained at 700 °C (a), very small peaks corresponding to phenol and aromatic products containing a diphenyl ether structure were observed but no products containing sulfone groups were distinguished due to the preferential elimination of SO_2 at this higher temperature. Moreover, the recovery of the observed products was relatively low, and a considerable amount of carbonaceous residue remained in the sample cup even after pyrolysis.

Conversely, in the pyrogram obtained using THM-GC at 300 °C, the peaks of the dimethyl ethers of constituents of the original PSF, BPA, and bis(4-hydroxyphemyl)sulfone (bisphenol S; BPS) were observed exclusively with almost quantitative recovery, without any residue in the sample cup. In this case, the strong electron-withdrawing nature of the sulfone groups might ,promote the THM reaction at the ether linkages in PSF, with almost complete retention of the sulfone structure. In a similar manner, another type of commercially available PSU—poly(*p*-phenylenesulfonyl-*p*-phenylene) [poly(ether sulfone) (PES)]—was also thoroughly decomposed into the dimethyl ether of BPS through the THM reaction [19].

REFERENCES

1. J. M. Challinor, *J. Anal. Appl. Pyrol.*, *61*: 3 (2001).
2. Y. Ishida, H. Ohtani, and S. Tsuge, *J. Anal. Appl. Pyrol.*, *33*: 167 (1995).
3. H. Ohtani, R. Fujii, and S. Tsuge, *J. High Res. Chromatogr.*, *14*: 338 (1991).
4. H. Ohtani, N. Sugimoto, M. Hirano, T. Yokota, and K. Katoh, *J. Anal. Appl.Pyrol.*, *79*: 323 (2007).
5. J. W. de Leeuw and M. Baas, *J. Anal. Appl. Pyrol.*, *26*: 175 (1993).
6. H. Sato, M. Hoshino, H. Aoi, T. Seino, Y. Ishida, K. Aoi, and H. Ohtani, *J. Anal. Appl. Pyrol.*, *74*: 193 (2005).
7. H. Sato, Y. Kiyono, H. Ohtani, S. Tsuge, H. Aoi, and K. Aoi, *J. Anal. Appl. Pyrol.*, *68/69*: 37 (2003).
8. Y. Ito, H. Ogasawara, Y. Ishida, H. Ohtani, and S. Tsuge, *Polym. J.*, *28*: 1090 (1996).
9. Y. Ishida, S. Kawaguchi, Y. Ito, S. Tsuge, and H. Ohtani, *J. Anal. Appl. Pyrol.*, *40/41*: 321 (1997).
10. K. Oba, Y. Ishida, Y. Ito, H. Ohtani, and S. Tsuge, *Macromolecules*, *33*: 8173 (2000).
11. K. Oba, H. Ohtani, and S. Tsuge, *Polym. Degrad. Stab.*, *74*: 171 (2001).
12. S. Tsuge, H. Ohtani, and K. Oba, *Macromol. Symp.*, *195*: 287 (2003).
13. K. Hayashida, H. Ohtani, S. Tsuge, and K. Nakanishi, *Polym. Bull.*, *48*: 483 (2002).
14. K. Oba, Y. Ishida, H. Ohtani, and S. Tsuge, *Polym. Degrad. Stab.*, *76*: 85 (2002).
15. H. Matsubara, A. Yoshida, H. Ohtani, and S. Tsuge, *J. Anal. Appl. Pyrol.*, *64*: 159 (2002).
16. H. Matsubara and H. Ohtani, *J. Anal. Appl. Pyrol.*, *75*: 226 (2006).
17. H. Matsubara and H. Ohtani, *Anal. Sci.*, *23*: 513 (2007).
18. H. Matsubara, A. Yoshida, Y. Kondo, S. Tsuge, and H. Ohtani, *Macromolecules*, *36*: 4750 (2003).
19. H. Ohtani, Y. Ishida, M. Ushiba, and S. Tsuge, *J. Anal. Appl. Pyrol.*, *61*: 35 (2001).

12 Index of Sample Pyrograms

Karen D. Sam

CDS Analytical, Inc., Oxford, Pennsylvania

Each analyst will optimize the chromatography of his Pyrolysis-GC system to provide the most pertinent information about the specific sample being analyzed. It is helpful, however, especially when first starting out, to have examples of typical analyses, giving some idea of what to expect in a pyrogram. This chapter includes capillary GC pyrograms of many different materials, all performed on readily available and typical columns. Although an analyst's samples will almost certainly be different in some way from the example materials shown here, it may be helpful to review results obtained on similar samples as a starting point in developing a specific pyrolysis method.

The examples are grouped roughly according to sample material type.

FIGURE NUMBER	PYROGRAM NUMBER	SAMPLE MATERIAL	SETPOINT °C	DETECTOR
		SYNTHETIC POLYMERS		
1	S-1	Kraton 1107	800	FID
		Kraton is a copolymer of styrene and isoprene.		
2	S-2	Polyester shirt thread	750	FID
3	S-3	Polychloroprene	750	FID
4	S-4	Polyethyl methacrylate	600	FID
5	S-5	Polymethyl methacrylate	600	FID
6	S-6	Polystyrene	750	FID
		Peak number 1 = monomer, 2 = dimer, 3 = trimer.		
7	S-7	Polyvinylchloride	600	FID
		Peak number 1 = Benzene, 2 = Toluene, 3 = Naphthalene.		
8	S-8	Polybutyl acrylate	750	MS
9	S-9	Polyvinylidine chloride	750	MS
10	S-10	Polyvinyl toluene	750	MS
11	S-11	Polystyrene butyl acrylate	750	MS
12	S-12	Polystyrene acrylonitrile	750	MS
13	S-13	Polyvinyl chloride	750	MS
14	S-14	Teflon	750	MS
15	S-15	Polyurethane	750	MS
16	S-16	Poly(4-tert-butylstyrene)	700	MS
17	S-17	Poly(vinyl alcohol-co-ethylene) 32 mol% ethylene	750	MS
18	S-18	Kraton 1161 PT	750	MS
19	S-21	Poly(styrene-co-methyl methacrylate) styrene 40 mol %	700	MS
20	S-22	Poly-a-Pinene	700	MS
21	S-23	Polylimonene	700	MS
22	S-24		600	MS

		Polyvinyl alcohol, 88% hydrolyzed DB1701 column		
23	S-25	Polyvinyl alcohol, 100% hydrolyzed DB1701 column	600	MS
24	S-26	Ethylene tetrafluoroethylene, ETFE	700	MS
25	S-27	Fluorinated ethylene propylene copolymer	700	MS
26	S-28	Polynorbornene	700	MS
27	S-29	Poly(vinylidine fluoride-co-hexafluoropropylene)	700	MS
28	S-30	Polyvinyl butyral	700	MS
29	S-31	Polyvinylpyrrolidone	700	MS

POLYOLEFINS

30	O-1	Polyethylene	725	FID
31	O-2	Polypropylene, isotactic	700	FID
32	O-3	Polyisobutylene	800	FID
33	O-4	Polybutadiene	800	FID
34	O-5	Polypropylene-l-butene copolymer (butene content 47%)	750	FID
35	O-6	Polyethylene in air	650	FID
36	O-7	Polyethylene in air	750	FID
37	O-8	Polypropylene in air Heating rate 100°/s	800	FID
38	O-9	Polyisoprene	750	FID
39	O-10	Poly-1-butene	600	FID
40	O-11	Polypropylene, atactic	750	MS
41	O-12	Polyethylene (25%) propylene	750	MS
42	O-13	Ethylene (52%) propylene rubber	750	MS
43	O-14	Polyethylene, high density	700	MS

NYLONS

44	N-1	Nylon 11	900	FID
45	N-2	Nylon 12	800	FID
46	N-3	Nylon 6/T	800	FID
47	N-4	Nylon 6/6	800	FID
48	N-5	Nylon 6/9	800	FID
49	N-6	Nylon 6/10	850	FID
50	N-7	Nylon 6/12	800	FID
51	N-8	Nylon 6	750	MS

BIOLOGICAL AND NATURAL MATERIALS

52	B-1	Amber	650	FID
53	B-2	Baltic amber Peak marked "S" is succinic acid	650	FID
54	B-3	Chitin from crab shells	450	FID
55	B-4	*E. Coli* bacteria	650	FID
56	B-5	Gelatin	750	FID
57	B-6	Starch	650	FID
58	B-7	Phenylalanine Peak number 1 = benzene, 2 = toluene, 3 = ethyl benzene.	700	FID
59	B-8	Tyrosine Peak number 1 = benzene, 2 = toluene, 3 = ethyl benzene, 4 = Phenol, 5 = methyl phenol	700	FID

60	B-9	Animal glue	500	FID
		From a 1500-year-old Egyptian artifact		
61	B-10	Human hair	750	FID
62	B-11	Lamb's wool	750	FID
63	B-12	Cotton thread	750	FID
64	B-13	Human fingernail	750	FID
65	B-14	Kerogen	800	FID
66	B-15	Dried linseed oil	700	FID
67	B-16	Oak leaf	750	FID
68	B-17	Straw	750	FID
69	B-18	Natural rubber	800	FID
		Polyisoprene, *cis* configuration.		
70	B-19	Oil shale	800	FID
		From rock sample heated at 60°C/min		
71	B-20	Oil shale	800	FID
		Total organic content 1.2%, pulse heated		
72	B-21	Beeswax	500	FID
73	B-22	Silk	675	FID
74	B-23	Tobacco	750	MS
75	B-24	Tobacco (mentholated)	750	MS
76	B-25	Sucrose	750	MS
77	B-26	Beeswax	750	MS
78	B-27	Cupuacu butter	750	MS
79	B-28	Collagen	600	MS
80	B-29	Flax fiber	650	MS
81	B-30	Flax wool fiber blend	650	MS
82	B-31	Hemp wool fiber blend	650	MS
83	B-32	Nettle fiber	650	MS
84	B-33	Nettle wool fiber blend	650	MS
85	B-34	Wool fiber	650	MS

MANUFACTURED GOODS

86	G-1	Unprinted newspaper	650	FID
87	G-2	White magazine paper	650	FID
88	G-3	White bond paper	750	FID
89	G-4	Bathroom cleaner product	650	FID
		Peak marked 1 = C_{19} , 2 = C_{16} hydrocarbons		
90	G-5	Disinfectant cleaner product	650	FID
		Peak number 1 = benzene, 2 = toluene,		
		3 = C12,		
		4 = C_{14}, 5 = C_{16}, 6 = C_{18} hydrocarbons		
91	G-6	Dishwashing liquid	700	FID
		Peak number 1 = $_{Clot}$ 2 = C11,		
		3 = C_{12} hydrocarbons.		
92	G-7	Ink, black, ball-point pen	650	FID
93	G-8	Ink, black, ball-point pen	650	FID
		Pyrolyzed intact with paper on which it		
		had been written		
		Numbered peaks are from the paper,		
		lettered peaks from the ink		
94	G-9	Ink, blue, ball-point pen	650	FID
		Pyrolyzed intact with paper on which it		
		had been written		
		Numbered peaks are from the paper,		
		lettered peaks from the ink		

95	G-10	Printing ink	700	FID
		Formulation included linseed oil, wax, petroleum resins		
96	G-11	Petroleum resin	700	FID
97	G-12	Petroleum resin	700	FID
98	G-13	Kodak photocopy	650	FID
		Paper and toner material pyrolyzed together; peak number		
		1 = Methyl methacrylate, 2 = Styrene.		
99	G-14	Xerox photocopy	650	FID
		Peak number 1 = styrene, 2 = butyl methacrylate		
100	G-15	Mascara	750	FID
		Peaks indicated with arrows result from beeswax		
101	G-16	Mascara with acrylate	750	FID
102	G-17	Shirt thread	750	FID
		50/50 cotton/polyester blend fabric.		
103	G-18	n-Tetracontane	650	FID
104	G-19	Silicone grease	900	FID
105	G-20	Epoxy paint	750	MS
106	G-21	Rayon	750	MS
107	G-22	Packaging	750	MS
		PVA, PVC, styrene, MMA		
108	G-23	Automobile paint	750	MS
109	G-24	Automobile tire rubber	750	MS
110	G-25	Chewing gum	750	MS
111	G-26	Suede, dyed brown with a phthalate	600	MS
112	G-27	Medication blister packaging	700	MS
113	G-28	Cash register paper	600	MS
114	G-29	Porkhide chew toy, smoked, with terephthalate	600	MS
115	G-30	Suede	600	MS
116	G-31	Uncured epoxy	700	MS
117	G-32	Kapton tape	700	MS
		POLYESTERS		
118	P-1	Mirel® polyhydroxyalkanoate polymer	700	MS
119	P-2	Polyhydroxybutyrate	700	MS
120	P-3	Poly[butylene terephthalate-co-poly(alkylene glycol) terephthalate]	700	MS
121	P-4	Vectran LCP (liquid-crystal polymer)	700	MS
122	P-5	Polyethylene terephthalate	700	MS
123	P-6	poly(3-hydroxybutyric acid-co-3-hydroxyvaleric acid)	700	MS

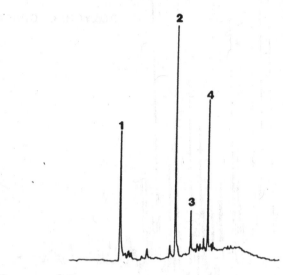

PYROGRAM S-1

FIGURE 12.1   S-1 Kraton 1107. Kraton is a copolymer of styrene and isoprene.

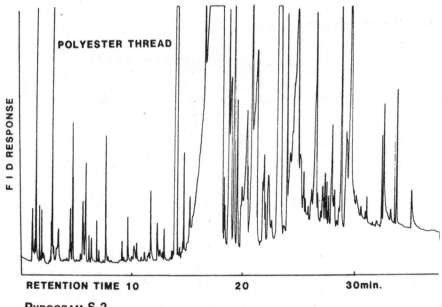

PYROGRAM S-2

FIGURE 12.2   S-2 Polyester shirt thread.

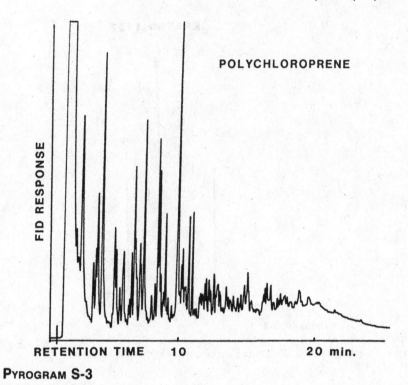

FIGURE 12.3   S-3 Polychloroprene.

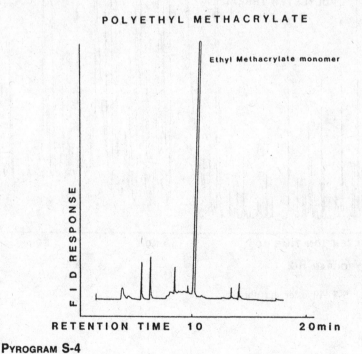

FIGURE 12.4   S-4 polyethyl methacrylate.

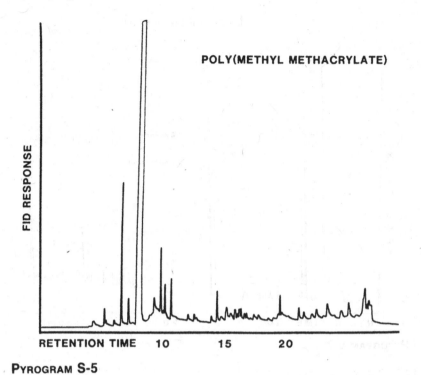

PYROGRAM S-5

FIGURE 12.5   S-5 polymethyl methacrylate.

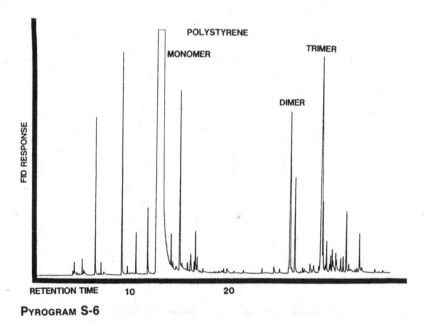

PYROGRAM S-6

FIGURE 12.6   S-6 polystyrene.

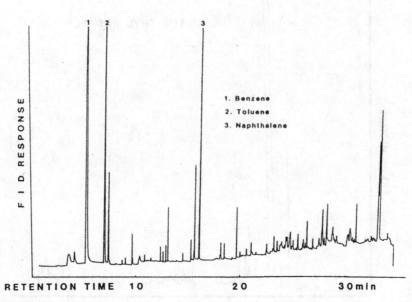

PYROGRAM S-7

FIGURE 12.7   S-7 Polyvinylchloride.

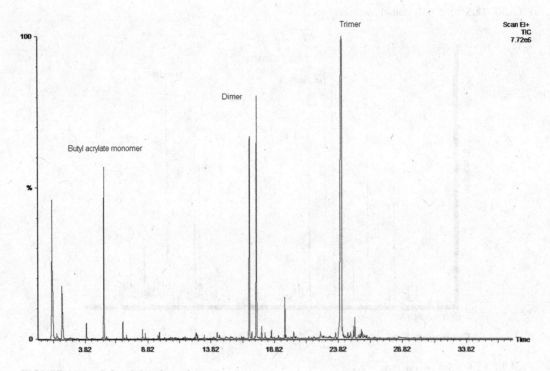

FIGURE 12.8   S-8 polybutyl acrylate.

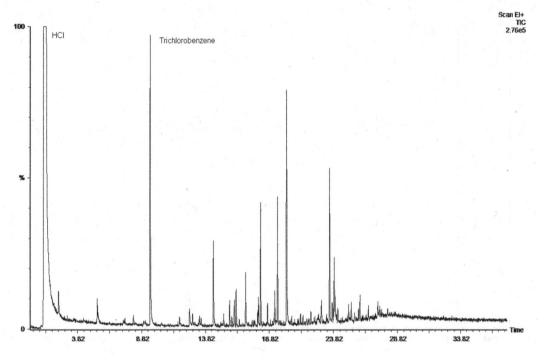

FIGURE 12.9   S-9 polyvinylidine chloride.

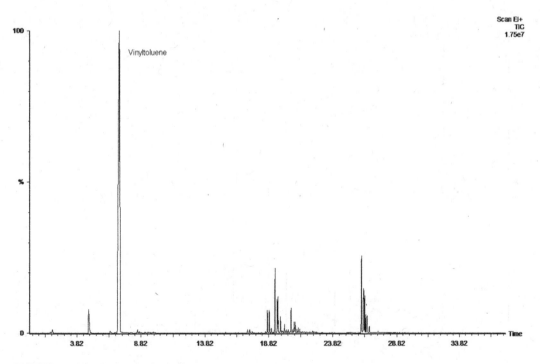

FIGURE 12.10   S-10 polyvinyl toluene.

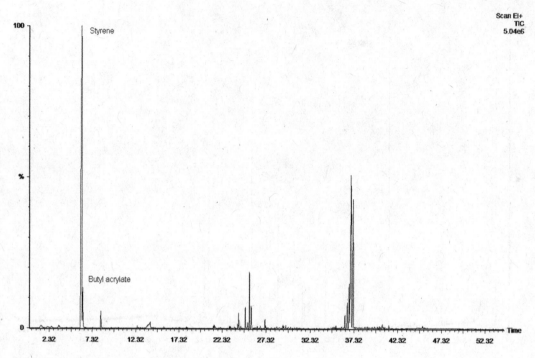

FIGURE 12.11 S-11 polystyrene butyl acrylate.

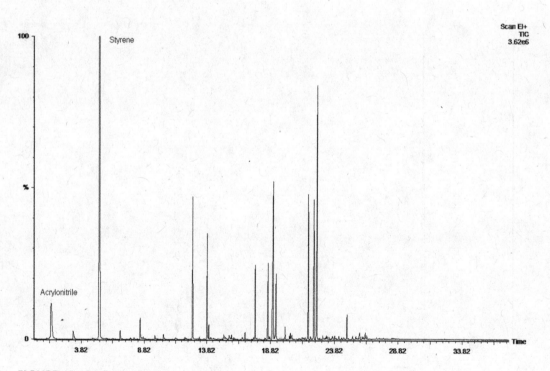

FIGURE 12.12 S-12 polystyrene acrylonitrile.

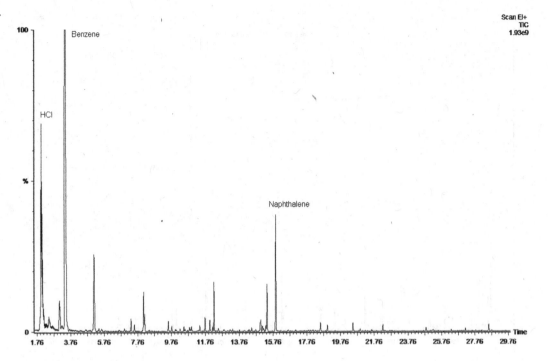

FIGURE 12.13 S-13 polyvinyl chloride.

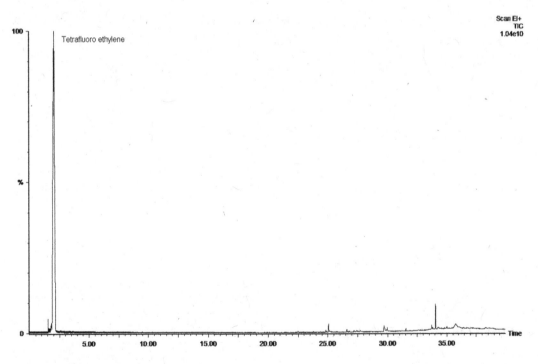

FIGURE 12.14 S-14 teflon.

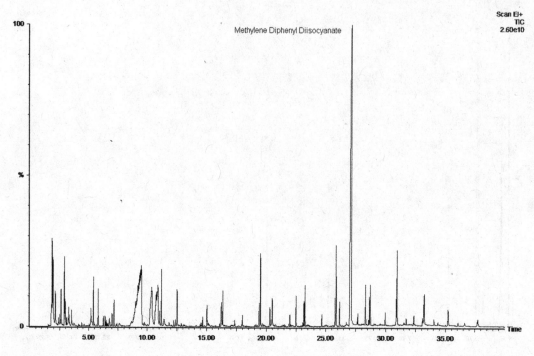

FIGURE 12.15 S-15 polyurethane.

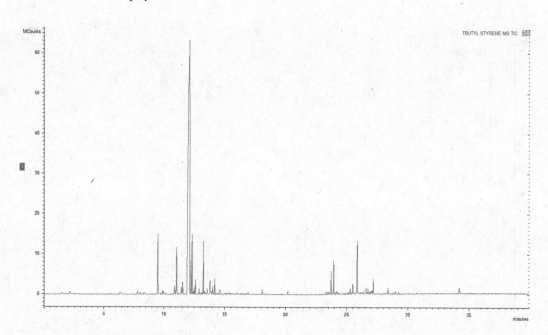

FIGURE 12.16 S-16 poly(4-*tert*-butylstyrene).

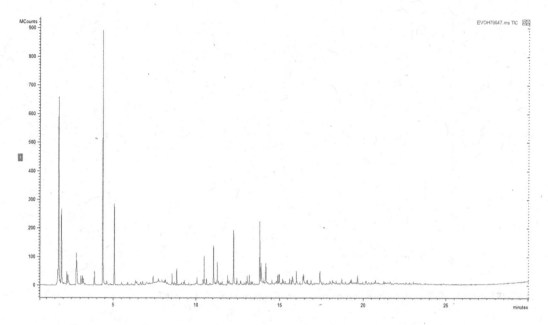

FIGURE 12.17   S-17 poly(vinyl alcohol-co-ethylene) 32 mol% ethylene.

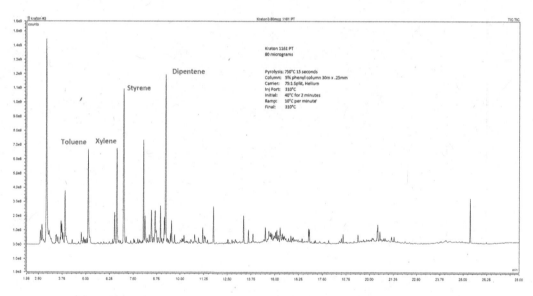

FIGURE 12.18   S-18 Kraton 1161 PT.

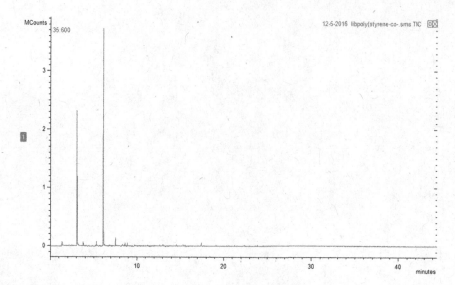

FIGURE 12.19   S-21 poly(styrene-co-methyl methacrylate) styrene 40 mol%.

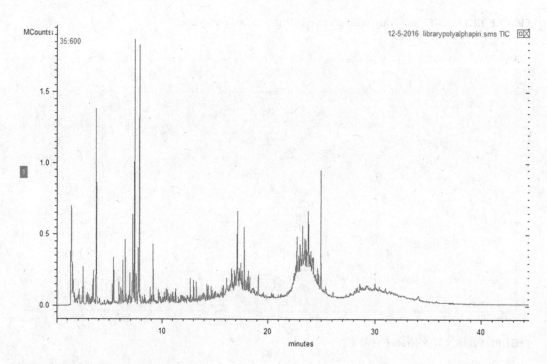

FIGURE 12.20   S-22 poly-a-pinene.

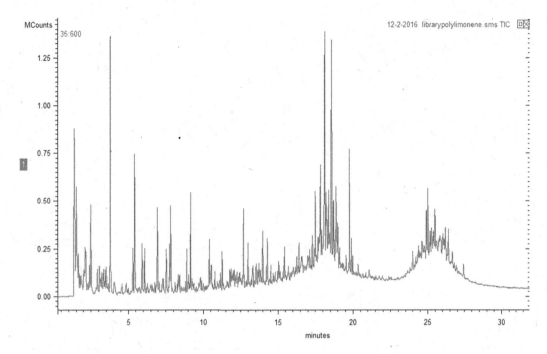

FIGURE 12.21 S-23 polylimonene.

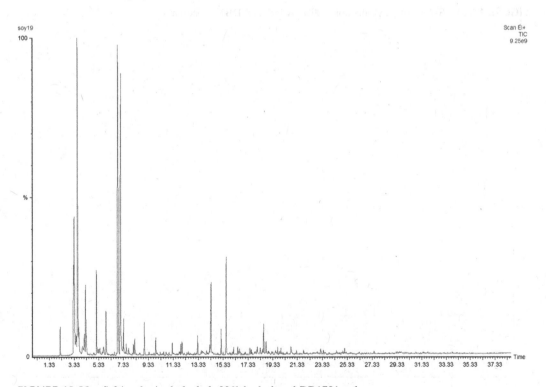

FIGURE 12.22 S-24 polyvinyl alcohol, 88% hydrolyzed DB1701 column.

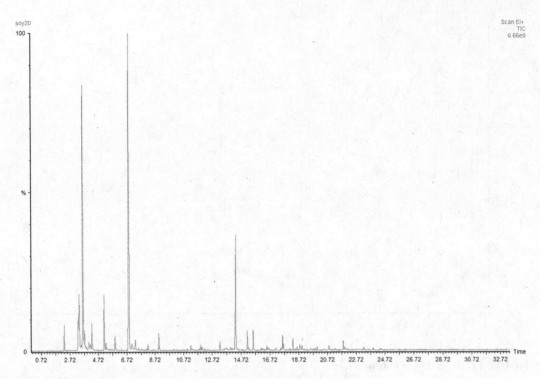

FIGURE 12.23 S-25 Polyvinyl alcohol, 100% hydrolyzed DB1701 column.

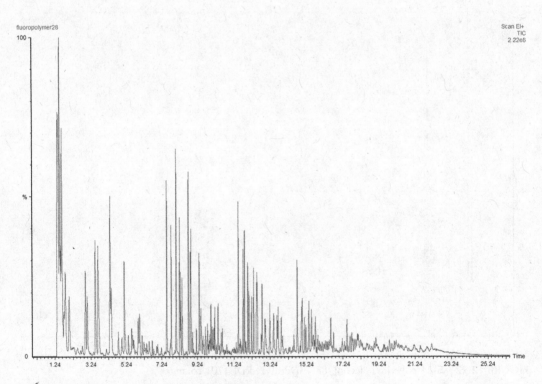

FIGURE 12.24 S-26 Ethylene tetrafluoroethylene, ETFE.

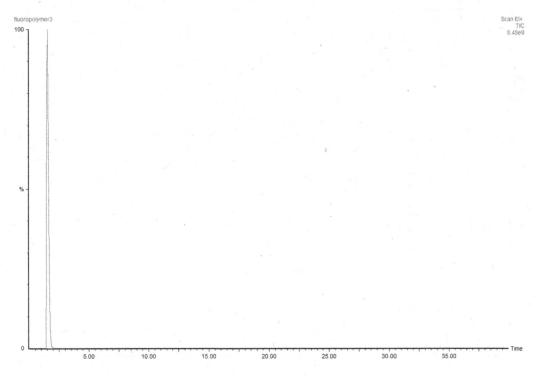

FIGURE 12.25   S-27 Fluorinated ethylene propylene copolymer.

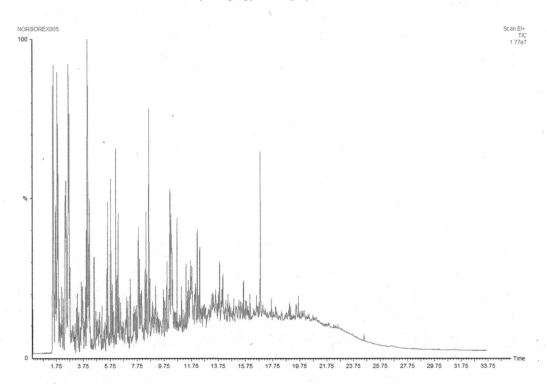

FIGURE 12.26   S-28 polynorbornene.

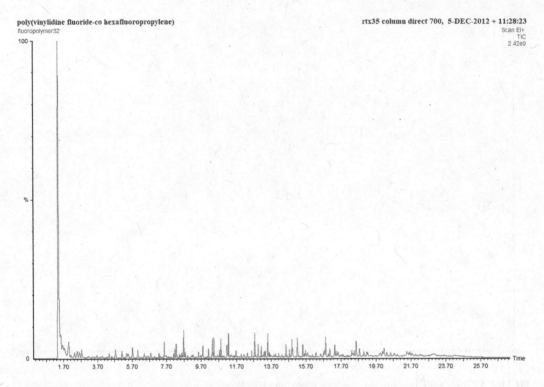

FIGURE 12.27   S-29 poly(vinylidine fluoride-co-hexafluoropropylene).

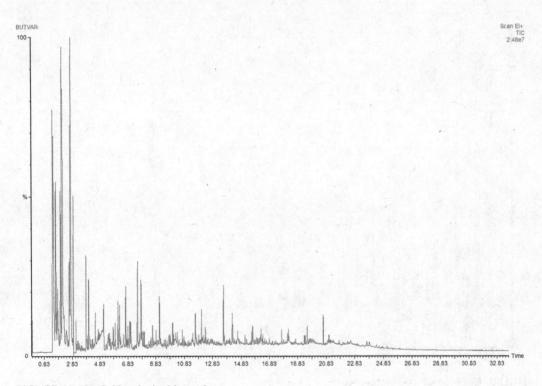

FIGURE 12.28   S-30 polyvinyl butyral.

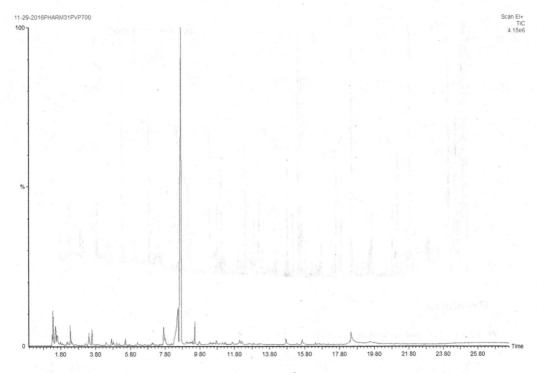

FIGURE 12.29 S-31 polyvinylpyrrolidone.

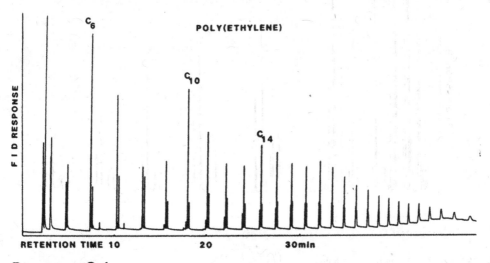

FIGURE 12.30 O-1 polyethylene.

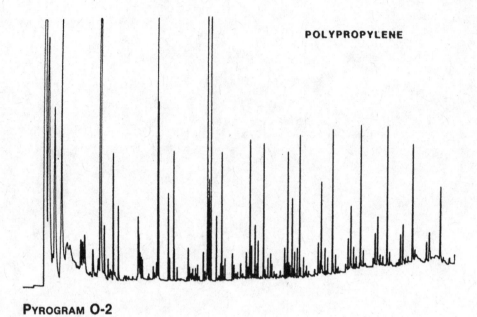

PYROGRAM O-2

FIGURE 12.31   O-2 polypropylene, isotactic.

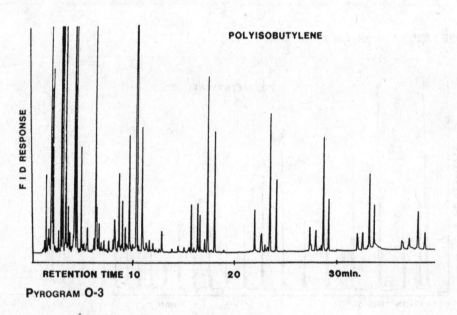

PYROGRAM O-3

FIGURE 12.32   O-3 polyisobutylene.

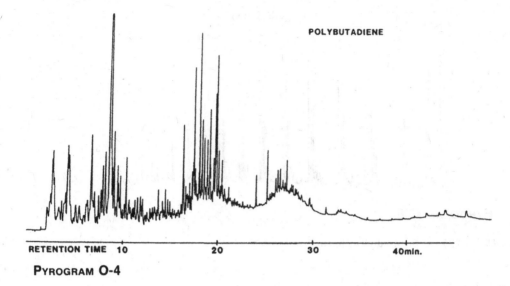

POLYBUTADIENE

RETENTION TIME 10   20   30   40min.

PYROGRAM O-4

FIGURE 12.33   O-4 polybutadiene.

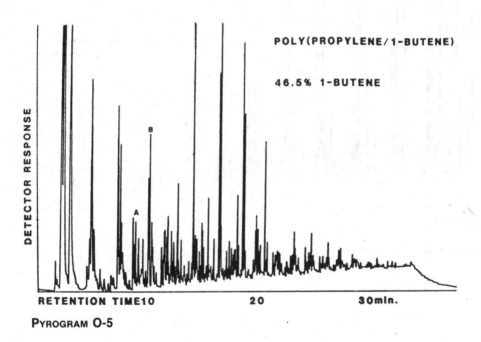

POLY(PROPYLENE/1-BUTENE)

46.5% 1-BUTENE

B

A

DETECTOR RESPONSE

RETENTION TIME10   20   30min.

PYROGRAM O-5

FIGURE 12.34   O-5 polypropylene-l-butene copolymer (butene content 47%).

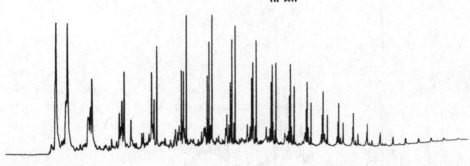

PYROGRAM O-6

FIGURE 12.35   O-6 polyethylene in air.

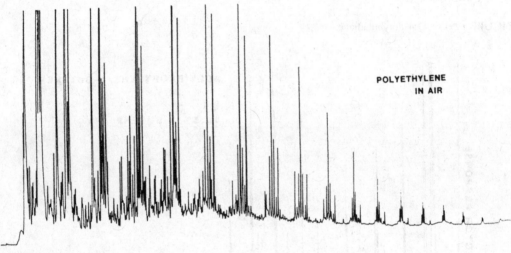

PYROGRAM O-7

FIGURE 12.36   O-7 polyethylene in air.

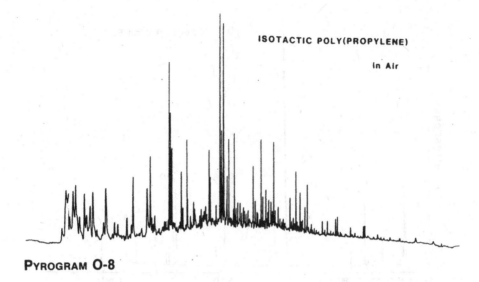

PYROGRAM O-8

FIGURE 12.37   O-8 polypropylene in air.

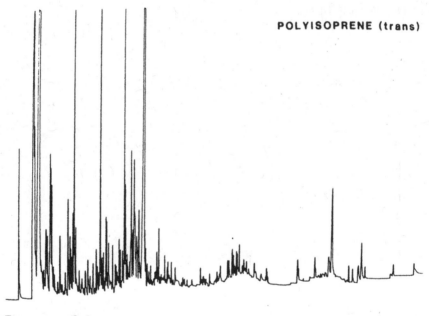

PYROGRAM O-9

FIGURE 12.38   O-9 polyisoprene.

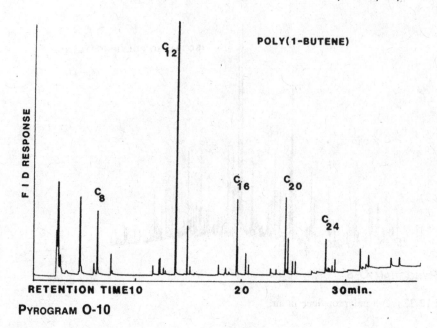

PYROGRAM O-10

FIGURE 12.39   O-10 poly-1-butene.

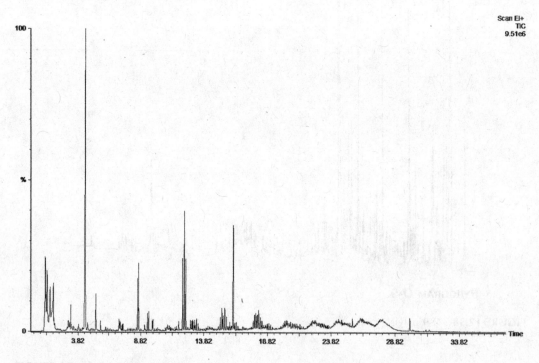

FIGURE 12.40   O-11 polypropylene, atactic.

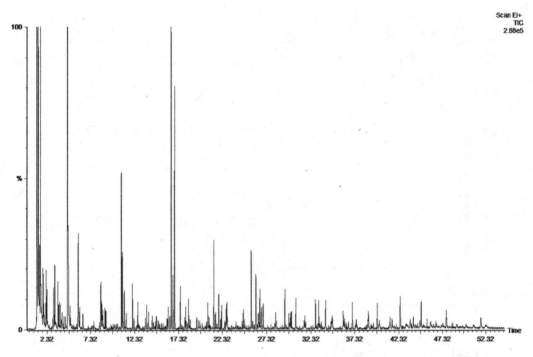

FIGURE 12.41   O-12 polyethylene (25%) propylene.

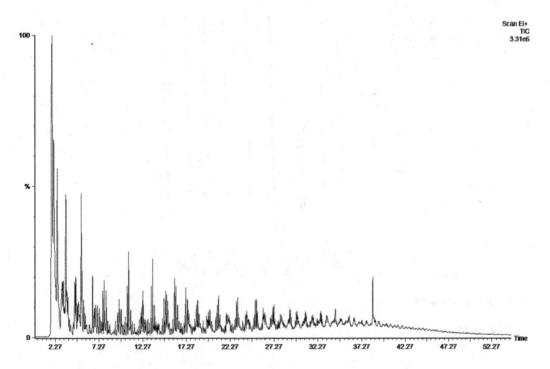

FIGURE 12.42   O-13 ethylene (52%) propylene rubber.

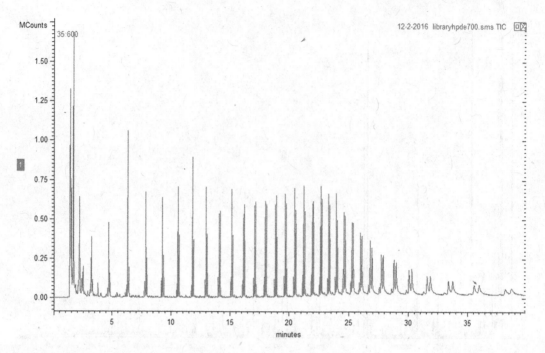

FIGURE 12.43   O-14 polyethylene, high density.

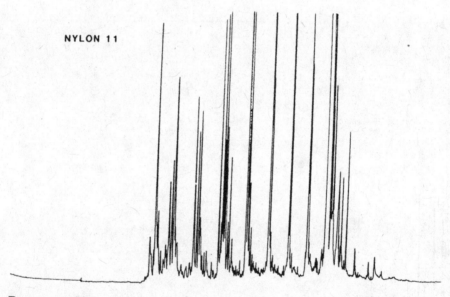

FIGURE 12.44   N-1 nylon 11.

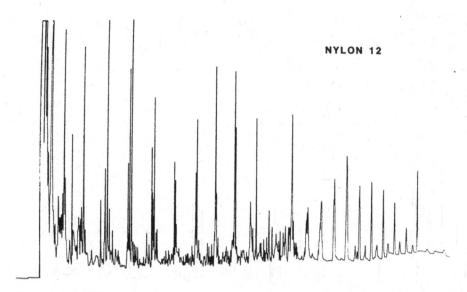

NYLON 12

Pyrogram N-2

FIGURE 12.45 N-2 nylon 12.

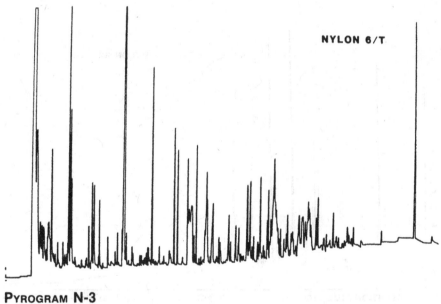

NYLON 6/T

Pyrogram N-3

FIGURE 12.46 N-3 nylon 6/T.

SAMPLE: NYLON 6/6

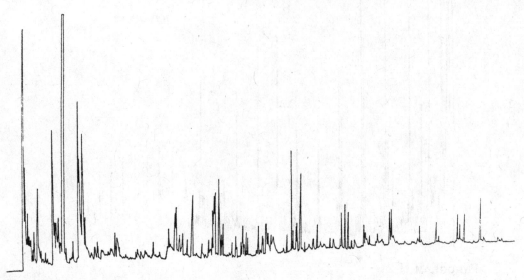

PYROGRAM N-4

FIGURE 12.47   N-4 nylon 6/6.

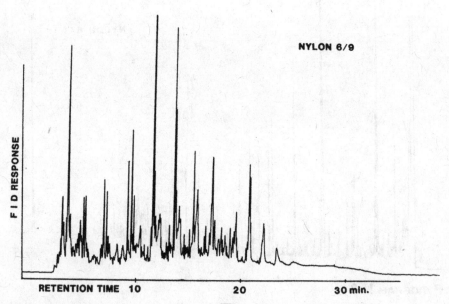

NYLON 6/9

PYROGRAM **N-5**

FIGURE 12.48   N-5 nylon 6/9.

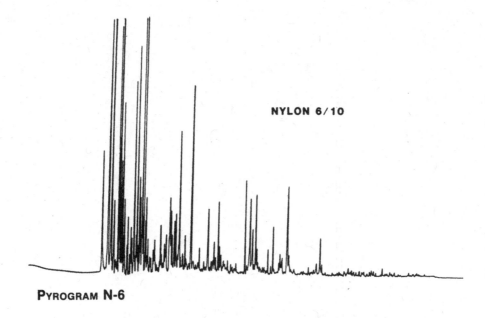

NYLON 6/10

PYROGRAM N-6

FIGURE 12.49 N-6 nylon 6/10.

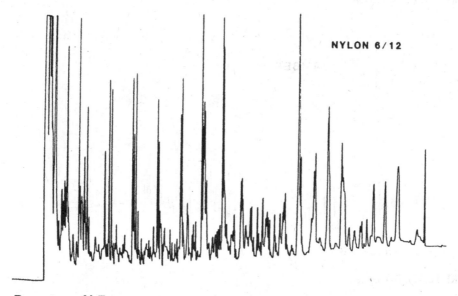

NYLON 6/12

PYROGRAM N-7

FIGURE 12.50 N-7 nylon 6/12.

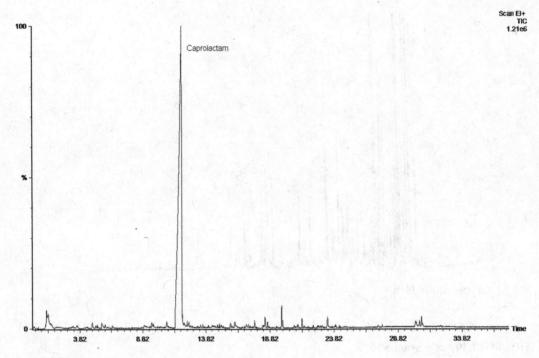

FIGURE 12.51   N-8 nylon 6.

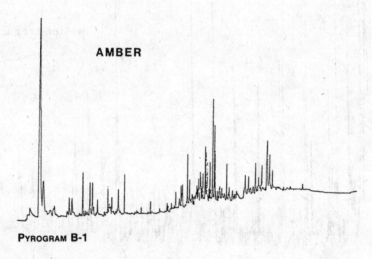

FIGURE 12.52   B-1 amber.

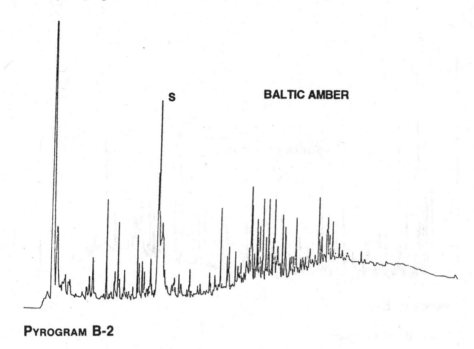

PYROGRAM B-2

FIGURE 12.53   B-2 baltic amber – peak marked "S" is succinic acid.

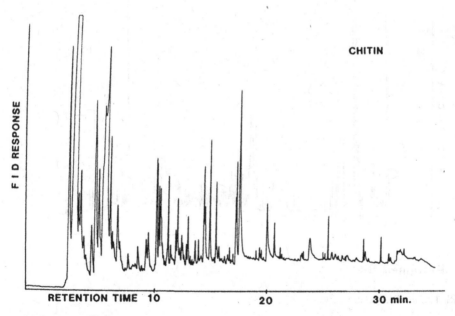

PYROGRAM B-3

FIGURE 12.54   B-3 Chitin from crab shells.

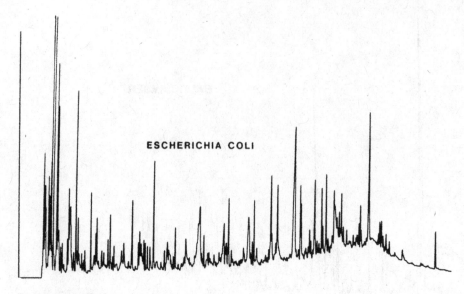

PYROGRAM B-4

FIGURE 12.55 B-4 *E. Coli* bacteria.

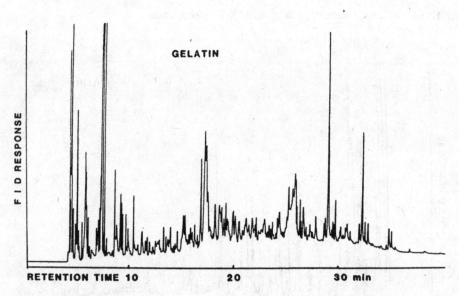

PYROGRAM B-5

FIGURE 12.56 B-5 gelatin.

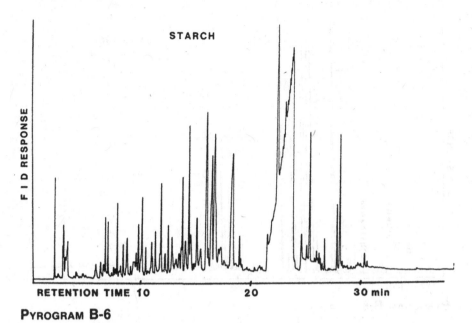

PYROGRAM B-6

FIGURE 12.57   B-6 starch.

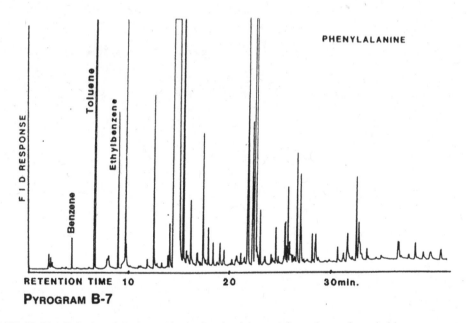

PYROGRAM B-7

FIGURE 12.58   B-7 phenylalanine peak number 1 = benzene, 2 = toluene, 3 = ethyl benzene.

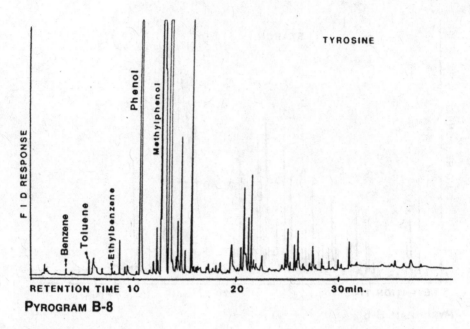

PYROGRAM B-8

FIGURE 12.59   B-8 tyrosine peak number 1 = benzene, 2 = toluene, 3 = ethyl benzene, 4 = phenol, 5 = methyl phenol.

ANCIENT ANIMAL GLUE

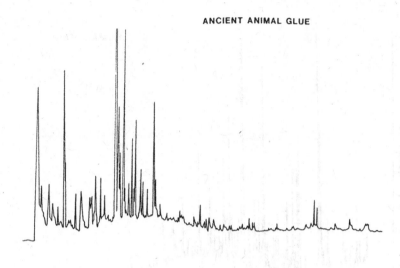

SARCOPHAGUS "GROUND"

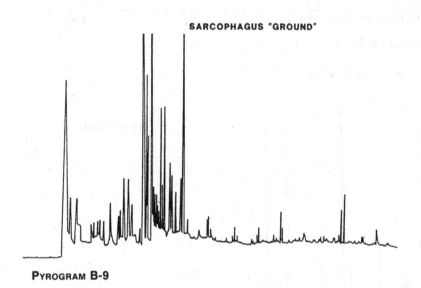

PYROGRAM B-9

FIGURE 12.60 B-9 animal glue From a 1500-year-old Egyptian artifact.

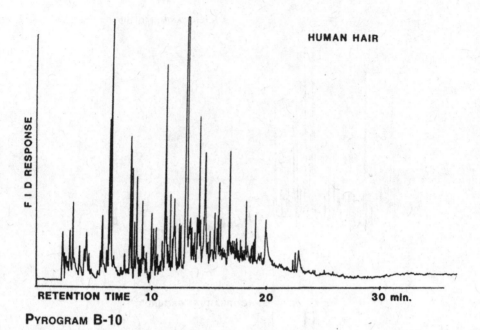

FIGURE 12.61   B-10 human hair.

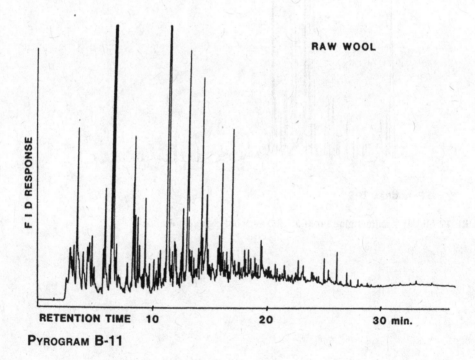

FIGURE 12.62   B-11 lamb's wool.

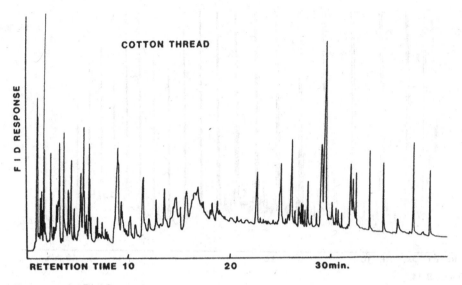

PYROGRAM B-12

FIGURE 12.63   B-12 cotton thread.

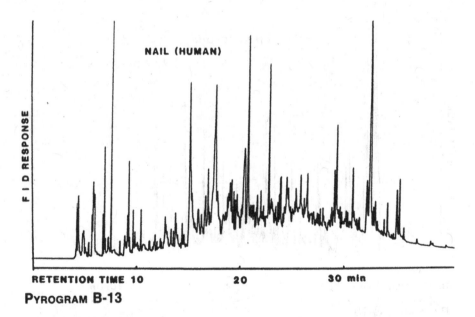

PYROGRAM B-13

FIGURE 12.64   B-13 human fingernail.

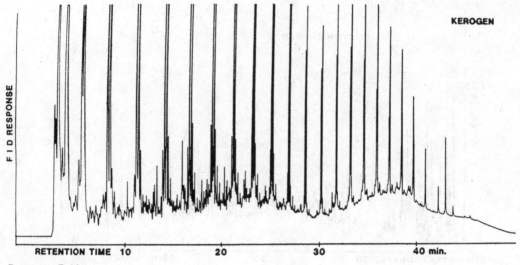

PYROGRAM B-14

FIGURE 12.65   B-14 kerogen.

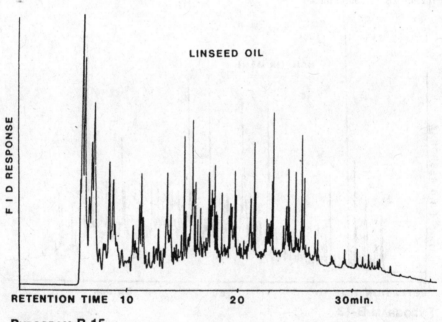

PYROGRAM B-15

FIGURE 12.66   B-15 dried linseed oil.

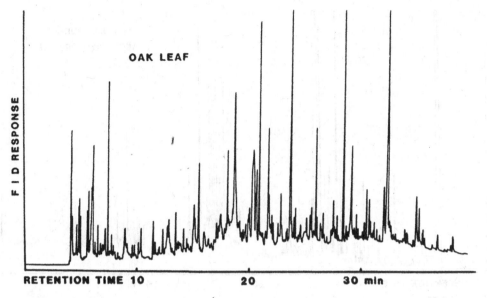

PYROGRAM B-17

FIGURE 12.67 B-16 oak leaf.

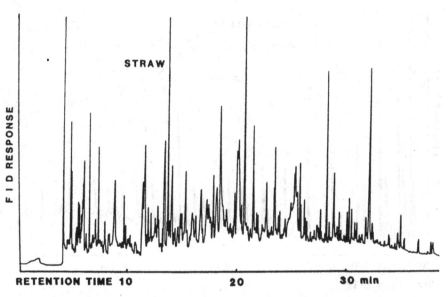

PYROGRAM B-16

FIGURE 12.68 B-17 straw.

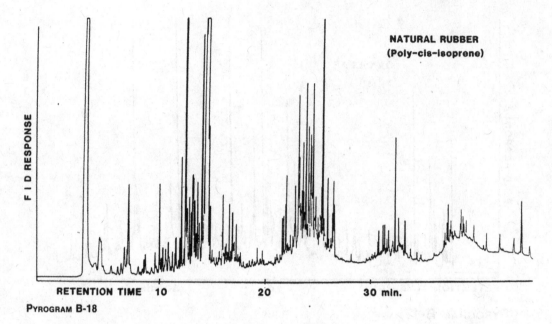

FIGURE 12.69 B-18 natural rubber polyisoprene, *cis*-configuration.

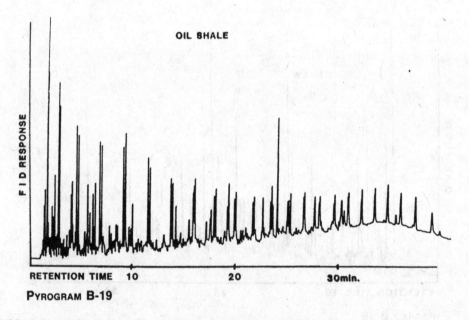

FIGURE 12.70 B-19 oil shale from rock sample heated at 60°C/minute.

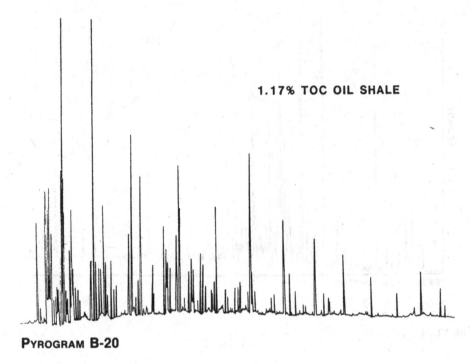

1.17% TOC OIL SHALE

PYROGRAM B-20

FIGURE 12.71   B-20 oil shale total organic content 1.2%, pulse heated.

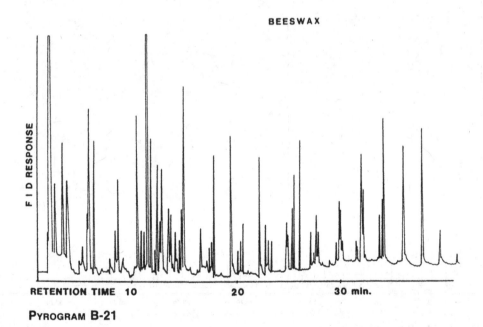

BEESWAX

PYROGRAM B-21

FIGURE 12.72   B-21 beeswax.

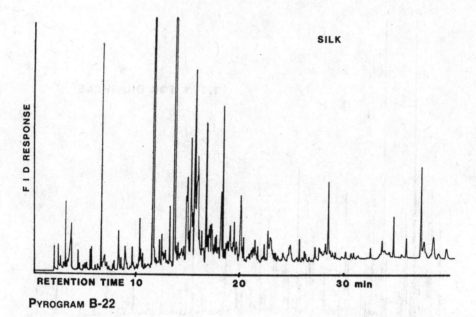

PYROGRAM B-22

FIGURE 12.73   B-22 silk.

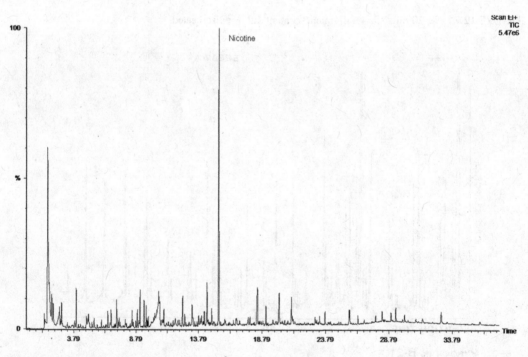

FIGURE 12.74   B-23 tobacco.

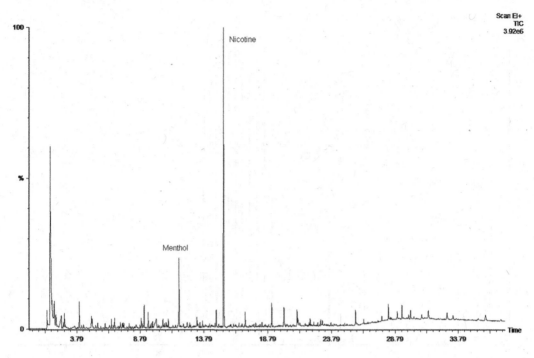

FIGURE 12.75   B-24 tobacco, mentholated.

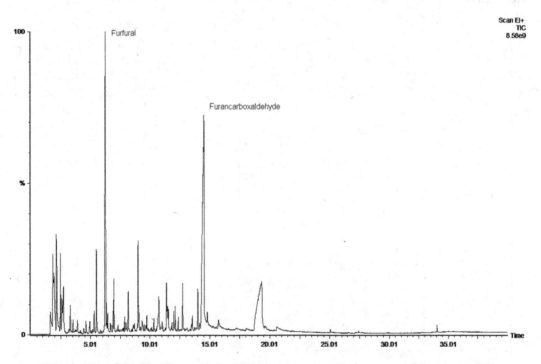

FIGURE 12.76   B-25 sucrose.

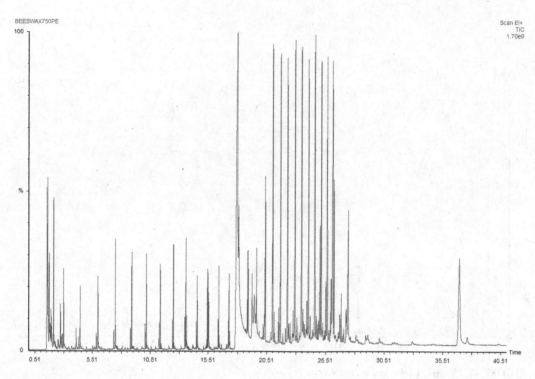

FIGURE 12.77   B-26 beeswax.

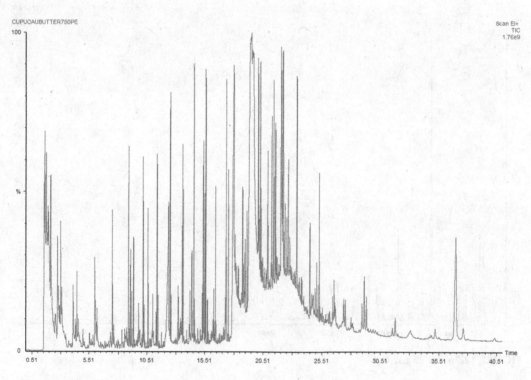

FIGURE 12.78   B-27 cupuacu butter.

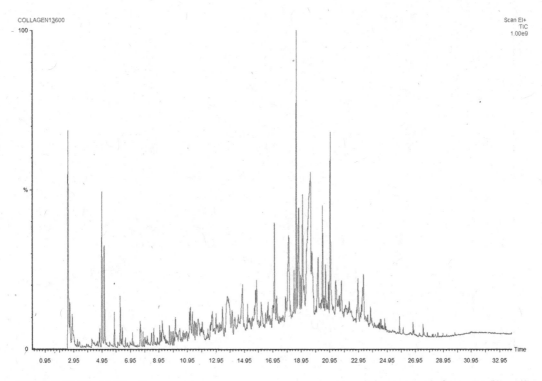

FIGURE 12.79   B-28 collagen.

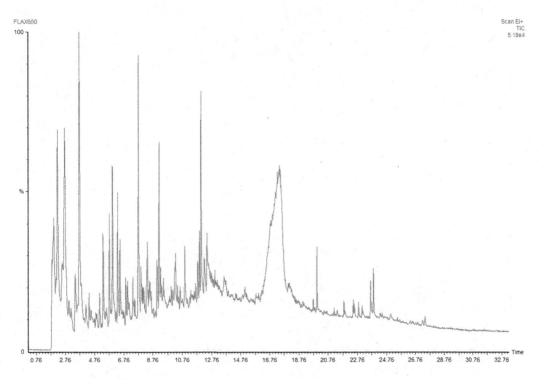

FIGURE 12.80   B-29 flax fiber.

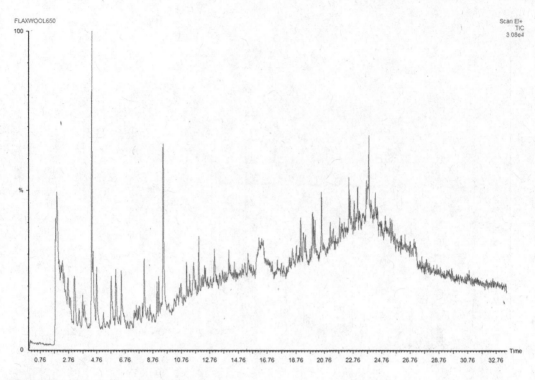

FIGURE 12.81   B-30 flax wool fiber blend.

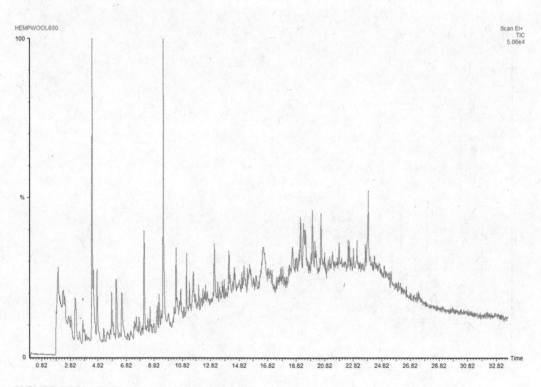

FIGURE 12.82   B-31 hemp wool fiber blend.

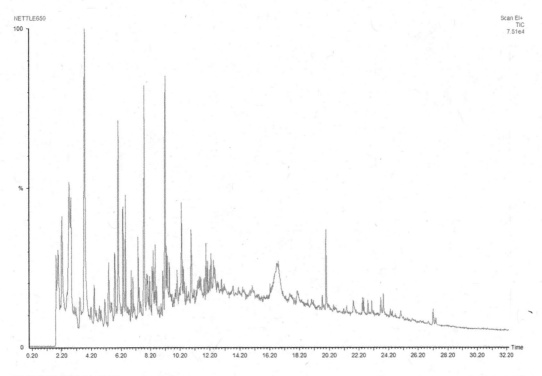

FIGURE 12.83   B-32 nettle fiber.

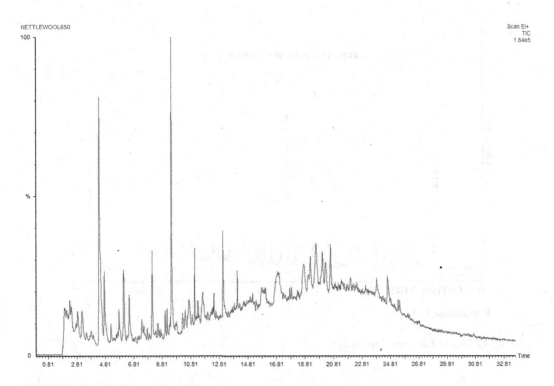

FIGURE 12.84   B-33 nettle wool fiber blend.

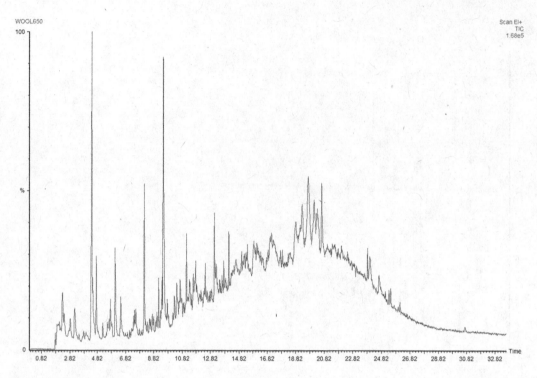

FIGURE 12.85 B-34 wool fiber.

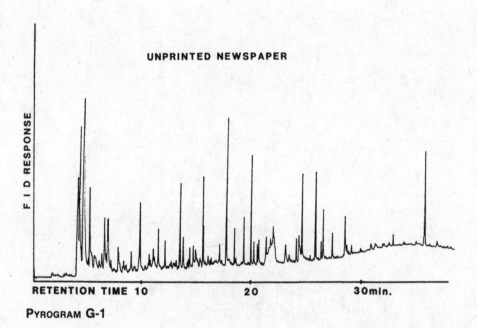

FIGURE 12.86 G-1 unprinted newspaper.

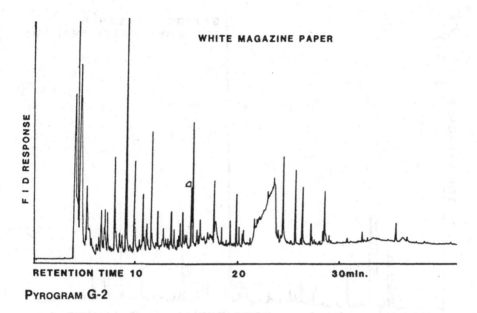

WHITE MAGAZINE PAPER

F I D RESPONSE

RETENTION TIME 10   20   30min.

PYROGRAM G-2

FIGURE 12.87   G-2 white magazine paper.

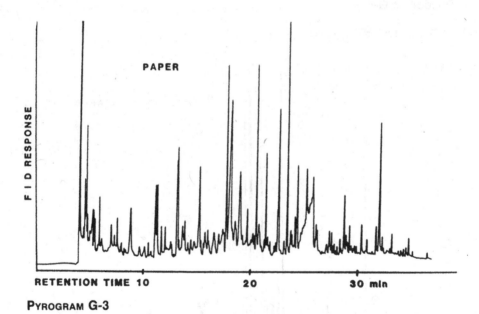

PAPER

F I D RESPONSE

RETENTION TIME 10   20   30 min

PYROGRAM G-3

FIGURE 12.88   G-3 white bond paper.

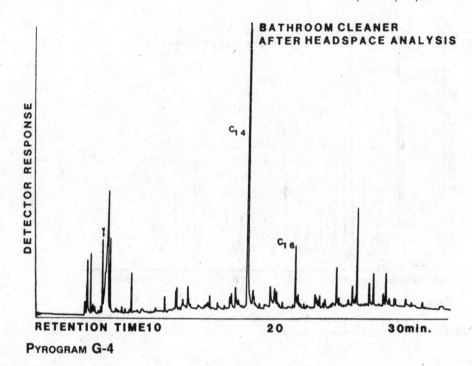

PYROGRAM G-4

FIGURE 12.89   G-4 bathroom cleaner product.

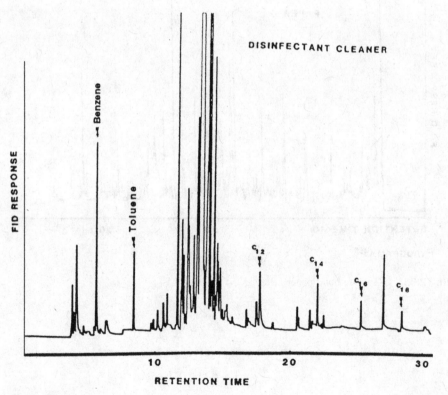

PYROGRAM G-5

FIGURE 12.90   G-5 disinfectant cleaner product. Peak number 1 = benzene, 2 = toluene, 3 = C_{12}, 4 = C_{14}, 5 = C_{16}, 6 = C_{18} hydrocarbons.

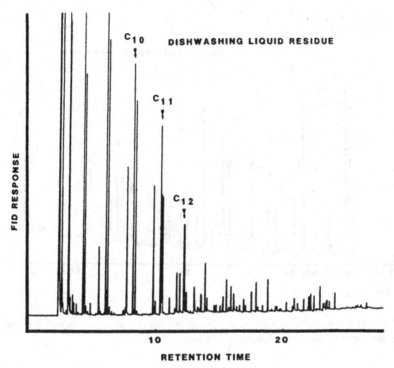

PYROGRAM **G-6**

FIGURE 12.91   G-6 dishwashing liquid. Peak number $1 = C_{10}$ $2 = C_{11}$, $3 = C_{12}$ hydrocarbons.

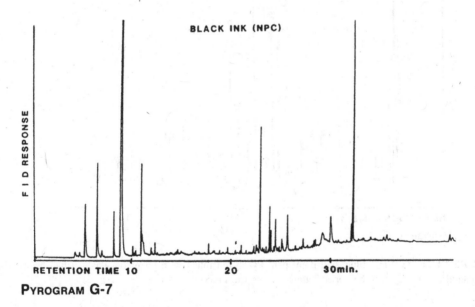

PYROGRAM **G-7**

FIGURE 12.92   G-7 ink, black, ball-point pen.

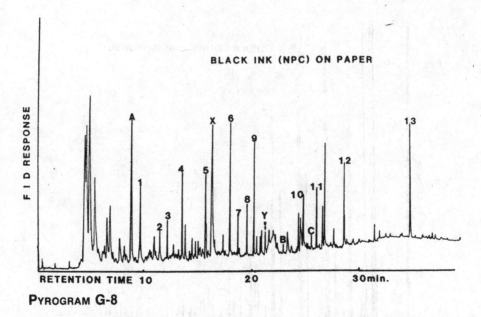

FIGURE 12.93   G-8 ink, black, ball-point pen. Pyrolyzed intact with paper on which it had been written. Numbered peaks are from the paper, lettered peaks from the ink.

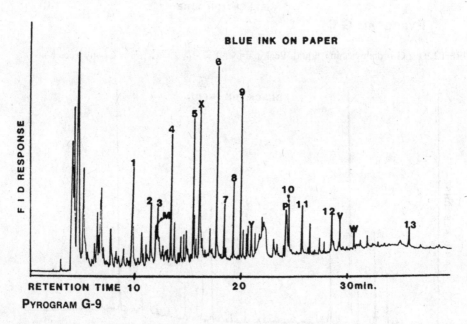

FIGURE 12.94   G-9 ink, blue, ball-point pen. Pyrolyzed intact with paper on which it had been written. Numbered peaks are from the paper, lettered peaks from the ink.

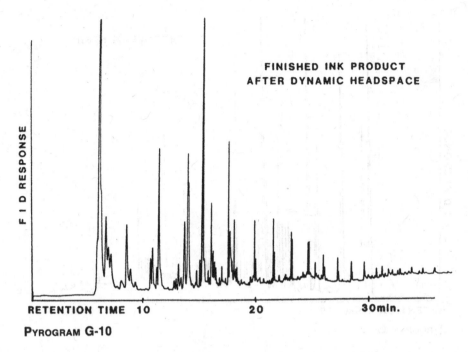

PyROGRAM **G-10**

FIGURE 12.95   G-10 printing ink. Formulation included linseed oil, wax, petroleum resins.

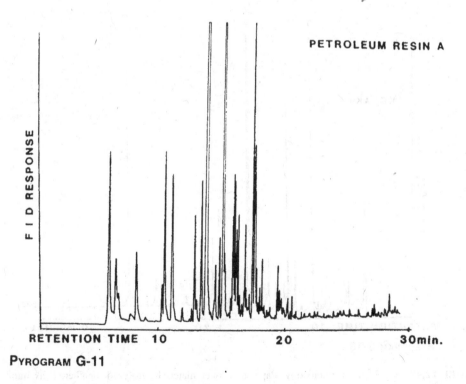

PyROGRAM **G-11**

FIGURE 12.96   G-11 petroleum resin.

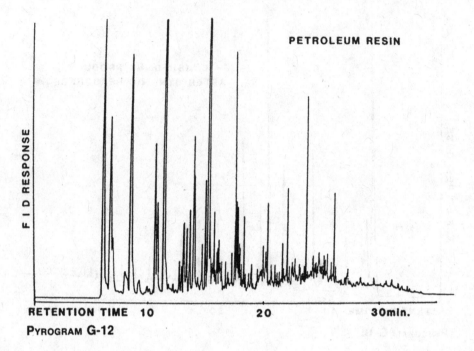

PYROGRAM G-12

FIGURE 12.97   G-12 petroleum resin.

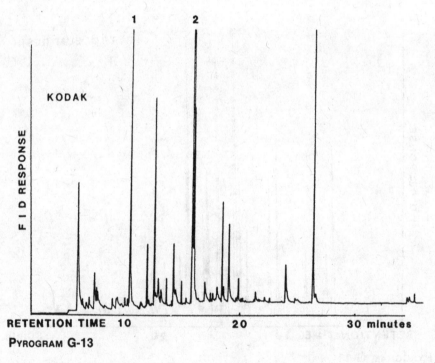

PYROGRAM G-13

FIGURE 12.98   G-13 Kodak photocopy. Paper and toner material pyrolyzed together. Peak number 1 = methyl methacrylate, 2 = styrene.

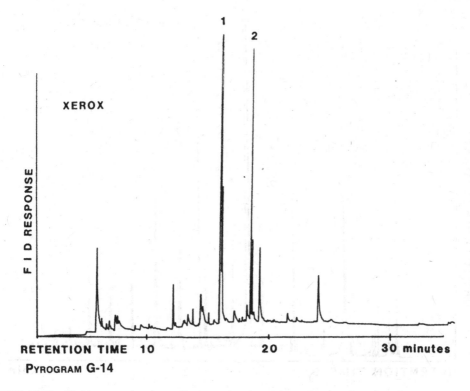

FIGURE 12.99 G-14 xerox photocopy. Peak number 1 = styrene, 2 = butyl methacrylate.

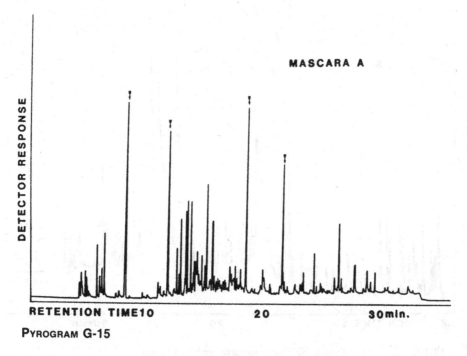

FIGURE 12.100 G-15 mascara. Peaks indicated with arrows result from beeswax.

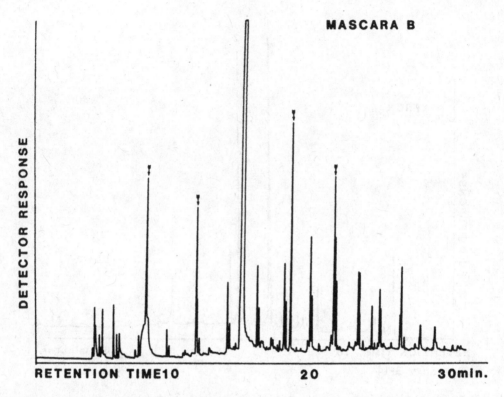

FIGURE 12.101   G-16 Mascara with acrylate.

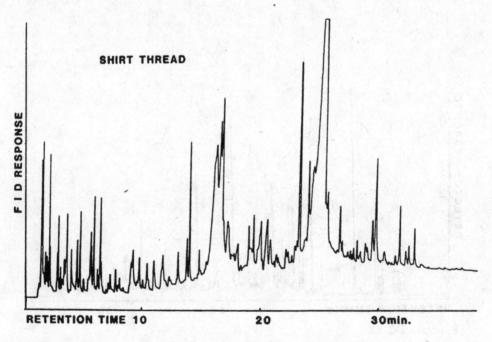

FIGURE 12.102   G-17 Shirt thread. 50/50 cotton/polyester blend fabric.

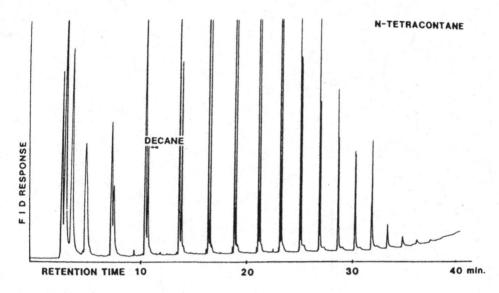

FIGURE 12.103 G-18 *n*-tetracontane.

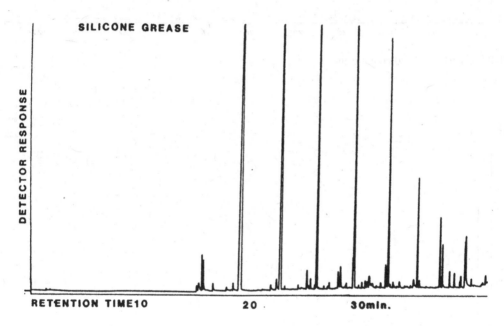

FIGURE 12.104 G-19 silicone grease.

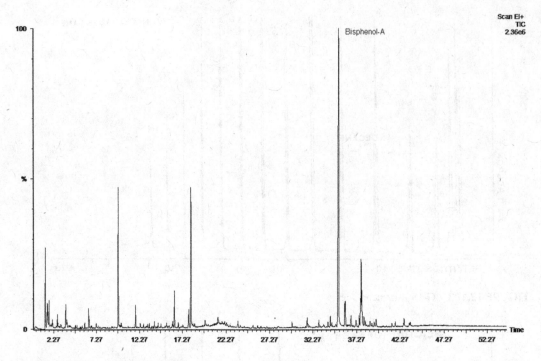

FIGURE 12.105 G-20 epoxy paint.

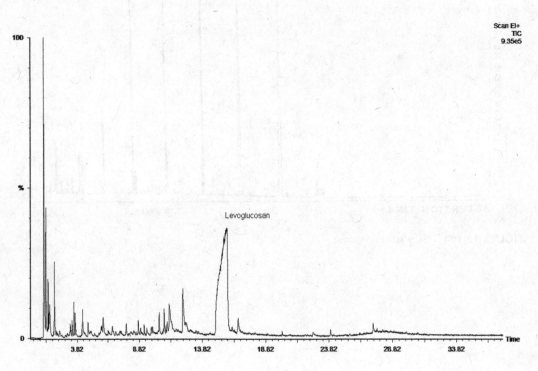

FIGURE 12.106 G-21 rayon.

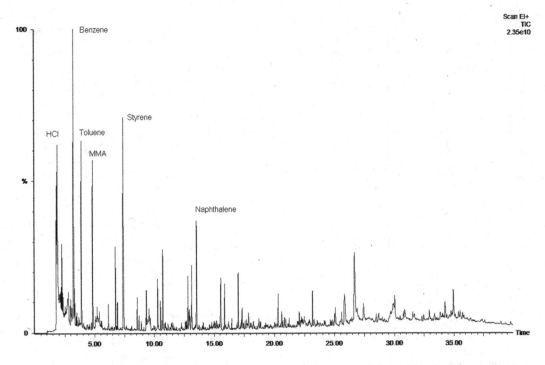

FIGURE 12.107   G-22 packaging. PVA, PVC, styrene, MMA.

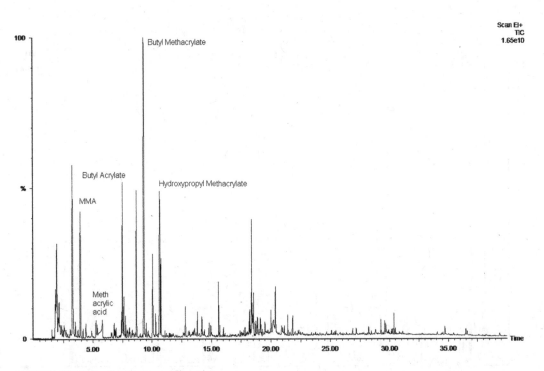

FIGURE 12.108   G-23 automobile paint.

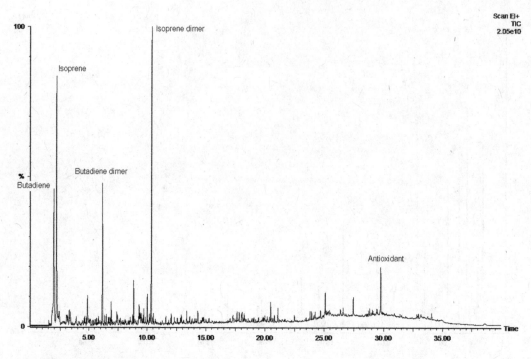

FIGURE 12.109   G-24 automobile tire rubber.

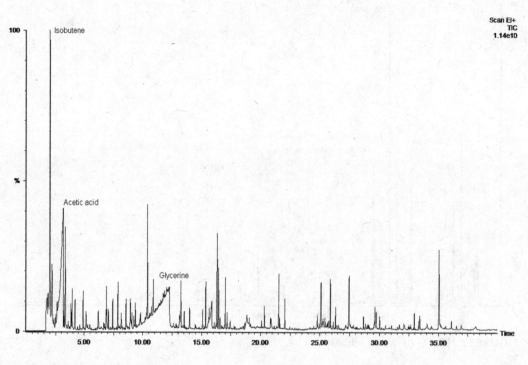

FIGURE 12.110   G-25 chewing gum.

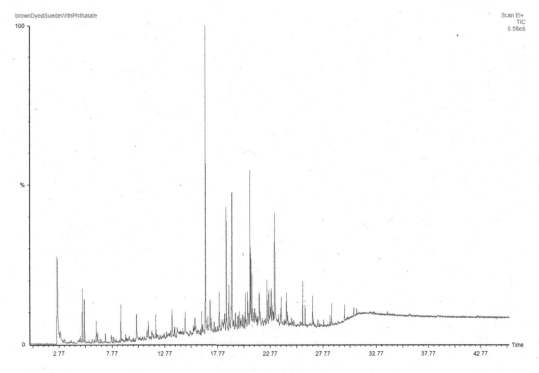

FIGURE 12.111 G-26 Suede, dyed brown, with a phthalate.

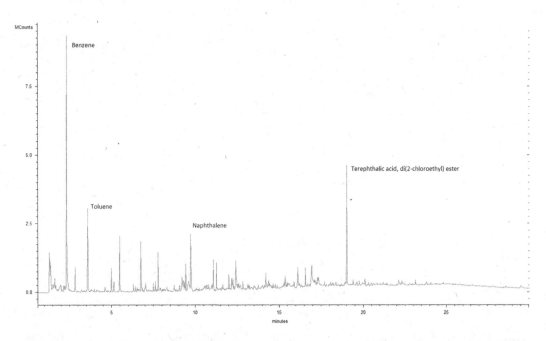

FIGURE 12.112 G-27 medication blister packaging.

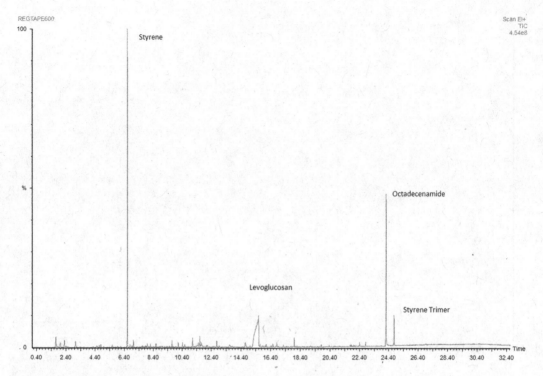

FIGURE 12.113   G-28 cash register paper.

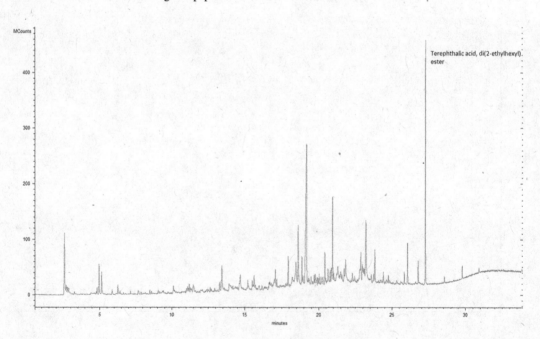

FIGURE 12.114   G-29 porkhide chew toy, smoked, with terephthalate.

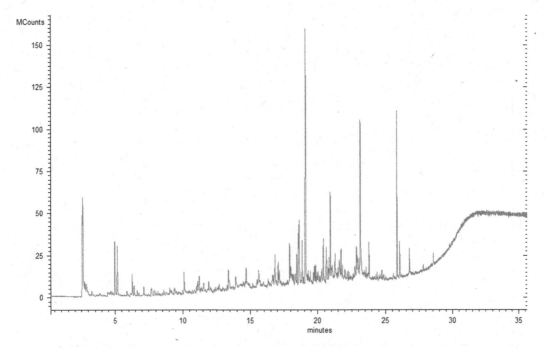

FIGURE 12.115   G-30 Suede.

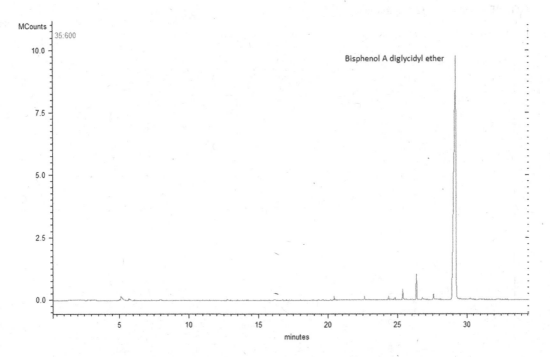

FIGURE 12.116   G-31 uncured epoxy.

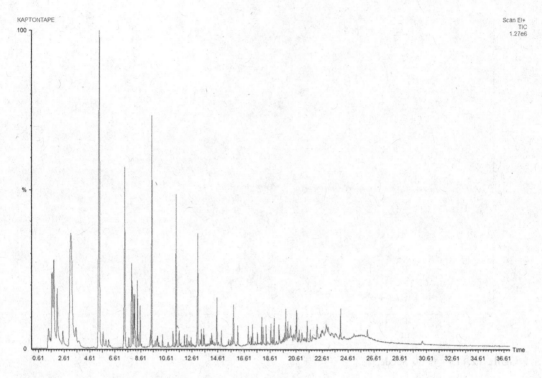

FIGURE 12.117   G-32 Kapton tape.

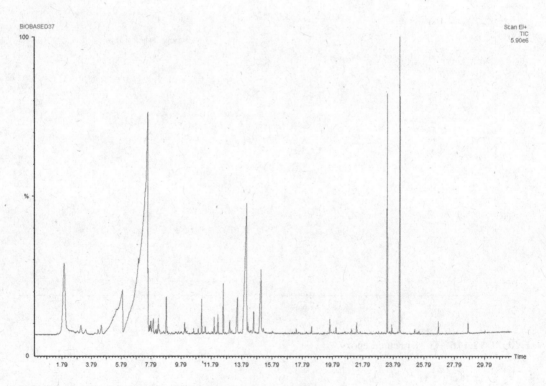

FIGURE 12.118   P-1 Mirel® polyhydroxyalkanoate polymer.

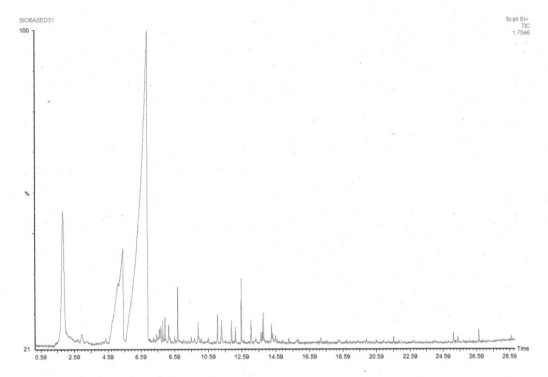

FIGURE 12.119   P-2 polyhydroxybutyrate.

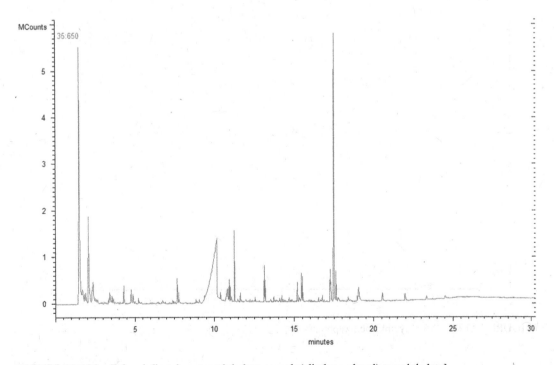

FIGURE 12.120   P-3 poly[butylene terephthalate-co-poly(alkylene glycol) terephthalate].

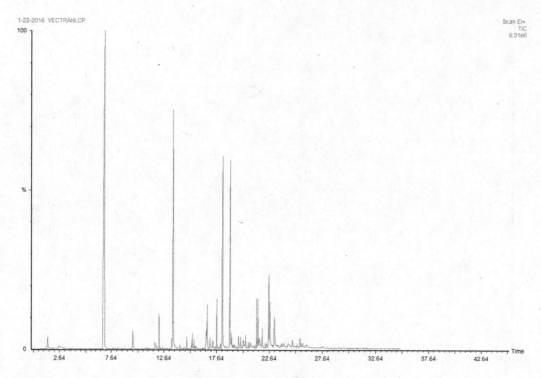

FIGURE 12.121   P-4 Vectran LCP (liquid-crystal polymer).

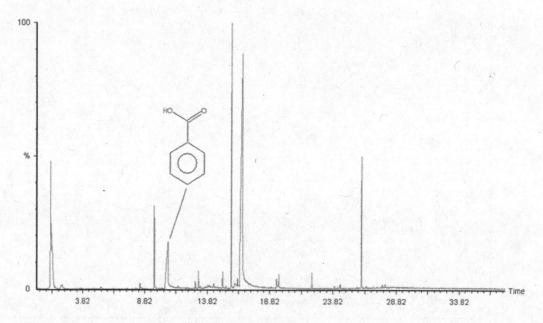

FIGURE 12.122   P-5 polyethylene terephthalate.

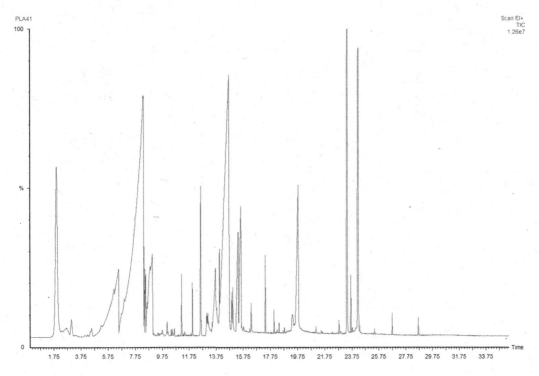

FIGURE 12.123 P-6 poly(3-hydroxybutyric acid-co-3-hydroxyvaleric acid).

Index

Note: *Italicized* page numbers refer to figures, **bold** page numbers refer to tables.

Printed in the United States
by Bookmasters

Printed in the United States
By Bookmasters